Computer Communications and Networks

The **Computer Communications and Networks** series is a range of textbooks, monographs and handbooks. It sets out to provide students, researchers, and non-specialists alike with a sure grounding in current knowledge, together with comprehensible access to the latest developments in computer communications and networking.

Emphasis is placed on clear and explanatory styles that support a tutorial approach, so that even the most complex of topics is presented in a lucid and intelligible manner.

More information about this series at http://www.springer.com/series/4198

Veljko Milutinovic · Milos Kotlar
Editors

Exploring the DataFlow Supercomputing Paradigm

Example Algorithms for Selected
Applications

Editors
Veljko Milutinovic
Indiana University
Bloomington, IN, USA

Milos Kotlar
School of Electrical Engineering
University of Belgrade
Belgrade, Serbia

ISSN 1617-7975   ISSN 2197-8433   (electronic)
Computer Communications and Networks
ISBN 978-3-030-13802-8   ISBN 978-3-030-13803-5   (eBook)
https://doi.org/10.1007/978-3-030-13803-5

Library of Congress Control Number: 2019931864

This Springer imprint is published by the registered company Springer Nature Switzerland AG
The registered company address is: Gewerbestrasse 11, 6330 Cham, Switzerland

Preface

Instead of having here a traditional preface, we decided to ask some specific questions and to answer them in a way that widens the dataflow horizons of interested readers.

1. Why is this book different from others?

When it comes to mapping of algorithms onto the dataflow architecture, this book contains algorithms not covered in previous books.

2. Where to find more algorithms?

When it comes to the mapping of algorithms onto the dataflow architecture, one could find more algorithms in the previous books of the same publisher.

3. What Alibaba said about artificial intelligence?

In a recent public memo, Alibaba of China said that FPGAs (dataflow) is better suited for artificial intelligence, compared to GPUs (control-flow).

4. What Intel said about artificial intelligence?

The recent claims of Intel state the same as above, meaning that the major ICT businesses worldwide are unisonic on the great potentials of the dataflow paradigm.

5. Why is the programming model of this book corresponding to the programming model of a future Intel chip?

The recent Intel's patent reflects the programming model used in dataflow machines of Maxeler Technologies, which is used in this book, and in the past Springer books of the same editors.

6. Where the editors of this book used the related materials in their teaching so far, worldwide?

The previous books of this publisher were used in official teaching and public lectures of one or both editors of this book; a partial list of universities, companies, research labs, and government institutions is a pretty long one: MIT, Harvard,

Boston, NEU, Dartmouth, University of Massachusetts at Amherst, USC, UCLA, Columbia, NYU, Princeton, NJIT, CMU, Temple, Purdue, IU, UIUC, Michigan, Wisconsin, Minnesota, FAU, FIU, Miami, Central Florida, University of Alabama, University of Kentucky, GeorgiaTech, Ohio State, Imperial, King's, Manchester, Huddersfield, Cambridge, Oxford, Dublin, Cork, Cardiff, Edinburgh, EPFL, ETH, TUWIEN, UNIWIE, Graz, Linz, Karlsruhe, Stuttgart, Bonn, Frankfurt, Heidelberg, Aachen, Darmstadt, Dortmund, KTH, Uppsala, Karlskrona, Karlstad, Napoli, Salerno, Siena, Pisa, Barcelona, Madrid, Valencia, Oviedo, Ankara, Bogazici, Koc, Istanbul, Technion, Haifa, Bersheba, Eilat, etc. Also at the World Bank in Washington DC, IMF, the Telenor Bank of Norway, the Raiffeisen Bank of Austria, Brookhaven National Laboratory, Lawrence Livermore National Laboratory, IBM TJ Watson, HP Encore Labs, Intel Oregon, Qualcomm VP, NCR, RCA, Fairchild, Honeywell, Yahoo NY, Google CA, Microsoft, Finsoft, ABB Zurich, Oracle Zurich, and many other industrial labs, as well as at Tsinghua University, Shandong, NIS of Singapore, NTU of Singapore, Tokyo, Sendai, Seoul, Pusan, Sydney University of Technology, University of Sydney, Hobart, Auckland, Toronto, Montreal, Durango, MontereyTech, Cuernavaca, UNAM, etc.

Bloomington, Indiana, USA   Veljko Milutinovic
March 2019   Milos Kotlar

Contents

Part I Theoretical Issues

1 **Method of Big-Graph Partitioning Using a Skeleton Graph** 3
Iztok Savnik and Kiyoshi Nitta

2 **On Cloud-Supported Web-Based Integrated Development
Environment for Programming DataFlow Architectures** 41
Nenad Korolija and Aleš Zamuda

Part II Applications in Mathematics

3 **Minimization and Maximization of Functions:
Golden-Section Search in One Dimension** 55
Dragana Pejic and Milos Arsic

4 **Matrix-Based Algorithms for DataFlow Computer Architecture:
An Overview and Comparison** 91
Jurij Mihelič and Uroš Čibej

5 **Application of Maxeler DataFlow Supercomputing to Spherical
Code Design** 133
Ivan Stanojević, Mladen Kovačević and Vojin Šenk

**Part III Applications in Image Understanding, Biomedicine,
Physics Simulation, and Business**

6 **Face Recognition Using Maxeler DataFlow** 171
Tijana Sustersic, Aleksandra Vulovic, Nemanja Trifunovic,
Ivan Milankovic and Nenad Filipovic

7 **Biomedical Images Processing Using Maxeler DataFlow
Engines** 197
Aleksandar S. Peulic, Ivan Milankovic, Nikola V. Mijailovic
and Nenad Filipovic

8   An Overview of Selected DataFlow Applications in Physics
    Simulations 229
    Nenad Korolija and Roman Trobec

9   Bitcoin Mining Using Maxeler DataFlow Computers 241
    Rok Meden and Anton Kos

Index 313

Contributors

Milos Arsic Faculty of Mathematics, University of Belgrade, Belgrade, Serbia

Uroš Čibej Faculty of Computer and Information Science, University of Ljubljana, Ljubljana, Slovenia

Nenad Filipovic Faculty of Engineering, University of Kragujevac, Kragujevac, Serbia;
Research and Development Center for Bioengineering (BioIRC), Kragujevac, Serbia

Nenad Korolija School of Electrical Engineering, University of Belgrade, Belgrade, Serbia

Anton Kos Faculty of Electrical Engineering, University of Ljubljana, Ljubljana, Slovenia

Mladen Kovačević Department of Electrical and Computer Engineering, National University of Singapore, Singapore, Singapore

Rok Meden Faculty of Electrical Engineering, University of Ljubljana, Ljubljana, Slovenia

Jurij Mihelič Faculty of Computer and Information Science, University of Ljubljana, Ljubljana, Slovenia

Nikola V. Mijailovic Faculty of Engineering, University of Kragujevac, Kragujevac, Serbia;
Research and Development Center for Bioengineering (BioIRC), Kragujevac, Serbia

Ivan Milankovic Faculty of Engineering, University of Kragujevac, Kragujevac, Serbia;
Research and Development Center for Bioengineering (BioIRC), Kragujevac, Serbia

Kiyoshi Nitta Yahoo Japan Corporation, Tokyo, Japan

Dragana Pejic Faculty of Mathematics, University of Belgrade, Belgrade, Serbia

Aleksandar S. Peulic Faculty of Engineering, University of Kragujevac, Kragujevac, Serbia

Iztok Savnik University of Primorska, Koper, Slovenia

Vojin Šenk Faculty of Engineering (a.k.a. Faculty of Technical Sciences), University of Novi Sad, Novi Sad, Serbia

Ivan Stanojević Faculty of Engineering (a.k.a. Faculty of Technical Sciences), University of Novi Sad, Novi Sad, Serbia

Tijana Sustersic Faculty of Engineering, University of Kragujevac, Kragujevac, Serbia;
Research and Development Center for Bioengineering (BioIRC), Kragujevac, Serbia

Nemanja Trifunovic School of Electrical Engineering, University of Belgrade, Belgrade, Serbia

Roman Trobec Parallel and Distributed Systems Laboratory, Department of Communication Systems, Jozef Stefan Institute (JSI), Ljubljana, Slovenia

Aleksandra Vulovic Faculty of Engineering, University of Kragujevac, Kragujevac, Serbia;
Research and Development Center for Bioengineering (BioIRC), Kragujevac, Serbia

Aleš Zamuda Faculty of Electrical Engineering and Computer Science (FERI), University of Maribor, Maribor, Slovenia

Part I
Theoretical Issues

Chapter 1
Method of Big-Graph Partitioning Using a Skeleton Graph

Iztok Savnik and Kiyoshi Nitta

Abstract We propose a new method of graph partitioning for big graphs that include a conceptual schema. The conceptual schema of a graph database, called a *schema graph*, is defined implicitly as part of the graph database itself. A graph database is stored in a distributed triple-store, i.e., a distributed database system for managing graph edges represented by triples. We define the statistics of a graph database on the basis of the schema graph. The statistics are gathered for all *schema triples*, i.e., the types of graph edges. The space of the schema triples is partially ordered by the is-more-general relationship that is defined through the class and predicate taxonomies. The graph partitioning method has two phases. A skeleton graph of the triple-store is computed in the first phase. The skeleton graph is composed of the set of schema triples that have the extensions of an appropriate size to serve as the fragments of the distribution. The edges of the skeleton graph are selected in a top-down manner, i.e., from the most general schema triple to more specific schema triples. The edges of the skeleton graph are clustered into n partitions in the second phase of the algorithm. The function DISTANCE that is used in the clustering algorithm is based on the statistics of the schema triples. The graph partitioning function maps each schema triple from the skeleton graph to its partition, stored on a separate data server. The partitioning function is well defined in that it maps the types of the triple-patterns to k fragments such that k corresponds to the size of the portions of the triple-store addressed by the triple-patterns. In other words, it maps the types of triple-patterns that address a large number of triples to multiple distributed fragments, and the types of triple-patterns that address few triples to a single fragment.

I. Savnik (✉)
University of Primorska, Koper, Slovenia
e-mail: iztok.savnik@upr.si

K. Nitta
Yahoo Japan Corporation, Tokyo, Japan
e-mail: knitta@yahoo-corp.jp

© Springer Nature Switzerland AG 2019
V. Milutinovic and M. Kotlar (eds.), *Exploring the DataFlow*
Supercomputing Paradigm, Computer Communications and Networks,
https://doi.org/10.1007/978-3-030-13803-5_1

3

1.1   Introduction

The development of a graph-based semantic web speaks to society's enormous interest in constructing a detailed knowledge base (graph) that includes properties of categories from all popular areas of human activity. Knowledge bases such as Knowledge Graph, Wikidata, YAGO, and Knowledge Vault currently include from 1,000 to up to 350,000 categories, up to 570,000,000 instances of categories, up to 35,000 relationship types, and up to 18,000,000,000 relation instances [9]. However, existing knowledge graphs are in their infancy in many respects, and more systematic use of intelligent tools for extracting knowledge from various data sources has just begun.

The need for triple-store systems capable of storing and managing from tera to peta (10^{15}) triples is obvious. The scaling of storage and query-processing systems to this amount of data is currently possible through the large-scale distribution of data into shared-nothing clusters. A triple-store is too large to be stored on a single database server. It must be partitioned into parts that have the appropriate size to be stored on low-cost commodity servers. The partitions have to be replicated to improve durability and availability and to speed-up access to data. Finally, the triples must be distributed in such a way that the queries benefit from the distribution.

Distributed database systems that are based on the relational model [25] provide various methods and tools to allow a user to distribute a database and speed-up query processing. First, the algorithms for distributing individual tuples and blocks can partition the contents of a table, or a group of tables, to a set of data servers. The mapping between the tuples and blocks and the data servers is implemented by various algorithms such as round-robin, hash-based, and range partitioning. Unfortunately, the distributed data servers that hold partitions of the database must deliver results (including the intermediate results) of queries to the coordinator, so the cluster network becomes a bottleneck. This kind of solution scales up to a few 100 data servers.

The most practical method used for partitioning in relational distributed systems is range partitioning [3]. A relation or a set of relations is split into partitions on the basis of a set of ranges that can be defined in different ways. Predicate-based distribution defines the ranges by using a set of predicates. However, the predicates used for the distribution function must be defined manually. The ranges can be specified using a special attribute on the basis of which the distribution is defined. Another partitioning method, list-based partitioning defines the ranges by listing the values of a special attribute for each particular interval.

The typical drawbacks of range partitioning in distributed relational systems are the need to know precisely the structure and semantics of the database as well as the common patterns used in the queries of a given application. The partitions must be carefully designed to serve as the basis for the efficient implementation of the range partitioning. Automatic partitioning is not possible at the semantic level because of the complexity of the structure of the database.

In contrast, Internet database systems such as Amazon's Dynamo [8] and Google's Bigtable [5], use a much simpler data model than relational systems. For this reason, the key-value pairs (or the column families in the case of Bigtable) can be distributed much more easily, especially because the keys serve as a universal, and in many cases the only, access method. Distribution based on hashing works well in modern Internet database systems. In particular, the consistent hashing [18] provides fast and uniform hashing of objects with their keys, tolerance to the failures of data servers, and scalability by allowing the addition of new data servers in real-time. The locality of data describing a particular key in Internet database systems is a common consequence of using simple sorted textual keys. Bigtable, for instance, stores columns on the basis of sorted textual keys, and rows with the same keys are stored on the same tablets.

Triple-stores have, in comparison to relational database systems, a significant advantage in the simplicity of their structure. Technically, a triple-store is composed solely of triples that have the same structure; the triples are used to represent the ground data, conceptual schema, and higher levels of the structural representation, including a variant of the predicate calculus [2, 24]. Triple-stores are much more appropriate than relational database systems for defining the automatic distribution of data to a set of data servers for this reason.

Unlike modern Internet database systems, triple-stores do not have one single type of key that could be used for accessing triples; rather, the triples can be accessed by using possibly very complex SPARQL queries [10] composed of triple-patterns. For this reason, the triples cannot be distributed merely in the manner implemented by Internet databases systems, especially if our goal is to exploit the semantics of a triple-store for defining its distribution in a way that is well organized for querying.

1.1.1 Proposed Method of Graph Partitioning

The approach to the automatic partitioning of the triple-stores proposed in this paper is based on the conceptual schema of the triple-store. By the conceptual schema of the triple-store we mean the taxonomy of the classes and predicates, and the definitions of the domains and ranges of all predicates. Here, we introduce the concept of a *schema triple* to serve as a type of the triples, i.e., the types of the graph edges. For example, the schema triple (person, worksAt, organization) is the type of triple (jim, worksAt, neo4j).

The schema triples that are defined in a given triple-store are organized in a partially ordered set by the relationship is-more-general. The relationship is-more-general is defined employing the class and the predicate taxonomies. Therefore, the types (schema triples) of the graph edges stored in a triple-store form a hierarchy from the most general schema triple to the most specific schema triples that are the leaves of the hierarchy. The interpretation of the most general schema triple includes the complete triple-store while the interpretations of the most specific schema triples include some subset of the triple-store. Moreover, the interpretation of the schema triple t_1, that is-more-general then some other schema triple t_2, subsumes the interpretation of t_2.

The partial ordering of the schema triples is used in the first phase of the partitioning algorithm as the framework for the computation of the *skeleton graph* of the triple-store. The skeleton graph is composed of the schema triples (edges) with interpretations that are of appropriate size to serve as fragments of the triple-store distribution. The size of the interpretation of the given schema triple is estimated by using the statistics of the triple-store. The skeleton graph can be seen as an abstraction of the graph stored in the triple-store.

The statistics of the schema triples are precomputed to be used to estimate the size of the interpretations of the schema triples. Given a triple t, the statistics are updated for all schema triples that represent the actual types of t. The set of actual types of t can be restricted by only taking into consideration the schema triples that are close to the stored types of the triple t.

The skeleton graph is partitioned into n subgraphs in the second phase of the partitioning algorithm by using some of the clustering algorithms. We can obtain the partitioning method that generates strongly connected subgraphs, depending on the definition of the function DISTANCE between a schema triple (edge) and a set of schema triples (subgraph). The extensions, or the interpretations, of the computed subgraphs, represent the actual partitions.

1.1.2 Contributions

The main contribution of the research presented here is a new method of partitioning large graphs that include a conceptual schema of the graph. The conceptual schema of a triple-store is given in the form of a class and a predicate taxonomy, and the characterization of predicates by the domain and range of the predicate. The graph partitioning method uses the conceptual schema of a graph database to partition the graph into n strongly connected components.

The skeleton graph is constructed from the conceptual schema and the statistics of the triple-store in the first part of the method. The skeleton graph is composed of schema triples that have appropriate size of their extensions. Furthermore, the edges of the skeleton graph must be complete in the sense that any schema triple t_s of a given triple-store can be expressed by a set of edges of the skeleton graph that determine the partitions, where the instances of t_s are stored. The strongly connected partitions are computed from the skeleton graph in the second part of the method by using one of the clustering algorithms.

The units of the distribution are the fragments that represent the extensions of the edges (schema triples) of the skeleton graphs. Therefore, the partitions of the original graph include fragments that correspond to the skeleton graph edges grouped by the clustering algorithm.

The proposed method is well defined in the sense that it consistently maps the triple-patterns, which represent the basic building blocks of SPARQL queries, to the set of partitions; a query that addresses a large portion of a triple-store is mapped to multiple partitions (a query that addresses complete triple-store is executed on all partitions) while a query that addresses a small number of triples needs to be executed within the context of one partition.

Note that the distribution method is guided by the actual structure of the querying space by using the types of triple-patterns as the characterization of the triples addressed by the triple-patterns. The structure of the querying space can be seen as the partially ordered set of schema triples (the types of triple-patterns), which implies the subsumption hierarchy of the interpretations (extensions) of the schema triples.

The major innovation of this research follows predominantly the creativity method Generalization [4].

1.1.3 Overview

The chapter is organized as follows. Section 1.2 presents the target system for the graph partitioning method. The architecture of the prototype triple-store big3store is also presented. The setup for the distributed triple-store system is detailed to obtain knowledge about the constraints imposed by the query distribution and execution environment.

Section 1.3 presents the formal view of the triple model and the algorithm for the computation of the statistics. The formalization presented in Sect. 1.3.1 defines the main concepts of the triple model used for the definition of the algorithms. We define the identifiers, the triples, the schema triples, the triple-patterns, and the semantic interpretations of the defined concepts. The identifiers (URI-s) and the triples from a triple-store are structured in a partially ordered set using the relationship is-more-general (is-more-specific). Finally, the algorithms for the computation of the statistics of a triple-store are described in Sect. 1.3.2. The definitions of the algorithms for the computation of the triple-store statistics follow the formal definition of the data model.

The proposed graph partitioning method is described in Sect. 1.4. First, a general overview of the proposed method is provided. The predicate-based, class-based, and edge-based partitioning methods are presented in Sects. 1.4.1.1–1.4.1.3. A detailed description of edge-based partitioning is given in Sects. 1.4.2–1.4.4. We explain how the skeleton graph of a triple-store is computed, how the skeleton graph is clustered to obtain graph partitioning and finally, how the localization function is constructed to return the set of partitions for a given schema triple. Finally, we describe the existing methods related to the proposed method.

Section 1.5 presents the initial empirical results. A comparison between two different distributions obtained by using the predicate-based partitioning method is presented. The first uses the random distribution of the fragments to the data servers and the second uses the clustering algorithm k-means to construct strongly connected partitions.

Finally, Sect. 1.6 concludes with remarks and provides some directions for further work.

1.2   Distributed Triple-Store Setup

Efficiently storing and querying huge volumes of data is currently possible by using a shared-nothing cluster architecture [32]. Efficient data servers with large amounts of random-access memory (RAM) and disk storage are commercially available as inexpensive commodity hardware. This allows heavy distribution and replication of data as well as large distributions of query processing on servers to form huge clusters.

We present a prototype triple-store system, big3store, which represents a practical environment for the development of the graph partitioning method. We expect that the reader is familiar with the Resource Description Framework (RDF) [27] and the RDF Schema [28] data model.

Section 1.2.1 explains the main ideas of the architecture of big3store by describing the big3store cluster and its components, front servers and data servers. The main concepts of the distributed query execution system are presented in Sect. 1.2.2. We describe the computational model used to express the queries, i.e., the logical algebra of RDF graphs and its counterpart, physical algebra. We present the Erlang processes that constitute the query execution system and their roles.

1.2.1   Architecture

Big3store is a *dataflow system* [3], where a triple-store is composed of an array of data servers arranged into columns and rows. The complete triple-store is partitioned and distributed into *columns* based on semantic information attached to triples via a triple-store schema. Each column stores a partition of a triple-store that is replicated to the column *rows*. The rows of a column, therefore, contain replicas of triple-store partitions assigned to the columns.

The query processing system of big3store is implemented in Erlang programming language [1]. A query is converted into a tree of processes that communicate by using asynchronous messages. The streams interconnect the query nodes that comprise query trees. A stream in big3store is based on blocks composed of RDF graphs, i.e., sets of triples. The implementation of streams is flexible in the sense that it provides detailed control of the stream parameters, such as the number of empty messages in circulation between each pair of processes, the size of stream chunks composed of multiple blocks, and the behavior and the size of queues.

Figure 1.1 outlines a cluster composed of two types of servers: *front servers* represented as the nodes of plane A, and *data servers* represented as the nodes of plane B. Data servers are configured in *columns* labeled from (a) to (f). A complete database is distributed into columns such that each column stores a portion of the complete database.

The portion of the database stored in a column is replicated into rows labeled from one to five. The number of replicas in a particular column can be dynamically

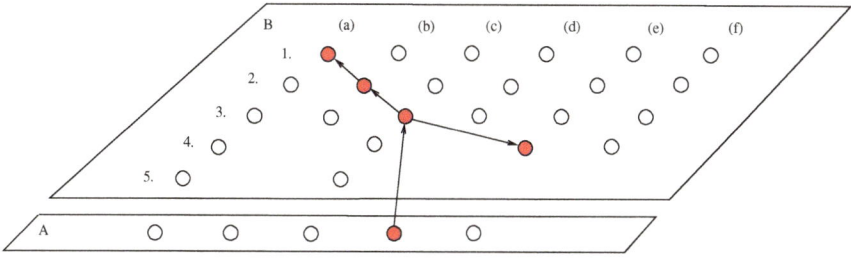

Fig. 1.1 Configuration of servers for a particular query

determined based on the query workload for each particular column. The more massive the load on a given column, the higher the number of row data servers chosen for replication. The specific row used for executing a query is dynamically selected based on the current load on servers in a column.

1.2.1.1   Front Servers

Front servers accept SPARQL queries initiated by the remote users and convert them into query trees. A query tree includes detailed information about access paths and methods for the implementation of joins used for processing the query. We refer to such a query tree as an *annotated query tree*.

The front server evaluates the annotated query tree as it does not need further optimization. The nodes of the query tree are localized by determining the columns of the data server array, where the queries represented by the query nodes are executed. The front server then constructs Erlang processes for each of the operations of the query nodes on the selected data server [29]. The task of the front server is also to initialize each of the previously created processes to construct a dataflow program composed of Erlang processes.

1.2.1.2   Data Servers

The data servers are passive in the sense that the front servers create all processes on the data servers, initialize them so that they implement a dataflow program for a given query tree, and initiate the evaluation of the dataflow program. After query execution is completed, the processes are terminated.

The data servers include the implementation of physical algebra operations together with the implementation of the access paths, i.e., methods of accessing indexed tables of triples in a local triple-store. The physical algebra operations comprise set operations for union, intersection, and difference; and implementations of the join operation, including the nested-loop join, the main-memory join, and variants of the hash-join. The logical operations select and project are included in each physical operation.

A non-distributed storage manager for storing triples and indices for accessing triples has to deal with similar problems to those faced by relational storage managers. We used a PostgreSQL database management system to store and manage tables of triples, which are referred to as *local triple-stores*. A local triple-store is a table that includes four attributes: triple id, subject, property, and object. We decided to add triple ids to a triple-store, because it allows more consistent and uniform storage of various data related to triples, such as named graphs, other groupings of triples, and properties of triples (reification). Each local triple-store maintains six indices for accessing subject, predicate, object (SPO) attributes, and an additional index for triple ids.

1.2.2 Distributed Query Execution System

The Erlang programming environment [1] is used for the implementation of big3store as an alternative to Hadoop-like systems [33]. It provides a remarkably simple and effective parallel programming model based on lightweight processes. Erlang processes use a "shared nothing" philosophy, where communication among processes is solely achieved by means of synchronous and asynchronous messages. A detailed presentation of the query execution system is given in [21].

1.2.2.1 Algebra of RDF Graphs

The logical algebra of RDF graphs is an abstract model used for the implementation of the query execution system. The logical algebra is defined using set semantics, in which the inputs and outputs of operations are sets of graphs. The denotational semantics of algebra and the implementation of algebra are described in [29]. The architecture for the query execution system based on the algebra allows for the use of pipelined and partitioned parallelism [7].

The logical algebra of RDF graphs is rooted in the relational algebra [6]. It reflects the nature of the RDF graph data model through the specific type of the access method, called *triple-pattern access method*. The operations of the logical algebra of RDF graphs are the selection, the projection, the join, and the set operations union, difference, and intersection. The logical algebra of RDF graphs is defined on sets. The arguments of algebraic operations and their results are RDF graphs.

The expressions of RDF graph algebra are graphs themselves; they are called *graph-patterns*. A graph-pattern is a set of triple-patterns, where the names of some nodes and edges are replaced with variables. Triple-patterns are the leaves of the graph-pattern, and inner nodes are the binary operations of the algebra defined on the basis of the equality of the variables from different triple-patterns.

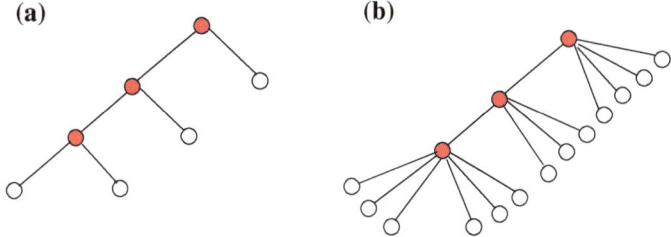

Fig. 1.2 **a** Left-deep query tree   **b** Left-deep query tree with multiple AM operations

The design of the physical algebra of RDF graphs follows the ideas used for implementing relational algebra in the framework of relational database management systems [11, 12]. The physical algebra has three types of operations: physical access methods (AM) AM, physical joins join, and physical set operations union, diff, and intsc.

A prototype triple-store system, big3store, currently uses left-deep query trees. The most important advantage of using left-deep trees is the *pipeline* that is formed by physical operations join. An example of a left-deep query tree with three join operations and four AM operations is given in Fig. 1.2a. The results of retrieving graphs from the outer query node of join operation can be used for index-based access to the inner query node. The graph that is constructed as the result of join operation is then sent to the parent query node, i.e., join query node. Consequently, there is no need to store intermediate results during query evaluation.

Triples-patterns that address huge numbers of instances are distributed to more data servers by using the semantic distribution algorithm. Therefore, physical operation AM, defined by using some triple-pattern, may be executed either on one or on several data servers. The number of servers depends on the size of the targeted set of triples. Consequently, left-deep query trees can have multiple AM query nodes as outlined in Fig. 1.2b.

A stream connects two Erlang processes that implement algebra operations. One process involves sending the data. The other process involves receiving the stream of graphs. The streams are defined by the following parameters that dictate the behavior of the streams.

First, each stream has a number of empty messages that are used as the carriers of the data. The sending part of the stream can only send a message if it owns one or more empty messages. If the sender runs out of empty messages, it has to wait. Second, the graphs are transferred between two processes by using blocks. Each block can hold a given number of triples. We can thus avoid network latency caused by the initiation of transfer when short messages occur. Third, the blocks between the local database management systems (DBMSs) and the access method are transferred in chunks that are composed of a predefined number of blocks.

Each stream has an outgoing queue that receives graphs from the local process and packs them in the blocks. A process at the other end of the stream has an input queue

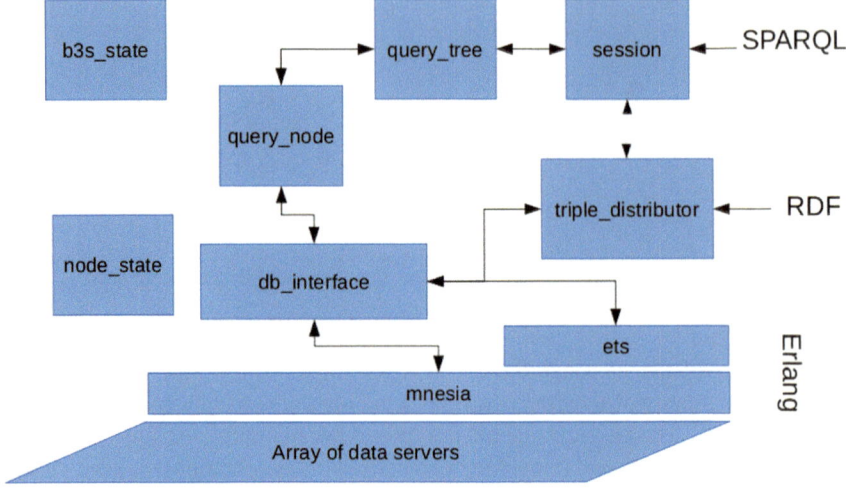

Fig. 1.3 Architecture for big3store query executor

that stores the blocks from the sending process and unblocks the graphs they contain. The stream parameters together with the parameters of the input/output queues can be used to control the behavior of the streams.

1.2.2.2 Query Execution System

The query execution system of big3store is composed of the modules outlined in Fig. 1.3. Each module includes the implementation of a particular type of process.

The state modules, `b3s_state` and `node_state`, are used for efficiently sharing big3store configuration data structures as well as for storing and querying the current state of the system, such as the number of processes running at each particular data server.

Each data server runs one instance of the PostgreSQL database system that serves as a *local triple-store*. A triple-store is attained utilizing a single table `triple_store` that is accompanied by six indices for all combination of SPO attributes. PostgreSQL provides transaction-based access to a local triple-store through a module `db_interface`. However, since a `db_interface` only provides very simple cursor-based access to a single table, the local triple-store can easily be replaced by another database engine. It could be even the file-based access method to RDF triples.

Module `triple_distributor` implements various schema-based algorithms for the distribution of triple-stores into a set of cluster columns.

Session processes are implemented in the module `session`. They serve as a user interface for interaction with users, initiate the creation of query-tree processes, control the execution of query trees, and collect the results from query execution. One session can spawn many query trees in parallel.

Module `query-tree` implements query-tree processes that run on front servers. The main task of query-tree processes is to prepare, schedule, and initiate the execution of queries in the form of query trees composed of query-node processes interconnected by means of streams. Therefore, each query-tree process controls one or more query node processes that constitute a query.

Physical algebra operations `AM`, `join`, `union`, `intsc`, and `diff` are implemented in `query-node` modules. Each physical algebra operation is realized as an independent Erlang query-node process that runs on one of the data servers. All operations are implemented as state machines that execute a particular protocol, i.e., an access method to the local triple-store, an indexed nested-loop join algorithm, a symmetric hash-join algorithm, a variant of the main-memory hash-join algorithm, or particular set operation.

The triple-pattern query node implements the access method that uses index-based access to the triple-store. After triples from the local triple-store have been obtained, the triple-pattern query node process checks them against a *selection list*. The selected triples are sent to the parent of the query node.

1.3 Formalization and Statistics

The conceptual schema of a triple-store is implicitly defined as the part of the triple-store itself—it is stored in the same manner as ordinary data. The conceptual schema of a triple-store is a graph, which is referred to as a *schema graph*. It is composed of the class nodes related using subclass relationships, and, by the definitions of the predicates linking the domain and range classes of the predicates. The schema of a triple-store can be very large. For example, Yago [16] includes approximately 500,000 classes.

The statistics of a triple-store is based on the triple-store schema that includes (a) the taxonomy of classes expressed using the predicate rdfs:subClassOf, (b) the definitions of the domains and the ranges of predicates expressed using the predicates rdfs:Domain and rdfs:Range, (c) the typing information defined for each particular ground triple using the predicate rdf:type, and (d) the definition of the hierarchy of the predicates expressed using the predicate rdfs:subPropertyOf.

A *schema triple* stands for the *type* of a triple. For instance, the schema triple (person, worksAt, organization) is the type of all triples that have the instance of the class person in the subject part, the predicate worksAt, and the instance of the class organization as the object part.

A *schema graph* is defined by a set of classes that stand for nodes of the schema graph and the set of schema triples that represent the edges of the schema graph. The *stored schema graph* is a schema graph defined through the triples that use the predicates rdfs:Domain and rdfs:Range, to state the domains and the ranges of the triple-store predicates. We assumed that all predicates used in the triple-store had the domains and the ranges defined.

The statistics are computed for each edge e of the schema graph. The statistical information stored for each edge includes (a) the size of the interpretation of the schema triple, e, i.e., the number of instances of a given schema triple, e, and (b) the number of distinct keys for each of the subsets of the components from e. The statistics are stored for all seven possible keys from a given schema triple to be able to accurately estimate the size of all possible types of triple-patterns.

Let us first define the formal framework that is used for the definition of the statistics of the triple-store as well as for the definition of the method of graph partitioning. We define formally a triple model used for the description of the concepts that constitute a conceptual schema of a triple-store. A complete formalization is presented in [29], while we present here solely the most important features of the formalization.

1.3.1 Formalization

The big3store triple-store is based on the RDF data model [27]. Let U be the set of URI-s, B be the set of blanks, and L be the set of literals. Let us also define sets, $S = U \cup B$, $P = U$, and $O = U \cup B \cup L$.

A *RDF triple* is a triple $(s, p, o) \in S \times P \times O$. An *RDF graph* $g \subseteq S \times P \times O$ is a set of triples. We state that an RDF graph g_1 is a *subgraph* of g_2, denoted as $g_1 \subseteq g_2$, if all triples of g_1 are also triples of g_2.

1.3.1.1 Identifiers

To abstract away the particulars of the RDF data model we define a set of identifiers $I = U \cup B \cup L$. The complete set of identifiers I is composed of the individual identifiers I_i, and the class identifiers I_c, i.e., $I = I_i \cup I_c$. The individual identifiers I_i can be further divided into individual identifiers that stand for the objects and things, and particular types of individual identifiers that stand for predicates. The set of predicate identifiers is named I_p. The predicates can have sub-predicates in the same manner as the classes have subclasses. While classes have instances, predicates do not have instances.

The interpretation of a given class identifier, c, is a set of individual identifiers that are the instances of c. The interpretation of c is denoted as $[\![c]\!]_g$. The interpretation of a class includes the interpretations of all its subclasses.

The relationship *is-more-specific* is defined on identifiers as a generalization of the relationships rdf:type, rdfs:subClassOf, and rdfs:subPropertyOf. The top of the partial ordering is composed of the most general classes together with the top class, $\top$. The bottom of the partial ordering of identifiers includes the individual identifiers from I_i. An identifier i_1 is-more-specific than another identifier i_2, written $i_1 \preceq i_2$, if i_1 is an instance of a class i_2, if a class i_1 is a subclass (sub-property) of a class (property) i_2, or, if i_1 is an instance of a subclass of a class i_2.

Another semantic interpretation is defined on top of the relationship is-more-specific. A *natural interpretation* maps identifiers to a set of all more specific identifiers. The natural interpretation of an individual identifier is the identifier itself. The natural interpretation of a class identifier is a set of all more specific identifiers.

The partial ordering relationship $\preceq$ and the natural interpretation function are defined *consistently*. The natural interpretation of a class c is denoted as $[\![c]\!]_g^*$. We have shown in [29] that an identifier i_1 is-more-specific then an identifier i_2, if and only if, the natural interpretation of i_1 is included in the natural interpretation of i_2. In short, $i_1 \preceq i_2 \iff [\![i_1]\!]_g^* \subseteq [\![i_1]\!]_g^*$.

1.3.1.2 Triples

The set of identifiers I is extended with the top class identifier $\top$. We assume in the sequel that $\top \in I \wedge i \in I \Rightarrow i \preceq \top$. The partial ordering relation, $\preceq$, is extended to triples. A triple, t_1, is-more-specific than a triple, t_2, if all components of t_1 are-more-specific than the corresponding components of t_2.

The interpretations of triples are based on the interpretations of the identifiers. The interpretation function $[\![\,]\!]_g$ is extended to triples. The interpretation of a triple $t = (s, p, o)$ is a set of triples $t' = (s', p', o')$ such that $s' \in [\![s]\!]_g$, $p' \in [\![p]\!]_g$ and $o' \in [\![o]\!]_g$. Similarly, the natural interpretation function $[\![\,]\!]_g^*$ is extended to triples.

The conceptual schema of a triple-store is now defined in the form of a *schema graph*. The arcs of the schema graph are called *schema triples*. The schema triples are composed solely of the class identifiers and the predicate identifiers. They stand for the types of the triples $t' \in g$. The schema graph, sg, defines the structure of ground triples from g. Each triple $t' \in g$ is an instance of at least one schema triple $t_s \in sg : t' \in [\![t_s]\!]_g$.

1.3.1.3 Triple-Patterns

A triple-pattern is a triple that can include variables in the place of the triple components. The interpretation of a triple-pattern, tp, is the set of triples, $t \in g$, such that t includes any value in place of variables, and has the values of other components equal to the corresponding tp components. The interpretation of tp is denoted as $[\![tp]\!]_g$.

The type of a triple-pattern tp is a schema triple t_{tp}, such that the interpretation of t_{tp} subsumes the interpretation of tp, or, $[\![tp]\!]_g^* \subseteq [\![t_{tp}]\!]_g^*$. Therefore, the triples addressed by a triple-pattern tp are restricted to the triples from $[\![t_{tp}]\!]_g^*$.

The partial ordering of the schema triples is used in the algorithm for the construction of the skeleton graph of a given triple-store. The types are organized in such a way so that the more general types include the interpretations of the more specific types. The skeleton graph includes the schema triples that have the interpretation of

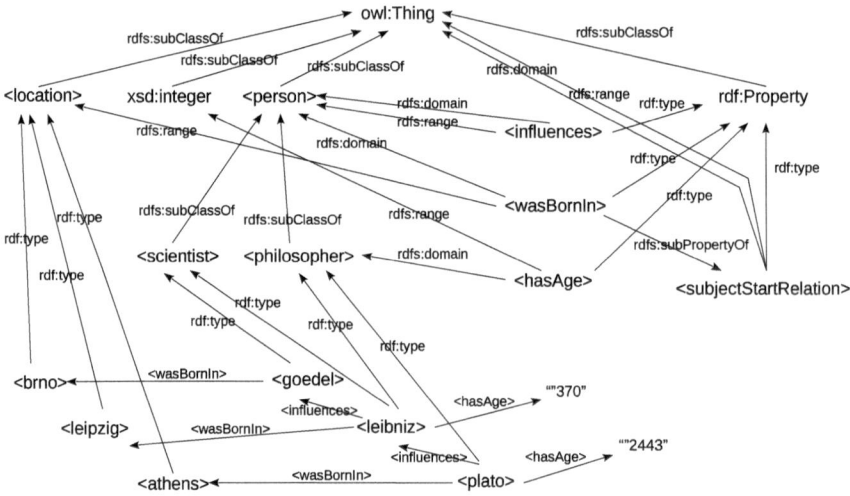

Fig. 1.4 Simple triple-store

the appropriate size to serve as the fragments of the distribution. The triple-patterns
that have the type more general than some schema triples from the skeleton graph
address more than one fragment.

Note that we only specify the semantic conditions for the definition of the type of
triple-patterns. The algorithm for the derivation of the type of triple-pattern is beyond
the scope of the presented research; it is defined in [30].

1.3.1.4 Example of Triple-Store

This example presents a simple triple-store composed of 33 triples. Four classes are
defined. The class person has subclasses scientist and philosopher. The class location
specifies the physical location of objects (Fig. 1.4).

The property, wasBornIn, describes the place of birth of a person. The property,
subjectStartRelation, specifies that the subject has started some relationship—it represents the generalization of the predicate wasBornIn. Finally, the property, hasAge,
that is defined for the class philosopher describes the age of a philosopher.

We will now present a few examples of the use of the natural interpretation function
$[\![\ ]\!]^*$ within the context of the above graph g. Let $\top$ stand for the most general class.
The natural interpretation of the schema triple, $t_{s_1} = (\top, \text{rdf:type}, \text{person})$, includes
four ground triples:

$$[\![t_{s_1}]\!]^*_g = \{(\text{plato,rdf:type,philosopher}), (\text{leibniz,rdf:type,philosopher}),$$
$$(\text{leibniz,rdf:type,scientist}), (\text{goedel,rdf:type,scientist})\}.$$

The natural interpretation of the schema triple, t_{s_2} =(person, wasBornIn, location), has three triples.

$$[\![t_{s_2}]\!]_g^* = \{(\text{plato, wasBornIn, athens}), (\text{leibniz, wasBornIn, leipzig}),$$
$$(\text{goedel, wasBornIn, brno})\}$$

1.3.2 Computing Statistics

The statistics are computed for each triple t of a given triple-store. First, the set of schema triples, which include the given triple, t, in their natural interpretations $[\![]\!]^*$ must be determined. Second, the corresponding counters of the computed schema triples are updated for all different keys that can be formed from t. We define three different procedures for the computation of the schema triples for a given triple t. Only the main ideas of the first two procedures, which are integrated into the third procedure for the computation of statistics, are presented here. The third procedure, which is used as a practical algorithm for the computation of the statistics, is presented in the sequel. The detailed presentation of the three procedures including the details about different ways of counting the triples as well as the experimental evaluation of the methods is given in [31].

The first procedure computes the statistics of the stored schema graph. For a given triple t with the predicate $t[P]$, the set of more general predicates, with respect to the relationship rdfs:subPropertyOf, is computed first. For each of the predicates obtained in this way, the domains and the ranges are determined by using the predicates rdfs:domain and rdfs:range. All possible schema triples are generated from the computed sets.

The second procedure computes the statistics for all possible schema triples. Given a triple t, the classes of the components $t[S]$ and $t[O]$ are computed first. These two sets are further closed by using relationship rdfs:subClassOf. In addition, the procedure closes the set, $\{t[P]\}$, by means of the relationship, rdfs:subPropertyOf. The Cartesian product of the three sets is used to enumerate all the schema triples that are used for recording the statistics.

The edges (schema triples) of the stored schema graph are used as reference points in the third procedure. The procedure combines the first and second procedures. First, it determines all possible schema triples of a given triple $t = (s, p, o)$ by computing the classes of components s and o, and the super-properties of the component p, in the same way as in the second procedure. Next, similarly to the first procedure, it uses the stored schema as the basis for the definition of the area composed of the schema triples that are taken into account for the computation of the statistics. The schema used for the statistics is defined as the intersection of the set of schema triples that are the types of t with the set of schema triples located some levels around the stored schema. Let us now present the third procedure for the computation of triple-store statistics given in Algorithm 1 in more detail.

Algorithm 1 Compute the statistics for a given triple t

1: **function** STATISTICS($t = (s, p, o)$: triple; u, l : integer)
2:    $g_p \leftarrow \{p\} \cup \{t_p | (p, \text{rdfs:subPropertyOf+}, t_p) \in g\}$
3:    $g_s \leftarrow \{t_s | \text{is\_class}(s) \wedge t_s = s \vee \neg \text{is\_class}(s) \wedge (s, \text{rdf:type}, t_s) \in g\}$
4:    $g_o \leftarrow \{t_o | \text{is\_class}(o) \wedge t_o = o \vee \neg \text{is\_class}(o) \wedge (o, \text{rdf:type}, t_o) \in g\}$
5:    $g_s \leftarrow g_s \cup \{c_s' | c_s \in g_s \wedge (c_s, \text{rdfs:subClassOf+}, c_s') \in g\}$
6:    $g_o \leftarrow g_o \cup \{c_o' | c_o \in g_o \wedge (c_o, \text{rdfs:subClassOf+}, c_o') \in g\}$
7:    $d_p \leftarrow \{c_s | (p, \text{rdfs:domain}, c_s) \in g\}$
8:    $r_p \leftarrow \{c_s | (p, \text{rdfs:range}, c_s) \in g\}$
9:    $s_s \leftarrow \{c_s | c_s \in g_s \wedge \exists c_s' \in d_p ((c_s \succeq c_s' \wedge \text{DIST}(c_s, c_s') \leq u) \vee$
10:                              $(c_s \preceq c_s' \wedge \text{DIST}(c_s, c_s') \leq l))\}$
11:    $s_o \leftarrow \{c_o | c_o \in g_o \wedge \exists c_o' \in r_p ((c_o \succeq c_o' \wedge \text{DIST}(c_o, c_o') \leq u) \vee$
12:                              $(c_o \preceq c_o' \wedge \text{DIST}(c_o, c_o') \leq l))\}$
13:    **for all** $c_s \in s_s, p_p \in g_p, c_o \in s_o$ **do**
14:        UPDATE- STATISTICS($(c_s, p_p, c_o), t$)

The parameters of the procedure STATISTICS are a triple, $t = (s, p, o)$, and two integer numbers u and l that define the area of the schema used for storing statistics. The number, u, specifies the number of levels above the stored schema and l specifies the number of levels below the stored schemata.

The first part of the algorithm, written in lines 1–6, determines all possible classes of s and o as well as all super-predicates of p. The second part of the algorithm, given in lines 7–8, determines the domains and ranges of the predicate, p, from the stored schema. Finally, the third part of the algorithm, given in lines 9–14, selects the schema triples l levels below and u levels above the stored schema from all possible schema triples that represent the types of (s, p, o).

We separately computed the classes to be used to enumerate the selected schema triples in lines 9–12. The set, s_s, is a set of classes, c_s, such that $c_s \in g_s$ and every c_s are either more general or more specific than some $c \in d_s$. Furthermore, the distance between c_s and c must be less than u when $c_s \succeq c$, or less than l when $c_s \preceq c$. The set s_o is defined similarly except that it includes the selected classes for the O component of the schema triples. The classes from s_s and s_o represent the domain and range classes of the schema triples selected for the schema of the statistics.

Finally, the statistic is updated in lines 13–14 for all schema triples that are generated by means of the Cartesian product of the sets, s_s, g_p, and s_o.

1.3.2.1   Retrieving Statistics of a Schema Triple

Let us suppose that we have the statistics of a triple-store computed by means of Algorithm 1. The statistics is based on the schema triples from u levels above the stored schema triples and l levels below the stored schema. The statistics are stored in the form of a key-value dictionary, where the key is a schema triple, and the value represents the number of its instances. The schema triples that represent the keys used for the definition of the statistics are referred to in the sequel as the *schema of the statistics* S_s.

Let tp be a triple-pattern and let t_{tp} be the type of tp. The query tp address the triples from the interpretation of t_{tp}, i.e., the triples $t \preceq t_{tp}$. When we want to retrieve the number of instances of a schema triple $t_{tp} = (s, p, o)$, we have two possible scenarios. The first is that the schema triple, t_{tp}, is stored in the dictionary. The statistics, in this case, is simply retrieved from the dictionary. This is the expected scenario, i.e., we expect that the statistics include all schema triples that can represent the types of triple-patterns.

The second scenario covers cases when the schema triple t_{tp} is either above S_s, below S_s, or, intervene with S_s. In these cases, the number of instances of the schema triple t_{tp}, i.e., the size of t_{tp}'s interpretation has to be approximated from the statistical data computed for the schema triples from S_s.

First, in the case that the type t_{tp} is above S_s, then the projection includes all schema triples from S_s that are more specific than t_{tp}. The interpretations of the projected schema triples from S_s include the triples addressed by the query tp. Second, in the case that the type t_{tp} is below S_s, then the projection includes one of the schema triples from S_s that are more general than t_{tp}. Note that all schema triples from S_s that are more general than t_{tp}, include in their interpretation also the interpretation of t_{tp}.

In all other cases, the t_{tp} intervene in some way with schema triples from S_s. Let us say that the schema triple from S_s is named t_s. The possible ways of intervening are as follows. First, one component of t_{tp} is more general than the corresponding component in t_s, and the other two are more specific. Second, two components of t_{tp} are more general than the corresponding components in t_ss, and the other one is more specific. In all these cases we can use the intervening schema triples from S_s to represent t_{tp}s.

1.4   Graph Partitioning Method

Parallel data processing is only possible if the data are partitioned into a range of servers. We would ideally expect that a query that addresses a large amount of data would be split into n queries running on different machines in parallel. Therefore, the data addressed by a query must ideally be split into n parts residing on n different machines. The method of distribution must know the structure of the space of queries to achieve such a distribution of data.

The structure of SPARQL queries is relatively simple if we compare them to the structure of queries based on predicate calculus, for example. The structure of triple-patterns can be observed through the types of triple-patterns. A type of a triple-pattern restricts the possible values of a triple-pattern to the interpretation of the type. The most general type of a triple-pattern may be (owl:Thing,rdf:Predicate,-

owl:Thing).[1] This schema triple is a type of a triple-pattern (?x, ?y, ?z)—it addresses the complete triple-store. All other triple-patterns have a type that is more specific than the most general triple-pattern type. More specific types have more specific components of the schema triples representing the types. For instance, the type, (owl:Thing,<startsExistingOnDate>, xsd:Date), is more specific than (owl:Thing,-rdf:Property, owl:Thing), while the type, (<person>,<wasBornOnDate>, xsd:Date), is more specific than (owl:Thing,<startsExistingOnDate>, xsd:Date). The relationship is-more-specific (is-more-general) is defined on the basis of class and predicate taxonomies.

The leading idea in the method of distribution is to use a partially ordered set of schema triples to direct the method of distribution. Schema triples are enumerated from the most general toward more specific schema triples (types). Schema triples that have extensions of the appropriate size are chosen as units of the distribution, which are referred to as *fragments*. Therefore, we chose fragments in such a way that more general schema triples (types) are represented by a set of more specific fragments (schema triples). The level of generality of a triple-pattern type dictates the number of fragments used for the distribution of the extension of a given type.

In presenting the method of graph distribution, we adopted the view of data distribution described by Ozsu and Valduriez [25]. The design of data distribution is comprised of two phases: (1) the partitioning of the database and (2) the placement (allocation) of the partitions to the distributed data servers. The result obtained from the partitioning phase is a set of fragments which are of an appropriate size to serve as the units of the distribution. The placement phase of the data distribution method determines for each fragment the partitions stored on the separate data servers where the fragments are allocated.

1.4.1 Semantic Distribution

A method of *semantic distribution* is an approach to distributing graphs that include the conceptual schema of the graph. The complete graph is partitioned by first constructing the schema graph and then partitioning the schema graph of a triple-store into n subgraphs. The statistics of a triple-store is used to estimate the size of the generated partitions.

We identify three types of elements from the schema graph that can be used as the basis for partitioning. First, we can use predicates to partition triple-store into n roughly equal sized subgraphs. We refer to this type of distribution as a *predicate-based* distribution. This was our first implementation because it is simple. The graph partitioning algorithm based on the predicates is presented in more detail in Sect. 1.4.1.1.

---

[1]The most general type depends on the design of the metadata included in a given triple-store (knowledge graph).

Second, the distribution of the triple-store can be defined on the basis of the S part of the triples. This gives rise to the object-oriented view of the triple-store and its distribution. The triples are partitioned by the type of the S part of the triple. The partitions are assigned to the selected classes of the triple-store ontology. The selected classes can be seen as a border in the class-based ontology where the size of the interpretations of the selected classes is appropriate. This type of distribution procedure is called *class-based* distribution. It is further described in Sect. 1.4.1.2.

The third type of semantic partitioning algorithm is based on the edges of the schema graph. This is the most general type of distribution procedure since it subsumes predicate-based as well as class-based methods of distribution. The taxonomy defined on the set of edges turns out to be more appropriate for the definition of the partitioning algorithm than the predicates or the classes are. As the partially ordered set (poset) of edges is defined by means of the posets defined on the S, P, and O components of the edges, the schema graph edges form a finer granularity for the space of triples ordered by the interpretations of the edges. Furthermore, the edges of the schema graph represent the possible types of triple-patterns. The interpretations of the edges are the targets of the queries. A description of the partitioning algorithm based on the schema graph edges is given in Sect. 1.4.1.3.

1.4.1.1   Property-Based Distribution

The method of predicate-based graph partitioning was the first graph partitioning approach implemented in the frame of big3store. The method has two phases. The fragments of the triple-store to be distributed are determined in the first phase. The fragments are grouped into partitions that are allocated in the columns of the data server arrays in the second phase.

Each predicate represents all the triples that have a given predicate as the P part of the triples. We define the *skeleton* used for the distribution method in the first phase. The skeleton is defined by enumerating the predicates from the most general predicate, rdf:Property, toward more specific predicates; the predicates that represent the set of triples of appropriate size are selected. All the predicates, p, that are more general (with respect to the subPropertyOf relationship) than the predicates used as units of distribution are represented by all more specific predicates than p that are also simultaneously the units of the distribution. The predicates that are more specific than the predicates from the skeleton are represented by any more general predicate that is also an element of the skeleton.

We have the set of fragments that form the skeleton for the distribution, and we have to define where the fragments are going to be allocated. This is done in the second phase of the method by constructing a mapping from the set of selected predicates to the partitions stored in the columns of the data server's array. We implemented two partitioning functions in the frame of big3store triple-store.

The first partitioning function was constructed by randomly dispersing the skeleton predicates to the columns. The second partitioning function was generated by using a variant of the k-means clustering algorithm. The criterion for clustering was

defined as the strength of the connection between a predicate and a group of predicates formed by the use of the k-means algorithm. The strength of the connections
was estimated by employing the statistics of the predicates.

Dataset YAGO Version 3 has 127 predicates that have been defined to be almost
flat. In the data set, 15 predicates are defined as sub-properties of some other predicates. It is expected that the hierarchy of the predicates used in the knowledge bases
will become richer and will have multiple hierarchical levels, similarly to the class
hierarchies in recent knowledge graphs.

There were problems with the "big" predicates, which stood for a large number
of triples. The number of triples may happen to exceed the available storage capacity
of a data server. Therefore, the instances of a single predicate must be manually split
into partitions to be stored in a list of columns. The function that implements the
partitioning method then maps a big predicate to a list of partitions (columns).

The localization of the queries is easy with the predicate-based partitioning
method. The location of a triple-store fragment that is addressed by a triple-pattern
is determined from the predicate used in the triple-pattern. A query is broadcast to
all data servers when the triple-pattern does not specify the predicate.

1.4.1.2   Class-Based Distribution

The method of class-based partitioning distributes the triples on the basis of the S
parts of triples. The triples can be seen as the properties of objects represented by
the identifiers from the S parts of the triples. We thus obtain an object-oriented view
of the triple-store; a triple-store is represented using the classes and their instances.
The classes are organized into an inheritance hierarchy by means of relationship
rdfs:subClassOf.

We first design the fragmentation of a triple-store similarly to that of the method of
predicate-based distribution. The fragmentation has to be complete in the sense that
every triple from the triple-store is an element of at least one fragment. Furthermore,
the fragmentation is led by the idea that the space for data fragmentation must follow
the space for the data addressed by the queries. Thus, the queries that address the
small number of triples should find them in one fragment. The queries that address
large amounts of data should be distributed to maximize parallel execution of queries.

The classes that have the size of their interpretation less than m triples are selected
to be the fragments. They are the units of distribution that are allocated and stored
in one of the columns of the array of data servers. The set of classes to serve as the
skeleton for triple-store partitioning is computed by starting at the root of the class
hierarchy and descending level-by-level until the appropriate classes are reached.

After the skeleton graph is computed to contain a set, S_g, of selected classes serving
as the units of distribution (fragments), we assign the fragments to the partitions
implemented as columns of the server array. The assignment function determines
the distribution of the triple-store and also constrains the query evaluation with the
given distribution configuration. Classes from the skeleton graph are clustered by
using some clustering algorithm. The function distance() defines the shape of the

partitions. To obtain locality based partitioning the function distance() may measure the distance between two classes estimated using the statistics of the triple-store.

The queries are localized by using the S parts of the types of triple-patterns that constitute the query. Let us say the type of the triple-pattern, tp, is (c_s, p, c_o). The location of tp is determined by applying the partitioning function to $c_s \in S_g$. The class, c_s, is above the classes that form the skeleton, S_g. The class, c_s, is projected onto the skeleton, S_g. The intersection is computed between the set of classes that are more specific than c_s and S_g. The selected classes are then mapped to columns using the partitioning function.

The drawbacks of class-based distribution are large intersections among the interpretations of the classes, which means that a triple may be stored in more than one fragment. This property of the distribution function can be seen as an advantage. The intersections among classes are replicated in all intersecting fragments. Consequently, the fragments are self-contained. There is no need to copy the common part from some other fragment.

Another problem appears with the class-based partitioning when classes with the huge numbers of instances do not have a rich classification hierarchy. Consequently, they can not be split into smaller classes. Similar to the "big" predicates described in the previous section, "big" classes must also be manually split.

1.4.1.3 Edge-Based Distribution

The method of predicate-based partitioning is defined on the poset of predicates. The method of class-based partitioning is defined on the poset of classes. The definitions of the posets are presented in Sect. 1.3.1.1. Edge-based partitioning can be seen as a generalization of the predicate- and class-based partitioning methods. It is defined on the poset of the schema triples that was introduced in Sect. 1.3.1.2. The leading idea underlying the edge-based partitioning algorithm is the same as for algorithms based on predicates and classes.

The skeleton of the graph database is computed in the first phase. It comprises the edges, S_g, from the hierarchy of schema triples that have the interpretations of an appropriate size. The edges represent a border in the poset of edges, where the specializations of the most general edge $(\top, \top, \top)$ turn the sizes of their interpretations to a number less than n. The number n defines the maximal number of triples in a fragment. Therefore, the fragments are the interpretations of the schema triples of the "right" size. The procedure for the computation of the skeleton graph is described in Sect. 1.4.2.

The edges of the skeleton graph are clustered into partitions in the second phase. The algorithm that is used to generate the partitions dictates the plan to evaluate queries by fixing the locations of the fragments (edges). The algorithms for clustering a set of schema triples are presented in Sect. 1.4.3. The assignment of the partitions to the schema triples is stored as the *partitioning function*.

It is important to be able to locate the partitions (columns) for arbitrary schema triple, t_s, efficiently. When $t_s \in S_g$ the columns are obtained by directly applying the partitioning function to t_s. When t_s is above the skeleton, S_g, i.e., t_s is more general than some elements of S_g, then t_s is projected onto the skeleton, S_g, yielding a set, S_{t_s}. The partitions that hold the triples from the interpretation of t_s are obtained by taking the union of the results of the partitioning function applied to the elements of S_{t_s}. Finally, when t_s is below S_g, then the instances of t_s are in any $t'_s \in S_g$ such that $t_s \preceq t'_s$. Therefore, we have to identify one such t'_s and then apply the partitioning function to t'_s. A more detailed explanation of the function that computes the partitions for a given schema triple, t_s, is given in Sect. 1.4.4.

If we fix the edges used in the partitioning algorithm to $(\top, P, \top)$, where $\top$ denotes the top class, we obtain the predicate-based method. If we fix the edges to $(S, \top, \top)$, we obtain the method of class-based partitioning. Since the method of edge-based partitioning generalizes predicate- and class-based partitioning, we will now on focus on the edge-based partitioning approach. The algorithm to compute the skeleton of the graph database is presented in the next section.

1.4.2   Computing Skeleton Graph

The skeleton graph, S_g, is a schema graph that is composed of edges that are "strong" in the sense that they represent a large number of instances (triples). However, the number of instances of the edges of the skeleton graph must not be too large—the size of the edges has to be appropriate to serve as the units of distribution that we call *fragments*. The set of fragments defined by S_g contain a complete triple-store.

The edges of the skeleton graph are very carefully chosen. First, all edges, e, that are above S_g are represented in S_g by a subset, $S_e \subseteq S_g$, such that all the elements of S_e are more specific than e. Therefore, *all* instances of e are contained in fragments that are represented by the set, S_e. Second, the edges, e, that are below the skeleton, S_g, are represented by an edge, $e' \in S_g$, such that $e' \succeq e$. Therefore, the interpretation of e is included in the fragment represented by e'. Based on the above observations, we can construct a function that given a schema triple, t_s, returns the set of fragments. The returned fragments are the interpretations of the schema triples from S_g. They contain the complete interpretation of t_s. The localization procedure based on the above observations is detailed in Sect. 1.4.4.

Algorithm 2 Compute skeleton graph: declarative definition

1: **function** COMPUTE- SKELETON(g : graph)
2:     $t \leftarrow$ the most general schema triple in g
3:     $s \leftarrow \{t' \mid t' \in g \wedge t' \preceq t \wedge \text{STATISTICS}(t') < \text{Threshold} \wedge$
4:         $\nexists t'' \in g(t' \preceq t'' \preceq t \wedge \text{STATISTICS}(t'') < \text{Threshold}\}$
5:     **return** s

Algorithm 2 gives a declarative definition of the function COMPUTE- SKELETON. Let the symbol "Threshold" denote the maximal number of triples in the units of distribution. The fragments constructed by the algorithm are the largest fragments that are smaller than "Threshold". Further, let STATISTICS(t_s) be a function that returns the number of instances (triples) of the schema triple, t_s. The set of edges, e, of the skeleton graph, S_g, represent the schema triples of the graph, g, that are more specific than the most general schema triple, t, and are of the appropriate size. Their interpretation is smaller than "Threshold" and there is no other schema triple, t', that would be more general and also simultaneously of appropriate size.

The above algorithm is an abstraction of the method of computing the skeleton graph of the graph g. Let us now give a procedural definition of the function, COMPUTE- SKELETON.

The procedural version of the algorithm, COMPUTE- SKELETON, is presented in Algorithm 3. The enumeration procedure used to generate the specializations of the stored schema triples is based on the following formal definition of the poset of the schema triples. The schema triple, (s, p, o), is more general than the schema triples, (s', p', o'), if all of the components, s, o, and o, are more general than their corresponding pairs, s', p', and o'. Formally, $(s, p, o) \succeq (s', p', o') \iff s \succeq s' \land p \succeq p' \land o \succeq o'$. The enumeration procedure is encoded in the functions COMPUTE- SKELETON, COMPUTE- SKELETON- PRED, and COMPUTE- SKELETON- EDGE.

The enumeration starts with the most general stored schema triples that represent the stored conceptual schema of the triple-store. The function, COMPUTE- SKELETON, retrieves the stored schema triples in lines 2–6. It first retrieves the predicates, p, finds the domain and range classes, (c_s's and c_o's), of predicates, p, and then applies the function, COMPUTE- SKELETON- PRED, to each of the generated stored schema triples, (c_s, p, c_o).

The order of enumeration is very important in selecting the actual fragments (edges) of the skeleton graph. We started with the specialization of the S parts of the schema triples in our enumeration procedure. Hence, we provided an advantage to the method of class-based partitioning that specialized the class parts of the schema triples in an object-oriented way. The depth-first search used for the enumeration procedure, after the S parts had been specialized, started to specialize the O part of the schema triples. After the specializations of the O parts of the schema triples had been exhausted, the enumeration continued with the specializations of the P parts of the schema triples. Therefore, the most general steps in the enumeration procedure were performed through the P part. The procedure continued with specialization of the O part at the next level. Finally, the specializations of the O parts were searched at the last level.

The function, COMPUTE- SKELETON- PRED, is used to iterate through all the sub-predicates of the predicates, p. It starts by checking if the given parameter, (c_s, p, c_o), is the schema triple of the appropriate size by using the function, COMPUTE- SKELETON- EDGE, in line 11. If the schema triple is of the appropriate size, then the function, COMPUTE- SKELETON- PRED, completes with the return value, *true*. Otherwise, the specializations of p are retrieved and the function, COMPUTE- SKELETON- PRED, is recursively called to explore for each particular (one step) specialization of p

---

Algorithm 3 Compute skeleton graph: algorithmic definition

1: **function** COMPUTE- SKELETON(g : graph) $\rightarrow$ bool
2:    done $\leftarrow$ true
3:    $s_p \leftarrow \{p|(p, \text{rdf:type}, \text{rdfs:Property}) \in g\}$
4:    **for all** $p \in s_p$ **do**
5:       $d_p \leftarrow \{c_s|(p, \text{rdfs:domain}, c_s) \in g\}$
6:       $r_p \leftarrow \{c_o|(p, \text{rdfs:range}, c_o) \in g\}$
7:       **for all** $c_s \in d_p, c_o \in r_p$ **do**
8:          done $\leftarrow$ done **and** COMPUTE- SKELETON- PRED((c_s, p, c_o))
9:    **return** done
10: **function** COMPUTE- SKELETON- PRED((c_s, p, c_o) : schema-triple) $\rightarrow$ bool
11:    **if** COMPUTE- SKELETON- EDGE((c_s, p, c_o)) **then**
12:       **return** true
13:    **else**
14:       $s_p \leftarrow \{p'|(p', \text{rdfs:subPropertyOf}, p) \in g\}$
15:       done $\leftarrow$ true
16:       **for all** $p' \in s_p$ **do**
17:          done $\leftarrow$ done **and** COMPUTE- SKELETON- PRED((c_s, p', c_o))
18:    **return** done
19: **function** COMPUTE- SKELETON- EDGE((c_s, p, c_o) : schema-triple) $\rightarrow$ bool
20:    **if** STATISTICS((c_s, p, c_o)) < Threshold **then**
21:       ADD- TO- SKELETON((c_s, p, c_o))
22:       **return** true
23:    **else**
24:       $s_s \leftarrow \{c_s'|(c_s', \text{rdfs:subClassOf}, c_s) \in g\}$
25:       $s_o \leftarrow \{c_o'|(c_o', \text{rdfs:subClassOf}, c_o) \in g\}$
26:       **if** done $\leftarrow |s_s| \neq \emptyset$ **then**
27:          **for all** $c_s' \in s_s$ **do**
28:             done $\leftarrow$ done **and** COMPUTE- SKELETON- EDGE((c_s', p, c_o))
29:       **if** done **then**
30:          **return** true
31:       **if** $|s_o| = \emptyset$ **then**
32:          **return** false
33:       **else**
34:          done $\leftarrow$ true
35:          **for all** $c_o' \in s_o$ **do**
36:             done $\leftarrow$ done **and** COMPUTE- SKELETON- EDGE((c_s, p, c_o'))
37:          **return** done

---

in lines 14–18. The results of all the recursive calls are combined by using the logical conjunction. Note that all specializations represent the candidates for the fragments and that all specializations must be selected as the fragments (units of distribution) to complete the function, COMPUTE- SKELETON- PRED, with the return value, *true*.

A call of the function, COMPUTE- SKELETON- EDGE, with the parameter, (c_s, p, c_o), has a fixed value, p. The function explores the subclass hierarchy of c_s and continues with the exploration of the specializations of c_o. The predicate, p, is fixed during the evaluation of COMPUTE- SKELETON- EDGE.

The function, COMPUTE- SKELETON- EDGE, starts by checking if the size of the interpretation of the given parameter schema triple, (c_s, p, c_o), is less than "Threshold". Here, the function completes with the return value, *true*, which means that a given schema triple is accepted as the fragment (edge) of the graph skeleton. When the statistics for the interpretation of the schema triple (c_s, p, c_o) is larger than "Threshold", the specializations of the given schema triples are determined and evaluated with the recursive calls of the function, COMPUTE- SKELETON- EDGE, in lines 24–34. As has already been discussed, the search of the function, COMPUTE- SKELETON- GRAPH, starts with the specializations of c_s and then continues with the specializations of c_o in a depth-first manner. Note that only one step specializations are used in one recursive evaluation of the function. Furthermore, note that all recursive evaluations of the function, COMPUTE- SKELETON- EDGE, on the generated specializations of (c_s, p, c_o) must succeed so that the current invocation of COMPUTE- SKELETON- EDGE succeeds with the return value, *true*.

We omitted some optimizations from the presentation of Algorithm 3. The enumeration procedure marks each of the schema triples that are in the process of specialization as *visited* to avoid repetition in the enumeration process. Before we invoke any of the recursive calls of the functions, COMPUTE- SKELETON- PRED and COMPUTE- SKELETON- EDGE, on the generated specializations (schema triples), we check if the generated schema triple has already been processed.

1.4.3 Clustering Skeleton Graph

The skeleton graph of a triple-store is composed of edges (schema triples) that have an extension of the selected size. The schema triples from the skeleton graph are the fragments that serve as the units of distribution. The fragments are allocated at the column defined by the distribution function that maps the fragments to the partitions (columns).

The clustering of the skeleton graph is based on the statistics of the triple-store. We clustered the edges of the skeleton graph since we relied on the statistics of the edges. The distance function estimates the strength of the connection between two edges. The edges may be directly connected. Here, the strength of the connection is defined to be the estimated size of the join between the extensions of the directly connected edges. If there is a path from the one edge to the other, the strength of the connection is estimated as the size of the query defined by the joined extensions of edges from the path. Finally, if there is no path from one edge to the other, the strength of the connection is zero.

The strength of the connection between an edge and a group of edges is the average of the strengths of the connections between an edge and all edges from the group. We define the strength of a connection of an edge with the empty set of edges that is the size of the extension of the edge. Finally, the strength of the connectivity of a group of edges is an average of the strengths among all possible pairs of edges.

Note that the path that connects the two edges can include the relationship rdfs:subClassOf or rdfs:subPropertyOf. These relationships are defined between the classes or between the predicates, and their interpretation is an empty set. Thus, they stand solely for the definition of the structural relationship between the classes that imposes inheritance. The more specific classes inherit the predicates of the more general classes. Similarly, the more specific predicates can stand in place of more general predicates. We have omitted from the path between two edges the schema triples defined with the predicates rdfs:subClassOf and rdfs:subPropertyOf. Apart from this, we have used the same procedure as when a path does not include the schema triples with those predicates.

Algorithm 4 Compute the distance between two schema triples

1: **function** DISTANCE(e_1, e_2 : schema-triple) $\rightarrow$ schema-triple-list
2:     $p \leftarrow$ SHORTEST- PATH(e_1, e_2)
3:     **if** p is empty **then**
4:         **return** 0
5:     $p \leftarrow$ REMOVE- TAXONOMY(p)
6:     **return** SIZEOF- PATH- QUERY(p)

Let us now present the algorithm for estimating the distance between two edges in more detail. The pseudocode is given in Algorithm 4. The function, DISTANCE, is based on the shortest path between the edges. The shortest path is used to compute the approximation of the strength of the connection between the edges in line 2. The function SHORTEST- PATH(e_1, e_2: schema-triple) $\rightarrow$ schema-triple-list, computes the shortest path between the edges, e_1 and e_2. It returns a list of edges (schema triples). Each edge is linked through some component with its predecessor and successor in the returned list.

The schema triples with the predicates, rdfs:subClassOf and rdfs:subPropertyOf, are removed from the path by using the function, REMOVE- TAXONOMY(p) in line 5. The size of the path query defined by the shortest path p is computed by function SIZEOF- PATH- QUERY(p). It estimates the size of the sequence of joins defined on the extensions of the edges from the path p. The joins in the path query are estimated by using the statistics of the triple-store. The function, DISTANCE(e_1, e_2), gives a positive value that increases as the edges, e_1 and e_2, are more strongly connected. Given the edges, e_1 and e_2, that are not connected, the function returns zero.

A variant of the k-means algorithm was used for the initial experiments with the clustering. However, any of the well-known clustering algorithms could have been used [34, 35]. The modified k-means algorithm starts with n empty partitions. Recall that the distance between an edge and an empty set is not zero, but the size of the extension of the added edge. The distances are computed for each of the edges from the skeleton graph to the existing clusters during the first iteration of the algorithm. The edge is inserted into the cluster with the strongest connection to the added edge.

We do not compute the new means in the further iterations but leave the partitions exactly as they are. The edges that are the elements of the partitions are the actual representations of the partitions. The distances of each edge of the skeleton graph to the partitions are computed in each new iteration. The edges are placed to the selected partitions. If one of the edges changes its partition, the iteration continues. If there are no more changes during the reassignment of the edges to the partitions, the iteration stops.

The algorithm for clustering the skeleton graph presented in Algorithm 5 is a variant of the k-means algorithm. Let us now describe the setup for the algorithm. First, the skeleton graph is represented as a set, S, of the schema triples that is passed as the parameter of CLUSTER- SKELETON. Further, the arrays, G and H, store n clusters that include edges (schema triples). The elements of G and H are indexed by $i \in C = \{1, \ldots, n\}$, where n is the number of partitions.

Algorithm 5 Clustering skeleton graph

1: **function** CLUSTER- SKELETON(S : schema-triple-list) $\rightarrow$ cluster-array
2:    INITIALIZE(G)
3:    **for all** $t \in S$ **do**
4:       $k \leftarrow i \mid \exists i \in C \land \neg$CLUSTER- FULL($G[i]$) $\land$
5:             $\nexists j \in C(j \neq i \land \neg$CLUSTER- FULL($G[j]$)$\land$
6:                DISTANCE($t, G[j]$) $<$ DISTANCE($t, G[i]$))
7:       CLUSTER- ADD($t, G[k]$)
8:    fix $\leftarrow$ true
9:    **while** fix **do**
10:      $H \leftarrow G$
11:      INITIALIZE(G)
12:      **for all** $t \in S$ **do**
13:         $k \leftarrow i \mid \exists i \in C \land \neg$CLUSTER- FULL($G[i]$) $\land$
14:            $\nexists j \in C(j \neq i \land \neg$CLUSTER- FULL($G[j]$)$\land$
15:               DISTANCE($t, H[j]$) $<$ DISTANCE($t, H[i]$))
16:         CLUSTER- ADD($t, G[k]$)
17:         **if** $t \notin H[k]$ **then**
18:            fix $\leftarrow$ false
19:    **return** G

The function, CLUSTER- SKELETON, uses a set of functions that are defined on clusters. The function, CLUSTER- FULL($G[i]$), returns true, if cluster, i, in array, G, is full. It returns false otherwise. Next, the function, CLUSTER- ADD($t, G[i]$)), adds a schema triple, t, to the cluster, $G[i]$. Finally, the function, DISTANCE($t, G[i]$), computes the distance between the schema triple, t, and a cluster of schema triples $G[i]$ by summing up the distances between t and all schema triples from $G[i]$.

The algorithm for clustering the skeleton graph starts by constructing the initial configuration for clusters in lines 2–7. First, all clusters stored in the array, G, are initialized to an empty set. Recall that the distance between an edge and an empty cluster is an extension of the edge. Therefore, the clustering algorithm behaves correctly when the initial clusters are empty. We find the most appropriate cluster, k, for each schema triple, $t \in S$, by using the function, DISTANCE, defined between a schema triple and a cluster. The sizes of the clusters are set in such a way that there is sufficient room to store the complete triple-store. The cluster, k, is chosen in lines 4–6 so that the distance between t and all elements of cluster k are minimal. In addition, cluster, k, must have enough room to store the complete fragments of t.

The initial configuration of clusters represented by the array, G, is refined in lines 8–18. The cluster membership is refined by recomputing the clusters. The closest partition, $i \in C$, for each $t \in S$ is determined by minimizing the function, DISTANCE, in lines 13–15. The variable fix is set to false when the cluster membership has changed. This means that the algorithm will recompute the mapping from the edges (fragments) of the skeleton graph to the partitions (clusters) again.

1.4.4 Triple-Pattern Localization

The type of the triple-pattern, tp, was defined in Sect. 1.3.1.3 as a schema triple, t_{tp}, such that the interpretation of tp was included in the interpretation of t_{tp}. This can be represented formally as follows:

$$\llbracket tp \rrbracket_g \subseteq \llbracket t_{tp} \rrbracket_g^*.$$

Given a partitioning of the triple-store defined by the mapping, P : schema-triple $\rightarrow$ partition-list, the locations where the triple-pattern is executed are determined by first inferring the type of tp. It maps obtained schema triples to the list of partitions.

The type of the triple-pattern, tp, is computed by employing the stored schema of the triple-store. First, we use the types of the individual identifiers, I_i, that are stored in a triple-store by means of the predicate, rdf:type. We use stored schema triples based on the predicates, rdfs:Domain and rdfs:Range, to determine the types of the S and O parts of tp, where the P part of tp is given. If the P part of tp is not given, then the inferred type is very general in most cases.

There are various interactions among the types of schema triples. For example, let $tp_1 = (?x, p_1, o_1)$ and $tp_2 = (?x, p_2, o_2)$ be triple-patterns. The type information of the individual identifiers and the stored schema triples based on rdfs:Domain and rdfs:Range are used to infer the two schema triples that represent the types of tp_1 and tp_2. Let us say that the schema triples are (t_1, p_1, t_{o_1}) and (t_2, p_2, t_{o_2}). If there is a subclass relationship among the classes, t_1 and t_2, we can use the more specific class as the type of both S components. A detailed definition of the type inference algorithm was not the subject of the present research. A description of the type inference algorithm for the graph-patterns is given in [30].

Algorithm 6 Retrieve partitions of a given schema triple

1: **function** RETRIEVE- PARTITIONS($t = (s, p, o)$: schema-triple) $\rightarrow$ partition-list
2:     **if** $\exists t \in S$ **then**
3:         **return** $\{P(t)\}$
4:     **else if** $\exists t_s \in S : t_s \preceq t$ **then**
5:         **return** $\{P(t_s) | t_s \in S \wedge t_s \preceq t\}$
6:     **else if** $\exists t_s \in S : t_s \succeq t$ **then**
7:         **return** $\{P(t_s) | t_s \in S \wedge t_s \succeq t\}$
8:     **else if** $\exists t_s \in S : t_s[1] \preceq s \wedge t_s[2] \succeq p \wedge t_s[3] \succeq o$ **then**
9:         **return** $\{P(t_s) | t_s \in S \wedge t_s[1] \preceq s \wedge t_s[2] \succeq p \wedge t_s[3] \succeq o \wedge \text{ONE}(t_s[2:3])\}$
10:     **else if** $\exists t_s \in S : t_s[1] \succeq s \wedge t_s[2] \preceq p \wedge t_s[3] \succeq o$ **then**
11:         **return** $\{P(t_s) | t_s \in S \wedge t_s[1] \succeq s \wedge t_s[2] \preceq p \wedge t_s[3] \succeq o \wedge \text{ONE}(t_s[1, 3])\}$
12:     **else if** $\exists t_s \in S : t_s[1] \succeq s \wedge t_s[2] \succeq p \wedge t_s[3] \preceq o$ **then**
13:         **return** $\{P(t_s) | t_s \in S \wedge t_s[1] \succeq s \wedge t_s[2] \succeq p \wedge t_s[3] \preceq o \wedge \text{ONE}(t_s[1:2])\}$
14:     **else if** $\exists t_s \in S : t_s[1] \preceq s \wedge t_s[2] \preceq p \wedge t_s[3] \succeq o$ **then**
15:         **return** $\{P(t_s) | t_s \in S \wedge t_s[1] \preceq s \wedge t_s[2] \preceq p \wedge t_s[3] \succeq o \wedge \text{ONE}(t_s[3])\}$
16:     **else if** $\exists t_s \in S : t_s[1] \preceq s \wedge t_s[2] \succeq p \wedge t_s[3] \preceq o$ **then**
17:         **return** $\{P(t_s) | t_s \in S \wedge t_s[1] \preceq s \wedge t_s[2] \succeq p \wedge t_s[3] \preceq o \wedge \text{ONE}(t_s[2])\}$
18:     **else**
19:         **return** $\{P(t_s) | t_s \in S \wedge t_s[1] \succeq s \wedge t_s[2] \preceq p \wedge t_s[3] \preceq o \wedge \text{ONE}(t_s[1])\}$

Given a schema-triple, t, the function, RETRIEVE- PARTITIONS, returns the list of partitions where the triples that are the interpretations of t are stored. The function, RETRIEVE- PARTITIONS, is presented in Algorithm 6. We assumed that the skeleton graph was stored globally as the list of schema triples, S. Furthermore, we assumed that the mapping between the schema triples of the skeleton graph, S, to the partition numbers was given as the globally defined function, P : schema-triple $\rightarrow$ partition-list.

The function, RETRIEVE- PARTITIONS, is defined in certain cases. The first case is given in lines 2–3, where the parameter schema triple, t, is an element of S. Hence, the partition of the schema triple, t, is defined as $P(t)$.

The schema triple, t, in the next two cases that are presented in lines 4–7 is either above or below the skeleton graph, S. A schema triple from S is more specific than t in the former case. Set S includes a schema triple that is more general than t in the latter case. If the schema triple, t, is above S, t is "projected" to S. The selected schema triples are the specializations of t that are simultaneously the elements of S. When

t is below S in the opposite case, the selected schema triples are the generalizations of t that are also the elements of S. The selected schema triples are mapped to a list of partitions by the use of the function P.

The schema triple, $t = (s, p, o)$, intervenes with S in six different ways in the following six cases that are given in lines 8–19. Let us consider that a schema triple, $t_s \in S$, was given. The first way of intervening is that s is more general than $t_s[1]$, while p and o are more specific than the corresponding components of t_s. The second way of intervening is that p is more general than $t_s[2]$, while s and o are more specific than the corresponding components of t_s. The remaining cases cover the remaining four types of intervention between t and S.

Let us now present the method for selecting schema triples from S. The position of t's components represented by $\preceq$ relationship was used in the determination process of the method. First, if a component of t, say c, is more general than the corresponding component of some schema triple from S, all schema triples from S include more specific components than c to store the interpretation of t. Second, if a component, c, of t is more specific than the corresponding component of the selected schema triples, t_s, from S, the interpretation of t is included in the schema triples, t_s, that include a single value that is more general than c.

We can observe that the queries retrieve the results for the six cases in lines 9, 11, 13, 15, 17, and 19:

(a) They use implicit universal quantification, when c is more general than the corresponding components of the selected schema triples from S.
(b) They restrict the components in c's place in the selected schema triples from S to a single value.

We use a function, ONE, to be able to restrict the selected schema triples to those that include a single value of the components instead of c. The range of the components that are required to have just one value in the resulting set of schema triples is specified by using a standard list of range operations (e.g., Python's list operations). For example, given a triple, t_s, the expression, $t_s[1 : 2]$, returns a list of components in the range from one to two.

1.4.5 Related Work

Relational approaches to the distribution of a database have been described in Sect. 1.1. The distribution of a triple-store is a simpler problem due to the simplicity and uniformity of the RDF graph data model in comparison to the relational data model. The proposed method of partitioning a triple-store can be compared to the range-based partitioning. However, ranges are in a triple-store defined through the poset of schema triples. The most general schema triple represents a range that encompasses complete triple-store. A range that is defined by a schema triple can be split into a number of more specific subranges that are defined by more specific schema triples.

The distribution in triple-store systems is often implemented using hash partitioning [13, 14, 22, 23, 37]. Most implementations use the hash function on either S, P, or O components of triples, depending on the type of storage structure that is distributed. For example, the S part of triples is used for hashing to implement the SPO index. This mechanism keeps closely related data stored on one data server.

Distributed hashing is further extended by Lee and Liu [19]. Semantic information is used for directing the hashing procedure. Triple groups are generated such that they are based on the combinations of S, P, and O parts of triples. Baseline hash partitions are generated from the triple groups by clustering them together in partitions in this phase. The use of the triple groups is appropriate for star-shaped queries but is not efficient for other types of queries. The hop-based triple replication was proposed for this reason. Semantic hash partitions are defined to maximize intra-partition query processing.

Semantic-aware partitioning of RDF graphs is proposed by Xu et al. [36]. The algorithm classRank computes top-k significant classes, which is inspired by the PageRank algorithm. The schema graph defined on selected classes is used as heuristics for the proposed graph partitioning algorithm. It has been shown that the algorithm outperforms the state-of-the-art graph partitioning methods.

The multilevel graph partitioning schemes were proposed by Karypis and Kumar [17]. The graph partitioning method was efficiently implemented in the programming environment METIS. The proposed class of graph partitioning algorithms reduces the size of a graph by collapsing the vertices to obtain an abstract graph. This graph is then partitioned and finally un-coarsened to enumerate the members of the graph partitions. The multilevel graph partitioning schemes have proved to be very efficient in practical applications, such as, finite element methods, linear programming, VLSI, and transportation.

Finally, spectral graph partitioning methods give excellent results in practice [15, 26]. A spectral graph partitioning is based on the eigenvalues and eigenvectors of the Laplacian matrix of a graph. The main problem with the method is its computational complexity. The execution time can be reduced by multilevel spectral bisection (MSB) algorithm that is an order of magnitude faster without any loss in the quality of the edge-cut.

1.5 Empirical Evaluation

This section describes empirical results obtained from the experiments using a prototype big3store implementation. We have implemented three distribution methods. The first was a random distribution method that stored each triple into a randomly selected column. It was prepared to obtain the baseline efficiency. The second was a property-based distribution method that stored triples that had the same predicate in the same column. The third was a manually controlled method of property-based distribution. While it was the same as the second approach, the mapping from properties to columns collects similar properties to the same column. We present the execution time for the fixed benchmark query set for each distribution method. We describe the benchmark queries in the following subsection.

1.5.1 Benchmark Environment

The system for executing benchmark tasks consists of five server machines that all had the same physical specifications. Each server had two 2.9 GHz Xeon E5-2960 central processing units (CPUs) (32 cores) and 256 GB of random-access memory (RAM). It invokes one Erlang interpreter process on a server for operating the front or data server. The benchmark configuration used in this research invokes a *front server* and a *data server* on one server machine. It invoked a *data server* on each of the other server machines. It used every data server for storing its column. There was no *rows* replication in the experiments shown below. We used all triples of the YAGO2s [16] data set. They were distributed into five columns and stored in five data servers.

We briefly describe the benchmark queries shown in Fig. 1.5. We present empirical results executing these queries on the whole YAGO2 dataset in the next subsection. Query Q1 tries to find all triples simply having property `<startedOnDate>`. It returns nine triples. Query Q2 tries to find all graphs of triples sharing the same subject that has `<startedOnDate>` and `<endedOnDate>` properties. It returns one graph. Query Q3 compares `<Slovenia>` and `<Japan>` by using the same predicate in triple-patterns. While it is similar to query Q2, query Q3 returns 241,596 graphs. This query produces a large number of intermediate results. Data servers send those results as inter-process messages each other.

Queries Q9, Q4, and Q5, respectively, correspond to YAGO queries A1, B1, and B2 from [20]. Because YAGO and YAGO2s [16] had different schema structures, those queries were rewritten to have similar meanings. Query Q9 returns graphs that describe scientists who were born in a city in Switzerland, and who has an academic adviser who was born in a city in Germany. It returns 15 graphs. Query Q4 returns pairs of actors who acted in the same film and live in the same city in England. It returns one graph. Query Q5 returns all married couples who were born in the same city. It returns 714 graphs.

Query Q6 returns the creation dates of all things that are classified as wordnet_language and created by Ericsson. It returns three graphs. Query Q7 tries to find graphs that describe Japanese computer scientists, who had created a programming language. It returns six graphs. Query Q8 tries to find graphs that describe developers who had created Erlang related software. It returns 35 graphs.

Query Q10 was constructed to test circular queries. It returns 33 graphs. While the current version of query Q10 is specific and can be executed quickly on the YAGO2 dataset, a more practical circular query could be constructed by removing the first two triple-patterns of `<Tim_Burton>` and `<Johnny_Depp>`.

```
SELECT * WHERE {
  ?sbj <startedOnDate> ?obj.
}

Query Q1

SELECT * WHERE {
  ?sbj <startedOnDate> ?obj1.
  ?sbj <endedOnDate>   ?obj2.
}

Query Q2

SELECT * WHERE {
  <Slovenia> ?prd ?obj1.
  <Japan>    ?prd ?obj2.
}

Query Q3

SELECT * WHERE {
  ?a1 <actedIn>      ?movie.
  ?a2 <actedIn>      ?movie.
  ?a1 <livesIn>      ?c1.
  ?c1 <isLocatedIn> <England>.
  ?a2 <livesIn>      ?c2.
  ?c2 <isLocatedIn> <England>.
}

Query Q4

SELECT * WHERE {
  ?p1 <isMarriedTo> ?p2.
  ?p1 <wasBornIn> ?city.
  ?p2 <wasBornIn> ?city.
}

Query Q5

SELECT * WHERE {
  <Ericsson> <created> ?pl.
  ?pl rdf:type <wordnet_language>.
  ?pl <wasCreatedOnDate> ?dt.
}

Query Q6
```

```
SELECT * WHERE {
  ?p rdf:type
  <wikicategory_Japanese_computer_scientists>.
  ?p <created> ?o.
  ?o rdf:type <wordnet_programming_language>.
}

Query Q7

SELECT * WHERE {
  <Erlang> <linksTo> ?pl.
  ?pl rdf:type <wordnet_software>.
  ?dev <created> ?pl.
}

Query Q8

SELECT * WHERE {
  ?p <hasGivenName>  ?gn.
  ?p <hasFamilyName> ?gn.
  ?p rdf:type <wordnet_scientist>.
  ?p <wasBornIn>        ?c1.
  ?c1 <isLocatedIn> <Switzerland>.
  ?p <hasAcademicAdvisor> ?a.
  ?a <wasBornIn>        ?c2.
  ?c2 <isLocatedIn> <Germany>.
}

Query Q9

SELECT * WHERE {
  <Tim_Burton> <directed> ?movie1.
  <Johnny_Depp> <actedIn> ?movie1.
  ?p1 <directed> ?movie1.
  ?p2 <influences> ?p1.
  ?p3 <actedIn> ?movie1.
  ?p3 <actedIn> ?movie1.
  ?p4 ?prd1 ?p3.
  ?p4 <actedIn> ?movie2.
  ?p1 ?prd1 ?p4.
}

Query Q10
```

Fig. 1.5 Benchmark queries

1.5.2 Benchmark Results on Different Distribution Algorithms

We have implemented three distribution algorithms in a big3store prototype system so far. The first one is a random distribution algorithm called *random*, the second one is a property-based distribution algorithm based on the vertical partitioning model called *p-based*, and the third one is a manually controlled property-based distribution algorithm called *mcp-based*.

Fig. 1.6 Query-tree processes of Q5 in p-based algorithm

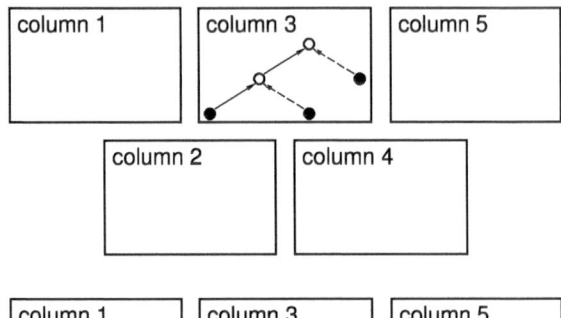

Fig. 1.7 Query-tree processes of Q9 in p-based algorithm

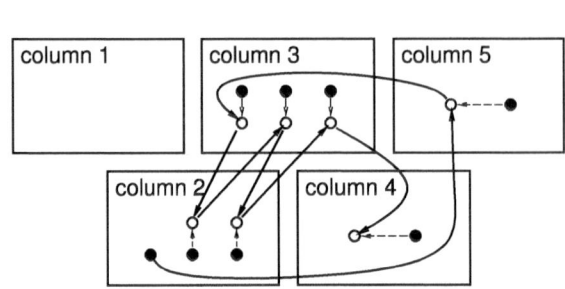

The p-based algorithm stores triples that have specific RDF predicates into the corresponding columns. In order to execute a triple-pattern with a constant predicate, the system runs query execution processes only on the data server that stores the given predicate. Figures 1.6 and 1.7 illustrate the query execution processes of big3store on the data partitioned by the p-based algorithm. The open circles stand for join operation processes and the closed circles denote pattern matching processes that access stored triples using indices. Only column 3 rectangle includes circles in Fig. 1.6. This means that big3store concentrates all query execution processes of query Q5 on column 3. There is no need for executing pattern match processes on columns 1, 2, 4, and 5, because triples having <isMarriedTo> and <wasBornIn> predicates are only stored in column 3 data server by the p-based algorithm. There is no circle placed in column 1 rectangle in Fig. 1.7. Each of column 2 and 3 rectangles includes three black circles and each of column 4 and 5 rectangles includes a black circle. This means that big3store invokes each pattern matching process of query Q9 only on one column. The initialization process of big3store query executor invokes a pattern matching process only on the corresponding column if the pattern has a constant predicate and p-based or mcp-based algorithms distributes the column data.

While the concentration of matching processes reduces the amount of message traffic between query execution processes, the other columns are idling in these query executions. Parallel executions of matching processes might improve the performance in some other queries. We implemented the random algorithm for comparing with the p-based and mcp-based algorithms. The random algorithm stores triples into columns at random. It is almost identical to hash-based distribution algorithms. The initialization process of big3store query executor invokes the processes of a matching

Table 1.1 Benchmark results executed in different distribution algorithms (in seconds)

Query	Random1	Random2	p-based	mcp-based
Q1	0.653	0.645	0.656	0.635
Q2	0.081	0.078	0.015	0.011
Q3	29.299	28.918	32.390	32.637
Q4	88.694	86.963	10.428	10.512
Q5	533.421	530.297	71.908	68.282
Q6	0.296	0.327	0.061	0.060
Q7	1.138	1.239	0.178	0.168
Q8	1.784	1.931	0.263	0.266
Q9	202.714	197.757	25.003	25.112
Q10	263.800	256.918	264.700	270.115

pattern on every column data servers if the random algorithm distributes the column data.

Table 1.1 lists the benchmark results executed on data columns distributed using random, p-based, and mcp-based algorithms. We have presented and explained queries Q1 to Q10 above. Column data distributed using the random 1 and 2 algorithms were different to each other. They used the same algorithm but different random seeds for selecting a column of each triple. The numbers show real execution time in seconds. There was no significant difference in query Q1. Because Q1 only returns 9 triples, message traffic cost did not affect the execution time. The p-based and mcp-based algorithms performed better than the random algorithms in query Q2. While the first triple-pattern of Q2 is identical to Q1, nested loops of the join operation produce much volume of inter-process messages in Q2. In executing queries Q5 and Q9 precisely described above, p-based and mcp-based significantly outperformed random. The results indicated that the p-based and mcp-based algorithms outperformed the random algorithm in most of the benchmark tasks.

1.6   Conclusion

We proposed a new method of graph partitioning. The method relies on the graph schema and statistics of a stored graph to construct a skeleton graph. The skeleton graph is constructed from the schema triples that have the interpretations of the appropriate size. Each of the skeleton graph's edges represents a fragment of the triple-store. The union of all fragments represents the complete triple-store. Then, the edges of the skeleton graph are clustered to obtain strongly connected partitions of the skeleton graph. The partitions of the original graph are defined as the extensions of the partitions of the skeleton graph.

There are a number of possible directions for further work. Some parameters that define the hardware environment related to the query distribution and execution have changed in the last two decades. First, a huge amount of main memory is often used instead of disk-based storage, which now plays a different role and is used to achieve durability in storage systems. Second, network bandwidth is continuously increasing, which raises questions about the definition of "strongly connected" partitions.

Consequently, it is reasonable to rethink the definitions of the partitions and distribution algorithms that were developed in the 1980s and 90s. They were designed in the context of relational distributed database systems. The graph data model is simple and provides a uniform representation of data at all semantic levels, so it might offer more appropriate solutions for developing automatic distribution methods in triple-stores.

Acknowledgements This work was partially supported by the Slovenian Research Agency ARRS Research program P1-0383.

References

1. Armstrong J (2013) Programming erlang: software for a concurrent world. Pragmatic Bookshelf
2. Baader F, Calvanese D, McGuinness D, Nardi D, Patel-Schneider P (2002) Description logic handbook. Cambridge University Press, Cambridge
3. Babu S, Herodotou H (2012) Massively parallel databases and mapreduce systems. Found Trendsin Databases 5(1):1–104
4. Blagojevic V et al (2016) A systematic approach to generation of new ideas for phd research in computing. Adv Comput 104:1–19
5. Chang F, Dean J, Ghemawat S, Hsieh WC, Wallach DA, Burrows M, Chandra T, Fikes A, Gruber RE (2008) Bigtable: a distributed storage system for structured data. ACM Trans Comput Syst 26(4)
6. Codd EF (1970) A relational model of data for large shared data banks. Commun ACM 13(6):377–387
7. DeWitt D, Gray J (1992) Parallel database systems: the future of high performance database processing. Commun ACM 36(6)
8. DeCandia G, Hastorun D, Jampani M, Kakulapati G, Lakshman A, Pilchin A, Sivasubramanian S, Vosshall P, Vogels W (2007) Dynamo: amazon's highly available key-value store. In: Proceedings of the twenty-first ACM SIGOPS symposium on operating systems principles, SOSP '07. ACM, New York, pp 205–220
9. Dong XL, Gabrilovich E, Heitz G, Horn W, Lao N, Murphy K, Strohmann T, Sun S, Zhang W (2014) Knowledge vault: a web-scale approach to probabilistic knowledge fusion. In: KDD-2014, KDD. ACM, New York
10. Garlik SH, Seaborne A, Prud'hommeaux E (2013) SPARQL 1.1 Query language. http://www.w3.org/TR/sparql11-query/
11. Graefe G (1993) Query evaluation techniques for large databases. ACM Comput Surv 25(2):73–169
12. Graefe G (2000) Dynamic query evaluation plans: some course corrections? IEEE Data Eng Bull 23(2):3–6
13. Harris S, Lamb N, Shadbolt N (2009) 4store: the design and implementation of a clustered rdf store. In: Proceedings of the 5th international workshop on scalable semantic web knowledge base systems

14. Harth A, Umbrich J, Hogan A, Decker S (2007) Yars2: a federated repository for querying graph structured data from the web. In: Aberer K, Choi K-S, Noy N, Allemang D, Lee K-I, Nixon L, Golbeck J, Mika P, Maynard D, Mizoguchi R, Schreiber G, Cudré-Mauroux P (eds) The semantic web. Springer, Berlin, pp 211–224
15. Hendrickson B, Leland R (1992) An improved spectral graph partitioning algorithm for mapping parallel computations. Technical report SAND92-1460, Sandia National Laboratories, Albuquerque
16. Hoffart J, Suchanek FM, Berberich K, Weikum G (2013) Yago2: a spatially and temporally enhanced knowledge base from wikipedia. Artif Intell 194:28–61. Artificial intelligence, wikipedia and semi-structured resources
17. Karypis G, Kumar V (1998) A fast and high quality multilevel scheme for partitioning irregular graphs. SIAM J Sci Comput 20(1):359–392
18. Karger D, Lehman E, Leighton T, Panigrahy R, Levine M, Lewin D (1997) Consistent hashing and random trees: distributed caching protocols for relieving hot spots on the world wide web. In: Proceedings of the twenty-ninth annual ACM symposium on Theory of computing, STOC '97. ACM, New York, pp 654–663
19. Lee K, Liu L (2013) Scaling queries over big rdf graphs with semantic hash partitioning. Proc VLDB Endow 6(14):1894–1905
20. Neumann T, Weikum G (2008) Rdf-3x: a risc-style engine for rdf. Proc VLDB Endow 1(1):647–659
21. Nitta K, Savnik I (2014) A distributed query execution method for rdf storage managers. In: Liebig T, Fokoue A (eds) International workshop on scalable semantic web knowledge base systems (SSWS2014), CEUR workshop proceedings, vol 1261, Aachen, pp 45–60
22. OpenLink software documentation team (2009). OpenLink virtuoso universal server: documentation
23. Owens A, Seaborne A, Gibbins N, mc schraefel (2009) Clustered tdb: a clustered triple store for jena. In: WWW2009
24. Owl 2 web ontology language (2012). http://www.w3.org/TR/owl2-overview/
25. Özsu MT, Valduriez P (1999) Principles of distributed database systems, 2nd edn. Prentice-Hall Inc., New York
26. Pothen A, Simon HD, Wang L, Bernard S (1992) Towards a fast implementation of spectral nested dissection. In: Supercomputing '92 proceedings. IEEE Computer Society Press, pp 42–51
27. Resource description framework (rdf) (2004). http://www.w3.org/RDF/
28. Rdf schema (2004). http://www.w3.org/TR/rdf-schema/
29. Savnik I, Nitta K (2016) Algebra of rdf graphs for querying large-scale distributed triple-store. In: Proceedings of the CD-ARES. Lecture notes in computer science, vol 9817. Springer, Berlin
30. Savnik I, Nitta K (2017) Type inference of the graph patterns. University of Primorska, Technical report (In preparation), FAMNIT
31. Savnik I, Nitta K, Krnc M, Skrekovski R (2017) Triple-store statistics based on conceptual schemata. Technical report submitted, FAMNIT, University of Primorska
32. Stonebraker M (1986) The case for shared nothing. Database Eng Bull 9(1):4–9
33. White T (2011) Hadoop: the definitive guide, 2nd edn. O'Reilly Media Inc., CA, USA
34. Xu R, Wunsch D (2005) II. Survey of clustering algorithms. Trans Neur Netw 16(3):645–678
35. Xu D, Yingjie T (2015) A comprehensive survey of clustering algorithms. Ann Data Sci 2(2):165–193
36. Xu Q, Wang X, Wang J, Yang Y, Feng Z (2017) Semantic-aware partitioning on rdf graphs. In: Chen L, Jensen CS, Shahabi C, Yang X, Lian X (eds) Web and big data. Springer International Publishing, Cham, pp 149–157
37. Zeng K, Yang J, Wang H, Shao B, Wang Z (2013) A distributed graph engine for web scale rdf data. Proc VLDB Endow 6(4):265–276

Chapter 2
On Cloud-Supported Web-Based Integrated Development Environment for Programming DataFlow Architectures

Nenad Korolija and Aleš Zamuda

Abstract Control-flow computer architectures are based on the von Neumann paradigm. They are flexible enough to support the execution of instructions in any order. Each instruction is fetched from the memory before it could be executed. Passing the data from the instruction that produces it to the instruction that requires it is done using registers or memory. DataFlow computer architectures are configured for execution of an algorithm, while data travel through the hardware. They are suitable for high-performance computing, where the same set of instructions should be run many times. Initialization of data and other processing is done by the processor based on control-flow. The Maxeler framework provides functionality for transforming any algorithm into a VHDL file, and further configuring the dataflow hardware. It also provides support for sending data from the control-flow host processor to the dataflow hardware, and bringing results back. Common programming languages are supported for the host processor, while dataflow hardware programming is done in MaxJ, which is an extended subset of the Java programming language. One can use an integrated development environment called MaxIDE, which is based on Eclipse. We present here a perspective overview of a cloud-supported web-based integrated development environment, WebIDE, which is a subset of MaxIDE, and enables users to develop and run programs for dataflow hardware even without owning dataflow hardware. The main concepts are explained, as well as differences in two integrated development environments. Then, our main focus is on the point of view of programmers, and the goal is to compare the MaxIDE and the WebIDE Maxeler framework, describing the technology needed to support the WebIDE Maxeler framework, providing that the MaxIDE already exists.

N. Korolija (✉)
School of Electrical Engineering, University of Belgrade,
Bulevar Kralja Aleksandra 73, 11120 Belgrade, Serbia
e-mail: nenadko@gmail.com

A. Zamuda
Faculty of Electrical Engineering and Computer Science (FERI),
University of Maribor, Koroška cesta 46, 2000 Maribor, Slovenia
e-mail: ales.zamuda@um.si

© Springer Nature Switzerland AG 2019
V. Milutinovic and M. Kotlar (eds.), *Exploring the DataFlow
Supercomputing Paradigm*, Computer Communications and Networks,
https://doi.org/10.1007/978-3-030-13803-5_2

2.1 Introduction

For many decades, control-flow computer architectures were dominantly used for servers, high-performance computers, and desktop computers. Executing instructions in any order requires control logic, as well as loading instructions that have to be executed in run time [1, 2]. Cache memories are introduced while the main memory was the bottleneck for many types of applications [3]. Shared memories were introduced to reduce the latency when multiprocessor architectures were introduced [4, 5]. DataFlow paradigm naturally solves this issue by treating program execution as a factory floor, where data travels through the preconfigured hardware [6–12]. However, it is justifiable only to instructions that are repeatedly executed over and over again. Therefore, dataflow hardware is usually combined with a processor based on control-flow [12, 13]. Many high-performance applications experience acceleration using the dataflow hardware [14–26].

Here, basic principles of control-flow programming will be described, followed by the dataflow programming. The Maxeler framework will be discussed, depicting how the control-flow processor and the dataflow hardware could operate together, directed by the host (control-flow processor) code, and the MaxJ code, where MaxJ is the language developed for configuring the dataflow hardware by programming. Example CPU code and MaxJ code will be presented. Then, the different architectures will be reviewed, the stand-alone approach, the cloud approach, and its web-based WebIDE on top of the cloud-supported approach [27].

2.2 The Control-Flow Hardware

Control-flow type of processors is based on the von Neumann architecture. This architecture assumes having one input device that feeds the central processing unit with data. The processed data are sent to the output device. The central processing unit processes the data by executing instructions using an arithmetic/logic unit. In order to be able to load and store operands, the central processing unit is connected to the memory unit.

In this type of computer architecture, instructions can be executed in any order. This gives programmers a possibility to develop any type of application that could be realized using the instruction set defined by the computer architecture. However, the ability to execute instructions in any order comes with a cost. The arithmetic/logic unit has to have the information about the instruction that it has to execute, before executing it. Having in mind the fact that an operand may need to be fetched from the main memory, and the result stored back to the main memory, fetching an instruction from the main memory brings additional penalties both in terms of the execution time and in the amount of cache memory spent on storing instructions for possible future use.

Programming the control-flow computer architecture is relatively simple. There is only CPU code that has to be written. For most of the programming languages, this code has to be compiled by the compiler, and then executed. In some cases, interpreters are available for executing the code without compiling it.

There are many problems that are characteristics of the computer architecture based on control-flow. They could be roughly divided into three categories, based on the type of problems they introduce: instruction execution inefficiency, excessive complexity, and unnecessary data transfers.

Before running an instruction, the instruction has to be fetched and decoded, and an auxiliary control unit has to control the arithmetic/logic unit. This imposes a much slower execution time than necessary for executing a single instruction. Also, this imposes more memory transfers than necessary solely for executing the instructions.

The first problem is only partially solved by introducing cache memories. However, this again comes with a cost. Today's processors based on the von Neumann architecture could spend more than 40% of the chip die solely on cache memories. This is only a part of the complexity issues introduced by the von Neumann architecture. An even bigger issue is the fact that the billions of transistors do exist on a modern processor based on this architecture, while only up to a few instructions could be executed at a single moment.

Another problem is that this architecture introduces a memory bottleneck, which is not only due to the fetching of instructions. When two instructions are being repeatedly executed one after the other, where the second one consumes the data produced by the first one, it would be more efficient if the result from the first instruction could be forwarded directly to the second instruction. However, this type of architecture assumes returning the result to the memory after executing the first instruction, and bringing it back for executing the second one.

2.3   The DataFlow Hardware

The dataflow computer architecture represents a configured hardware for executing algorithms, where data travels through the hardware. The process is similar to a factory floor, where each worker is responsible for a part of the process, while the product travels on the conveyor belt. Figure 2.1 depicts the kernel for processing the data that is given through streams.

However, the hardware has to be reconfigurable, so that it could be capable of executing various algorithms. Even in this case, there is a penalty in terms of time needed for recompiling the hardware. An even bigger shortage of this architecture is that the reconfigurable hardware is usually around one order of magnitude slower than the control-flow computer hardware. Another issue that raises with adding additional hardware is the probability of failure. However, it is expected that both components would work properly during the lifespan [28].

Since dataflow hardware is capable of executing only instructions for which it was configured, while configuring lasts many times more than execution of instructions,

Fig. 2.1 Example kernel
processing

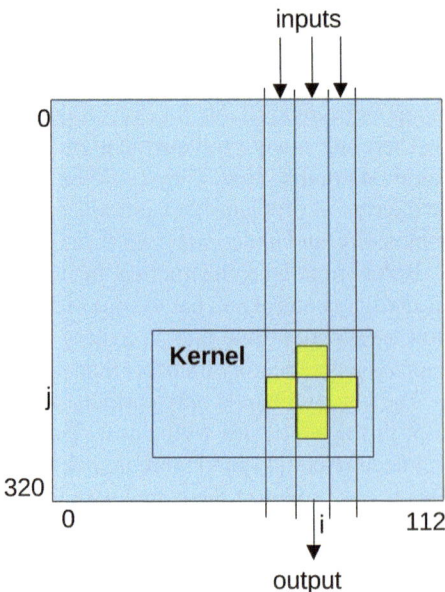

using dataflow hardware is justified only if instructions would be executed many times. Most of the algorithms for high-performance computing include repeatedly executing the same set of instructions. However, the initialization phase includes instructions that might be run only once. Configuring dataflow hardware for them is not justified. Therefore, dataflow hardware is usually combined with the control-flow hardware.

Having in mind the previously mentioned, programming dataflow hardware consists of programming the dataflow hardware for execution of instructions that would be repeated many times, and programming the host control-flow engine for initialization of the data.

2.4   The Maxeler Framework

This work is based on programming the Maxeler framework. This framework takes both the program for the control-flow type of processor and the dataflow hardware as inputs and configures the hardware according to specifications, schedules communications between these two, and controls the algorithm that would utilize the dataflow language. The code for the control-flow hardware will be referred to as CPU code in the further text, while the code for the dataflow hardware will be referred to as DFE code.

The Maxeler framework includes the toolchain that enables transforming both the CPU code and the DFE code into the system where a control-flow type of processor executes the initialization phase of the algorithm, streams data into the dataflow hardware, waits for the dataflow hardware to finish, and collects the results.

The CPU code supports some of the most commonly used programming languages, including C, C++, Java, Python, Fortran, and so on. Although Fortran is not so commonly used by programmers nowadays, it is still frequently used in many scientific simulators, e.g., in earthquake simulators.

The DFE code is supposed to be written in the MaxJ programming language. This language is similar to Java, except that it lacks a functionality that is not applicable to the dataflow hardware, e.g., the functionality related to GUI. The MaxJ also includes auxiliary elements. For example, in order to define a hardware variable, one could use the HWVar type of operand. Once some other variable gets copied into the hardware variable, there is no returning back. One could explain that as the following: if a binary variable value is stored into wires, the precise values of zero or one become voltages that are real physical property of any system. Voltages become lower and lower, as we move through the wire. At a certain stage, it becomes impossible to claim whether the value should be treated as zero or one.

The Maxeler framework application consists of a host code and a dataflow code. An example of CPU code is given in the following code:

```
package chap01_overview.ex2_movingaverage;
import
  com.maxeler.maxcompiler.v1.kernelcompiler.Kernel;
import
  com.maxeler.maxcompiler.v1.managers.
  standard.SimulationManager;
public class MovingAverageOverviewSimRunner {
        public static void main(String[] args) {
                SimulationManager m = new
                    SimulationManager(
                        "MovingAverageOverviewSim");
                Kernel k = new
                    MovingAverageOverviewKernel(
                      m.makeKernelParameters(), 6);
                m.setKernel(k);
                m.setInputData("x", 1, 5, 6, 7, 2, 0);
                m.setKernelCycles(6);
                m.runTest(); m.dumpOutput();
                double[] expected =
                        { 3, 4, 6, 5, 3, 1 };
                double[] got =
                        m.getOutputDataArray("y");
                //checkStream(expected, got);
                m.logMsg("stream is correct!");
    }
}
```

The corresponding dataflow code is given in the following code. This simple kernel is used for calculating average value on certain elements in the stream of data.

```
package chap01_overview.ex2_movingaverage;
import
```

```
    com.maxeler.maxcompiler.v1.kernelcompiler.Kernel;
import
    com.maxeler.maxcompiler.v1.
    kernelcompiler.KernelParameters;
import
    com.maxeler.maxcompiler.v1.
    kernelcompiler.types.base.HWVar;
public class MovingAverageOverviewKernel extends Kernel{
        public MovingAverageOverviewKernel(
                KernelParameters parameters, int N){
            super(parameters);
            HWVar x =
                io.input("x",
                    hwFloat(8, 24)); // Input
            HWVar x_prev =
                stream.offset(x, -1); // Data
            HWVar x_next = stream.offset(x, 1);
            HWVar cnt =
                control.count.simpleCounter(
                    32, N); // Control
            HWVar sel_nl = cnt > 0;
            HWVar sel_nu = cnt < N-1;
            HWVar sel_m = sel_nl & sel_nu;
            HWVar prev = sel_nl ? x_prev : 0;
            HWVar next = sel_nu ? x_next : 0;
            HWVar divisor =
                sel_m ? constant.var(
                    hwFloat(8, 24), 3) : 2;
            HWVar sum = prev+x+next;
            HWVar result = sum/divisor;
            io.output("y", result, hwFloat(8, 24));
    }
}
```

It should be noted that adding the extra hardware increases the probability of failure [28]. It also increases complexity in programming itself [29]. However, there are many open-source projects using the Maxeler framework. Some of them are available on the application gallery [30]. Using the application gallery, one could start a new project by modifying an existing one. There are even methods for transforming applications from control-flow architecture to the dataflow architecture [31–33]. New benchmarking methodologies are being developed to evaluate computing performance according to the recent advances in the technology [34] and support the transition to the dataflow paradigm [35]. New programming models are also developed [36]. Some of the research is not directly connected to the dataflow paradigm, but the principles could be used in transforming algorithms into the dataflow paradigm [37, 38].

Fig. 2.2 DataFlow
computer architecture: The
stand-alone approach

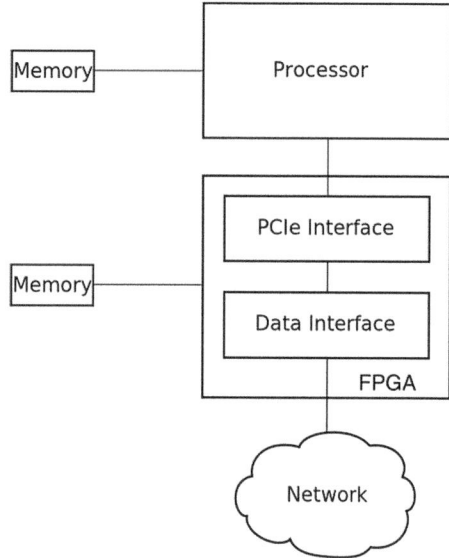

2.5   The MaxIDE Framework

Currently, Maxeler supports only Linux operating system. The code could either
be implemented and run from shell, or within the MaxIDE, or the WebIDE framework [27].

The MaxelerOS component is responsible for scheduling the communication
between the processor and the dataflow hardware.

The MaxIDE framework assumes that the dataflow hardware exists inside of
the computer where the host processor resides. Example architecture is depicted in
Fig. 2.2.

User aspects of the MaxIDE framework are described in the open literature [39].
Here, one could find a detailed description of how to install and use the framework.
Our main focus is on the point of view of programmers, and the goal is to compare the
MaxIDE framework and the WebIDE Maxeler framework, describing the technology
needed to support the WebIDE Maxeler framework, providing that the MaxIDE
already exists.

2.6   The WebIDE Framework

The WebIDE framework [27] assumes that the dataflow hardware exists in the cloud.
Example architecture is depicted in Fig. 2.3.

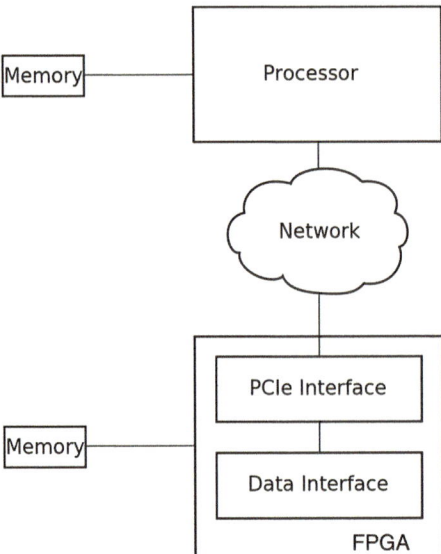

Fig. 2.3 DataFlow computer architecture: The cloud approach

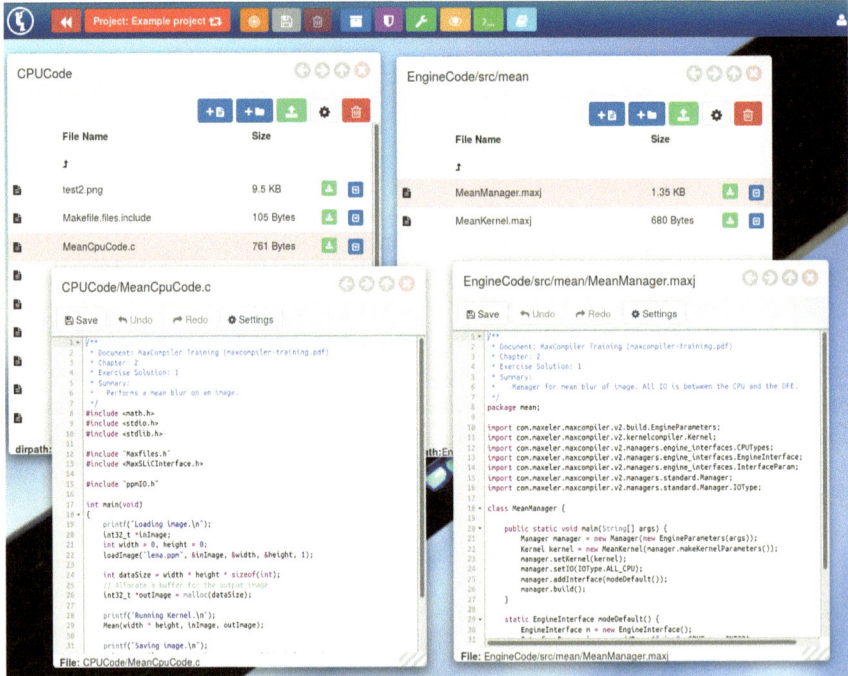

Fig. 2.4 Example WebIDE screenshot

The host processor code and the dataflow code are the same for both the MaxIDE framework and the WebIDE framework. Example screenshot of the framework is given in Fig. 2.4.

The integrated development environment is realized using AngularJS framework for the development of client-side application, and Bootstrap framework is used for the development of the user interface of the web application.

2.7   Conclusion

Comparing to a control-flow type of processor, the dataflow hardware along with the host processor based on control-flow can improve the computation in terms of total execution time, as well as power consumption. This applies to high-performance applications. We have presented the Maxeler framework, followed by the MaxIDE framework, and finally, WebIDE framework. We have shown that the user can efficiently use the dataflow hardware even if he does not possess one. The WebIDE is capable of executing the same host processor code, and the same dataflow code on the host processor based on control-flow and the dataflow hardware, which reside in the cloud. Future work on WebIDE includes integrating it with the Eclipse-integrated development environment. Authors believe this could bring dataflow computing to many programmers that are currently not familiar with programming dataflow hardware. Based on the cloud-supported WebIDE, also the Maxeler application gallery could be extended with new algorithm implementations such as artificial intelligence and analytics.

References

1. Milutinovic V (1996) Surviving the design of a 200MHz RISC microprocessor. IEEE Computer Society Press, Washington DC
2. Milutinovic V (ed) (1988) High-level language computer architecture. Computer Science Press, New York
3. Tartalja I, Milutinovic V (1997) The cache coherence problem in shared-memory multiprocessors: software solutions. IEEE Computer Society Press, Washington DC
4. Tomasevic M, Milutinovic V (1996) A simulation study of hardware-oriented DSM approaches. IEEE Parallel Distrib Technol 4(1)
5. Milutinovic V, Stenstrom P (1999) Special issue on distributed shared memory systems. Proc IEEE 87:399–404
6. Milutinovic V, Hurson A (2015) Dataflow processing, 1st edn. Academic Press, Cambridge, pp 1–266
7. Feynman RP, Hey AJ, Allen RW (2000) Feynman lectures on computation. Perseus Books, New York
8. Milutinovic V (1985) Trading latency and performance: a new algorithm for adaptive equalization. IEEE Trans Commun
9. Milutinovic V, Splitting temporal and spatial computing: enabling a combinational dataflow in hardware. In: The ISCA ACM tutorial on advances in supercomputing, Santa

10. Flynn M, Mencer O, Milutinovic V, Rakocevic G, Stenstrom P, Valero M, Trobec R (2013) Moving from PetaFlops to PetaData. Communications of the ACM. Margherita Ligure, Italy, 1995, pp 39–43
11. Trifunovic N, Milutinovic V, Salom J, Kos A (2015) Paradigm shift in big data supercomputing: dataflow vs. controlflow. J Big Data 2:4
12. Hurson A, Milutinovic V (2015) Special issue on dataflow supercomputing. Advances in computers, vol 96
13. Milutinovic V, Salom J, Veljovic D, Korolija N, Markovic D, Petrovic L (2017) Maxeler AppGallery revisited. Dataflow supercomputing essentials. Springer, Cham, pp 3–18
14. Stojanović S, Bojić D, Bojović M (2015) An overview of selected heterogeneous and reconfigurable architectures. In: Hurson A, Milutinovic V (eds) Dataflow processing, vol 96. Advances in computers. Academic Press, Waltham, pp 1–45
15. Kos A, Rankovic V, Tomazic S (2015) Sorting networks on Maxeler dataflow supercomputing systems. Adv Comput 96:139–186
16. Meden R, Kos A (2017) Bitcoin mining using Maxeler dataflow computers. Electrotech Rev 84(5):253–258
17. Umek A, Kos A (2016) The role of high performance computing and communication for real-time biofeedback in sport. Math Probl Eng
18. Kos A, Milutinović V, Umek A (2018) Challenges in wireless communication for connected sensors and wearable devices used in sport biofeedback applications. Future Gener Comput Syst
19. Ranković V, Kos A, Tomaz S, Milutinovic V (2013) Performance of the bitonic mergesort network on a dataflow computer. In: 21st Telecommunications forum (TELFOR), 2013. IEEE, Belgrade, pp 849–852
20. Milutinovic V, Salom J, Veljovic D, Korolija N, Markovic D, Petrovic L (2017) Polynomial and rational functions. Dataflow supercomputing essentials. Springer, Cham, pp 69–105
21. Korolija N, Milutinovic V, Milosevic S (2007) Accelerating conjugate gradient solver: temporal versus spatial data. In: The IPSI BgD transactions on advanced research
22. Ngom A, Stojmenovic I, Milutinovic V (2001) STRIP-a strip-based neural-network growth algorithm for learning multiple-valued functions. IEEE Trans Neural Netw 12:212–227
23. Milutinovic V (1989) Mapping of neural networks on the honeycomb architecture. Proc IEEE 77:1875–1878
24. Trobec R, Jerebic I, Janežič D (1993) Parallel algorithm for molecular dynamics integration. Parallel Comput 19:1029–1039
25. Korolija N, Djukic T, Milutinovic V, Filipovic N (2013) Accelerating Lattice-Boltzman method using Maxeler dataflow approach. Trans Int Res 9(2):5–10
26. Friston S, Steed A, Tilbury S, Gaydadjiev G (2016) Construction and evaluation of an ultra low latency frameless renderer for VR. IEEE Trans Vis Comput Graph 22(4):1377–1386
27. Milutinović V, Salom J, Trifunovic N, Giorgi R (2015) Using the WebIDE. Guide to dataflow supercomputing. Computer communications and networks. Springer, Cham
28. Huang K, Liu Y, Korolija N, Carulli J, Makris Y (2015) Recycled IC detection based on statistical methods. IEEE Trans Comput-Aided Des Int Circuits Syst 34(6):947–960
29. Popovic J, Bojic D, Korolija N (2015) Analysis of task effort estimation accuracy based on use case point size. IET Softw 9(6):166–173
30. Trifunovic N, Milutinovic V, Korolija N, Gaydadjiev G (2016) An AppGallery for dataflow computing. J Big Data 3(1):5
31. Korolija N, Popović J, Cvetanović M, Bojović M (2017) Dataflow-based parallelization of control-flow algorithms. Creativity in computing and dataflow supercomputing, vol 104. Advances in computers. https://doi.org/10.1016/bs.adcom.2016.09.003 Print ISBN 9780128119556
32. Milutinovic V, Salom J, Veljovic D, Korolija N, Markovic D, Petrovic L (2017) Mini tutorial. Dataflow supercomputing essentials. Springer, Cham, pp 133–147
33. Milutinovic V, Salom J, Veljovic D, Korolija N, Markovic D, Petrovic L (2017) Transforming applications from the control flow to the Dataflow paradigm. Dataflow supercomputing essentials. Springer, Cham, pp 107–129

34. Milutinović V, Furht B, Obradović Z, Korolija N (2016) Advances in high performance computing and related issues. Math Probl Eng
35. Kos A, Tomažič S, Salom J, Trifunovic N, Valero M, Milutinovic V (2015) New benchmarking methodology and programming model for big data processing. Int J Distrib Sens Netw 11:5
36. Mencer O, Gaydadjiev G, Flynn M (2012) OpenSPL: the Maxeler programming model for programming in space. Maxeler Technologies, UK
37. Knezevic P, Radnovic B, Nikolic N, Jovanovic T, Milanov D, Nikolic M, Milutinovic V et al (2000) The architecture of the Obelix-an improved internet search engine. In: Proceedings of the 33rd annual Hawaii international conference on IEEE system sciences, Hawaii, p 7
38. Kovacevic M, Diligenti M, Gori M, Milutinovic V (2004) Visual adjacency multigraphs-a novel approach for a web page classification. In: Proceedings of SAWM04 workshop
39. Trifunovic N et al (2016) The MaxGallery project. Advances in computers, vol 104. Springer, Berlin

Part II
Applications in Mathematics

Chapter 3
Minimization and Maximization of Functions: Golden-Section Search in One Dimension

Dragana Pejic and Milos Arsic

Abstract In this chapter, we will showcase a new approach to the calculation of the extrema of functions in one dimension by implementing the golden-section search algorithm using the dataflow paradigm. This paradigm has been around for quite some time already, but it was only recently, with the increased need to compute large datasets, that its use was brought to attention to many scientists around the globe. BigData has always been present to an extent, but with the ever-growing industry and the increasing speed in which information multiplies by the minute, the models that follow these changes have been expanding as well. Many fields use mathematical models as a way of explanation of various systems. These models are usually composed of equations whose number is counted in thousands. Too often it is needed to calculate the extrema of functions that those equations represent. Doing this the traditional way by using the control-flow paradigm shows that the majority of execution time is spent on calculation. In this chapter, we would like to show that this process can be sped up, and thus, leave more time to perform other actions regarding the exploration of the modeled systems.

3.1 Introduction

Mathematics is the backbone of many fields of science regardless of the fact whether they are natural or social. It is safe to say that mathematics has succeeded what almost no other science did and that is to become an integral part of numerous disciplines whose progress cannot be imagined without the use of results of mathematics. This is all thanks to the widespread use of mathematical models.

Since the modeling of devices and phenomena is essential to both engineering and science, there are very practical reasons for doing mathematical modeling. Even

D. Pejic (✉) · M. Arsic
Faculty of Mathematics, University of Belgrade, Belgrade, Serbia
e-mail: draganapejic93@gmail.com

M. Arsic
e-mail: milosharsic.matf.bg.ac.rs@gmail.com

© Springer Nature Switzerland AG 2019
V. Milutinovic and M. Kotlar (eds.), *Exploring the DataFlow Supercomputing Paradigm*, Computer Communications and Networks, https://doi.org/10.1007/978-3-030-13803-5_3

though these models can be executed in many ways such as differential equations, dynamical systems, logical models and many more, the one that is the most convenient for the use in computer science is the model that is used to maximize or minimize a certain output. In other words, the results of computer science are most widely used to solve the various optimization problems. The optimization problem usually consists of a large number of equations that need to be solved and the value of their variables determined in one way or the other.

However, due to the sheer number and the complexity of the equations, finding their solution is not an easy task. The majority of the time is not spent on setting up the model but on executing simple operations such as addition and multiplication. This makes the approach to the problem time inefficient despite the precision that can be achieved.

Doing these calculations by hand is almost impossible, but thanks to the rapid growth of computer science that started in the last century and continued into the new millennium, translating the equations into computer programs has proved to be an easier task. Executing the code on the machine did speed by the evaluation of the variables, but it did not solve the problem of the time inefficiency.

The time that it takes to solve the system of equations has drastically decreased but it still presents the significant percentage of the total time it takes to execute the computer program. The reason for this lies in the fact that mathematical algorithms can be translated the easiest into programs that belong to the control-flow programming paradigm. This paradigm is not the most suitable for processing large quantities of data that these algorithms require.

With the growth of technology and the speed in which information multiplies, the systems that mimic natural processes or behavior have grown as well. What used to be a couple of dozens of equations have become hundreds of thousands.

In the same way, information which used to be stored on traditional media has become digital too. According to the estimation of Lyman and Varian [1], the new data stored in digital media devices have already been more than 92% in 2002, while the size of these new data was also more than five exabytes. Years have passed since their research and amount of data has certainly increased over the last decade.

Even if the amount of data used in finding solutions to the optimization problem is just a small percentage of total data, it is still larger than it used to be in the past and it can be considered BigData.

Fisher et al. [2] point out that BigData means that the data will be unsuitable to be handled and processed by most current information systems or methods because data will not only become too big to be loaded into a single machine, but it also implies that most traditional data mining methods or data analytics developed for a centralized data analysis process may not be able to be applied directly to BigData.

The problems with dealing with BigData have birthed a new programming paradigm—dataflow. Although the advances of computer systems and internet technologies have witnessed the development of computing hardware following the Moore's law for several decades, the problems of handling the large-scale data still exist. One of the possible solutions for these problems offers dataflow programming.

In this chapter, we will present the basics of this paradigm that despite being introduced in the 70s [3] has not reached its potential or attracted the attention of a wider audience until many years later. The reason for this is the lack of enabler technology which exists today in the form of both software (OpenSPL) and hardware (FPGA). In a way, it can be said that the dataflow paradigm is both old and new. We will focus on the hardware dataflow approach and show how it can be used in solving the optimization problem.

Perhaps this problem is not the best to explain the concept of BigData, but it is one that troubles many scientists in various fields.

Since we are incapable of covering every topic in which tools of mathematical optimization are used, we will focus on an algorithm that is a part of many bigger systems and that is often crucial for solving them or, at least, making them simpler. That is the golden-section search.

In this chapter, we will explain the theoretic discoveries that this algorithm is based on as well as introduces some mathematical concepts that are important when it comes to solving the given task of finding the extrema of a function, and thus, show the unbreakable bond between mathematics and the real-world problems. We will try to connect two opposite sides and bring awareness to fact that many behaviors and phenomena can be observed from a different angle if we dare to change our approach.

This attempt will be carried throughout all sections of the chapter as we present the algorithm, show the source code for its implementations in two very different programming paradigms and offer explanations of the aforementioned code. In the end, we will compare the results we got after testing both programs on similar datasets.

By connecting mathematics and computer science, many things have become both easier and faster. Tasks that required hours now only need minutes if not seconds.

One thing that always intrigued people was finding the extrema of various functions that simulate natural forces or social behavior. The golden-section search algorithm presents one of the more efficient ways of solving this problem. It can be tailored to find either the maximum or the minimum of a given function.

In this chapter, we will present the golden-section search algorithm that evaluates the minimum of an unimodal function. We will translate mathematical formulae into code by using one of the languages that belong to the control-flow paradigm and test it on a different number of functions to show how much time is consumed simply on calculating the values inside various loops of the algorithm.

Control-flow paradigm is based on the Von Neumann architecture which is the most common computer architecture. But there is another one that is worth exploring. That is the Feynman architecture. We will do this by relying on the programming paradigm that is based both on the Feynman architecture and on reconfigurable FPGA chips and that is becoming more and more important as the amount of digital data increases. That is the dataflow paradigm. Comparing the two paradigms cannot be one easily in the general sense so we will do it on a more approachable level and that is through implementing the golden-section search in both these paradigms.

In the ever-changing world, any moment that we can save on calculation and that we can redirect to more important things such as examination of the problem is worthy. Thus we will try to decrease the time it takes for the algorithm to be executed by rewriting the program originally implemented in the control-flow paradigm. We will shift paradigms and implement the golden-section search algorithm by using the dataflow paradigm and all advantages it has to offer.

The chapter will briefly shift its focus from the algorithm to the dataflow accelerator provided by Maxeler technology, which is renowned as one of the leading corporations in this technological field, as we discuss the accelerator architecture, its memory organization, the way in which the accelerator communicates with a host computer, and the paradigm overall. The following sections will provide an insight in the way that our chosen algorithm can be accelerated using the dataflow paradigm and hint at the advantages that make this approach more time efficient compared to the control-flow implementation.

Once we show the detailed implementation using the dataflow accelerator and the source code of the algorithm, we will test the program again on the same sets of functions.

The results that we get from these two testings we will compare to show that time can be saved if we decide to implement the golden-section search algorithm in a paradigm that might not be the first one that comes to our mind.

We will also compare other factors in the two paradigms such as speed, power dissipation, and physical size. The collected results show that dataflow implementation gives a better performance in compared to the control-flow implementation.

However, this result does not make the control-flow inferior to the dataflow paradigm. There will be instances of algorithms that are more suitable for one or the other paradigm, but not both. What makes golden-section search algorithm more suitable for dataflow paradigm is not the size of the datasets that we tested both implementations on, but its iterative nature.

In the heart of the golden-section search algorithm is nothing but simple calculation that requires a lot of time. Thus we propose a new paradigm which presents an alternative method for parallel processing and show that in the case of this particular algorithm, the dataflow implementation shows better results that the control-flow implementation.

It is worth mentioning that this approach can be applied on every application which has the BigData problem or spends the significant amount of its execution time on calculation that happens inside numerous loops that make the majority of its code. However, it should be noted that all the benefits of the dataflow paradigm previously mentioned depend on the problem at hand, the structure of the algorithm that is used for solving it and quantity of data which have to be processed. In order to achieve a significant improvement of the execution time, it is of the crucial importance that the application meets some requirements that will be provided later on in this chapter.

More details of the theoretical backgrounds and on the possible future applications can be found in the relevant references that had the major impact on the research of the authors of this text [4–11].

3.2 Existing Solutions

The golden-section search is one of the methods used in solving optimization problems, but it is not the only technique worth mentioning. This algorithm belongs to a not so small group of algorithms that were discovered over time to help find the solutions of the problems that require computing the extrema of various functions.

As such, it is often used not as the main algorithm for solving a problem, but as a part of a different sub-procedures which goal is to bring the researchers one step closer to finding a correct value of the solution or, at least, an approximation of it that will have the required precision.

Optimization problems are not strictly based on numerical values. They can be formulated using other methods of mathematics, but models that are made that way are not the most suitable for solving problems that occur in computer science or other fields that use the achievements of applied mathematics. Therefore we will define an optimization problem in the form which is the most used and which gathered the interest of many scientists who were in need of a model that would solve practical problems they stumbled upon while doing their research.

To put it simply, an optimization problem consists of a set of independent variables or parameters, and often includes conditions or restrictions that define acceptable values of the variables. Such restrictions are called the constraints of the problem. The other essential component of an optimization problem is the objective function which is a single measure of "goodness" [12].

The objective function depends in some way on the variables and the goal is to find the solution of an optimization which, generally speaking, is a set of allowed values of the variables for which the value of the objective function is "optimal" according to the given restrictions. Sometimes "optimal" means "within the given bounds" or "satisfying a required precision". Usually optimization is finding either the maximum or minimum of the objective function.

Given the existence of constraints or the lack of them, optimization problems are divided into two classes

1. constrained optimization, where there are a priori limitations on the allowed values of independent variables and all values that don't conform with them are disregarded
2. unconstrained optimization, where such limitations don't exist and any solution that is found can be and usually is considered the right one.

We will consider the latter and view the objective function as the problem that needs solving.

But classification of the unconstrained optimization problems can be continued. Two general classes can be identified [13]

1. one-dimensional unconstrained problems
2. multidimensional unconstrained problems.

Problems of the second class are either very hard or impossible to solve. But luckily, there are ways to reduce them to the problems of the first class. Transforming the multidimensional unconstrained problems to one-dimensional problems is not always precise and some things can be lost, but it gives us the opportunity to solve a problem that is equivalent or similar to the original one and to find at least an approximation of the solution. Problems of the first class are easier to solve and there are more techniques for doing that, which gives us the freedom to pick the one that will give us the best result. That is the reason why efficient one-dimensional optimization algorithms are required, if we want to construct equally efficient multidimensional algorithms.

Since the multidimensional constrained problems can be translated into one-dimensional problems without losing much of their substance, in the following paragraphs we will consider only one-dimensional problems and discuss the ways they can be solved.

The optimization problem that most frequently occurs is finding the extreme value of the objective function, but since the tasks of maximization and minimization are closely related to each other because the minimum value of the function f is the maximum of the function $-f$, in the continuation of this section we will only consider the minimization problem.

The problem of finding numerical approximations to the minimum of functions, using hand computation, has a long history. Recently considerable progress has been made in the development of algorithms suitable for use on a digital computer, but the background of these algorithms remains to be a mathematical concept.

Depending of the amount of the given data, different approaches and algorithms for solving the minimization problem can be used. The requirements of these algorithms are that they need to be quick, cheap and require small amount of memory since they are run on a computer and resources are limited.

Often the evaluation of the function (and also its partial derivatives with respect to all variables, if the chosen algorithm requires them) is what eats up time and increases the execution time of the program. It is false to assume that this slows down the algorithm, but since the majority of time is spent on calculation, to the naked eye, this seems to be the case. If this happens, the aforementioned requirements are replaced with a simple "evaluate the function as few times as possible."

Even with this changed requirement, the number of algorithms that can be used for solving a particular problem does not decrease. This is because there is no perfect optimization algorithm that is tailored for just one group of problems. In regards to the provided data, one method can be better than another so the best practice is to try more than one algorithm in comparative fashion.

However, depending on the problem that is being dealt with, the initial choice of method and subsequently of algorithm can be made easier. The following considerations can be helpful if it is required to fine a solution of the one-dimensional minimization problem (minimize a function of one variable):

- Depending on the objective function and its properties, it is needed to choose between methods that need only evaluations of the function to be minimized and methods that also require evaluations of the derivative of that function. It is worth mentioning that algorithms that use the derivative are somewhat more powerful than those which only use the values of the function, but sometimes this isn't enough to compensate for the additional calculations of derivatives. Given some thought, examples that favour one approach over the other can be easily constructed. However, whenever it is possible to compute derivatives, one must be prepared to try using them.

- If the chosen method does not require calculation of the derivative, one of the best solutions is to first bracket a minimum using one of the various techniques that have been developed and then use a different method, for example Brent's method, to effectively calculate the minimal value. Both the routine for bracketing minimum and Brent's method can be found in [14]. If it happens that the objective function has a discontinuous second (or lower) derivative, then the parabolic interpolations of Brent's method are of no advantage. In this case, the simplest way to find the minimum is to use the golden-section search method.

- If we found Brent's method suitable, it can be further modified so that it makes limited use of the first derivative information. This information can be also used to construct high-order interpolating polynomials. But the tendency of polynomials to give extremely wrong interpolations at early stages, especially when it comes to function that might have sharp, almost "exponential" features, simply cannot be overlooked in favour of the improvement in convergence very near a smooth, analytic minimum. This improvement often can compensate for the errors that can occur when high-order interpolating polynomials are used.

Besides the mentioned methods, there are many others for solving the minimization problems. That still does not change the fact that, even given the previous advice for picking the correct one, all of them require some degree of calculation. Too often this process is too time-consuming and tedious to be done by hand, so the logical thing that comes to mind is translating these methods into programs that can be run on a digital computer.

Algorithms that are intuitively correct and easy to be comprehended can be tricky to implement in standard programming languages while the ones that require a certain amount of mathematical knowledge to be understood show better performance when translated into a series of commands. It is important to find an algorithm that is both—easy to understand and easy to implement.

One of the simplest but at the same time most efficient algorithms to find the minimum of a function is the golden-section search. The steps of the algorithm can be done by hand, but also quickly translated into the code that can be easily run on a machine.

That is one of the reasons why the use of this algorithm transcended the field of mathematics. Nowadays it found its home in many other sciences, natural or not. Some of the researches are hard to imagine without it. Once translated into lines of code, the algorithm is often used as a subfunction in a program in which finding the solution of a minimization problem is nothing more than one small piece in the grand scheme of things.

To illustrate all the ways in which the golden-section search algorithm can be applied, let us mention that is used for vertical profile optimization for the flight management systems [15], for finding a good shape parameter for meshless collocation methods [16], and for deriving a variant of the Nelder–Mead algorithm [17] to name a few. The mentioned applications make use of the golden-section search in various ways that can be quickly and almost directly translated into executable code. The fault with the code made this way is that it is straightforward. This means that it is made without caring about the time it will take for it to be executed nor the memory that will be used. Simply put, it is just an array of instructions that is executed in the order that it is written, which is one of the qualities of the control-flow paradigm in which the algorithm is most often implemented.

Due to its wide use in various fields, often as a part of important research projects, the execution time of the golden-section search algorithm is important. If it is slow, then it increases the execution time of the whole project that it is a part of, especially if the amount of data that needs to be processed is large enough to be considered BigData. Lately this has been the case since the partnerships between scientific centers located in the various parts of the world allows the collection of large quantities of data.

The golden-section search algorithm is efficient but not very fast when it is implemented using the control-flow paradigm and used for processing BigData. The implementations mentioned above are just that—implementations based on the control-flow paradigm, done in different programming languages. Regardless of the language used, the control-flow paradigm concentrates on the software level acceleration because each operation consists of fetching and decoding instructions, fetching data, calculating and saving results, which is essentially slow.

In the next section, we will introduce a new paradigm for solving BigData problems by using hardware accelerators, which relies on FPGA chips. By presenting the essence of the dataflow paradigm, the way in which it accelerates algorithms will be explained as well. Later in this chapter, we will provide the implementation of the Golden Section Search algorithm which presents a hardware acceleration that exploits the parallelism, but does not contribute to the added costs.

3.3   Essence of the DataFlow Paradigm

Some will say that the traditional control-flow paradigm has been the dominant programming paradigm over the last decades since the most popular languages belong to it. The others will agree and even add that things have not changed and that control-flow is still important today. One of the reasons for this is that the majority of digital computers are based on the Von Neumann architecture which makes the execution of the applications written in languages of this paradigm easy, but not necessarily fast. The other is that until a few years ago, all computers around us belong to the category of control-flow computers.

Fig. 3.1 A control-flow
program in action

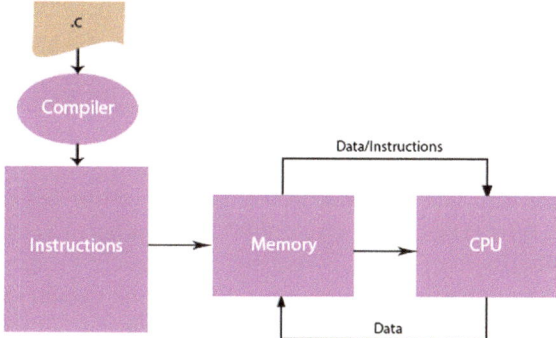

In the case of control-flow computers, the program is written to control the flow of data through the hardware. A program's source code is transformed into a list of instructions for a particular processor ("control-flow core"). This list is then loaded into the memory and the instruction move through the processor and occasionally read or write data to and from memory. Figure 3.1 shows how a control-flow program is executed.

All instructions have to be fetched, decoded, and executed. During the execution phase, the computer has to compute the addresses of data to be fetched, to fetch the data, to compute and store the result. This process is repeated over and over again until the end of the program is reached. The continuous push-pull of data through the memory hierarchy and the synchronization across multiple cooperating threads can be extremely time-consuming, which makes the execution process slow.

In order to improve the efficiency of this paradigm and decrease the time that is spent on writing and fetching data from memory, modern processors contain many levels of caching, forwarding and prediction logic, where the closest level of cashing to the processor has the shortest access time. Despite all of this, there are situations where requested data has not been store in cache memory and has to fetched from lower levels. This too is a time-consuming operation.

All these improvements do not change the fact that the programming model is inherently sequential and performance depends on the latency of memory accesses and the time for a CPU clock cycle.

Just like everything else, the trends in computer science are changing. It can be understood that this is a consequence of the presence of BigData in all kinds of research. Given this amount of that, the slowness of the control-flow paradigm is painfully evident. Since everybody expects the execution time to be within certain time bounds and that the results are achieved quickly, scientists have turned to the new-old programming paradigm and that is dataflow. It has been proven that this paradigm is the most suitable for BigData and it offers superior speedups (in some isolated cases, even up to 2000), brings the size reduction and power savings.

This paradigm has been present since the '70s when the scientists had proposed dataflow computers for the first time [18], but due to the lack of technology, this proposal had not become reality. With the development of technology, the dataflow computers are no longer a farfetched ideal but reality and in the recent years they emerged on the market as a viable alternative for general purpose computing.

What has not changed is that one still writes a program, but it is no longer a program to control the flow of data through the hardware. In the case of dataflow computer, the program is written so that it can configure the hardware (in space) in a way that input data, when it arrives, can flow through the computer hardware in only one way—the way how the computer hardware has been configured. The result is generated much faster since the dataflow is not driven by a program, but by a voltage difference between the input and the output of the hardware.

In a dataflow program, we describe the operations and data choreography for a particular algorithm. In a DataFlow Engine (DFE), data streams from memory into the processing chip where data is forwarded directly from one arithmetic unit ("dataflow core") to another until the chain of processing is complete [19]. Each dataflow core can compute only one simple arithmetic operation that enables one to combine lots of cores in one DFE, which can be seen in Fig. 3.2. After a dataflow program has processed its streams of data, the dataflow engine can be reconfigured for a new application in less than a second. Figure 3.2 shows how a control-flow program is executed.

As seen above, in the case of the control-flow programs. the compilation goes until the machine level code and the execution process is performed at the machine code level. Things are quite different in the dataflow paradigm since the compilation reaches the levels much below the machine code; it goes till the levels of wires and gates. This means that the process is executed at the gate transfer level.

As indicated before, the dataflow paradigm brings the benefits of speed, power, and size. But for these benefits to be achieved, some conditions have to hold. We explain them below.

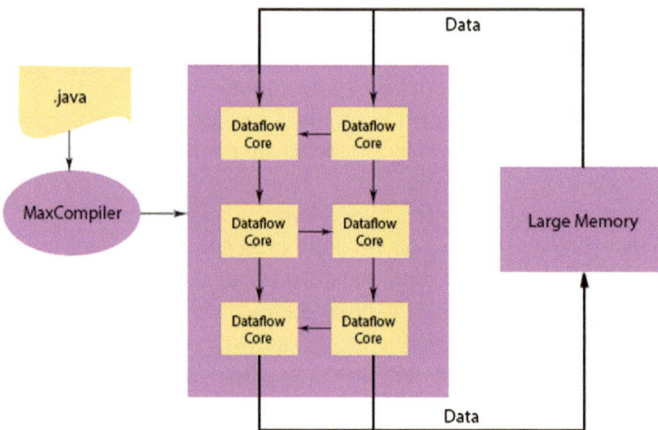

Fig. 3.2 A dataflow program in action

Since the dataflow technology goes below the machine level, it can be said that it migrates the execution of loops from software to hardware. This is an obvious method to make the loop execution faster since the large amount of time is spent on repeated execution of the instructions nestled inside the loop. In ideal cases, the loop execution time is reduced to almost zero. The decrease of time depends on the level of data reusability inside the migrated loops. It has been proved that the remarkable acceleration can be achieved if the parallel part of the application (loops) takes up 95% of the whole application. To harvest the other benefits such as power dissipation and size, it is advised that the serial part of the application continues to run on the control-flow paradigm and takes less than 5% of the application, while the parallel part migrates to the dataflow accelerator [20].

To put it simply, dataflow computers are accelerators. We continue to run the old program on the host machine of the control-flow type until the time for a time-consuming loop to be executed comes. When this happens, the execution is passed to the dataflow accelerator that is connected to the host machine via a PCI Express bus (or for larger systems, via InfiniBand).

It is worth mentioning that Maxeler's custom solutions achieve a speed up from 20 to 200 times over the control-flow's cutting edge technology, as well as the power dissipation reduction up to 80% and space saving over 95% regarded to datacenter space. And that is the reason why we will consider Maxeler card that usually serves as an accelerator for the CPU.

Essentially, the host CPU puts data into the accelerator and gets the result. But it is not that simple because the loop has to migrate from host CPU to the accelerator and then the data has to be moved as well. This is done by data streams. The host CPU is able to transfer more than one stream simultaneously to the same DFE. Since more DFEs can be connected together via high-bandwidth links, this enables parallel computing inside the accelerator system.

Programs that have only one loop are rare, which can be an inconvenience since one Kernel file needs to be written for every migrated loop. But no matter how many loops are migrated, in addition to the Kernel files, we only need to write one Manager file.

Kernel file presents an implementation of the migrated loop for DFE while the Manager code is responsible for moving data from the host to the accelerator, moving results from the accelerator to the host, and moving data in-between the kernels. Both files are written in MaxJ which is a superset of standard Java extended with new functionalities.

Since the host program is written in a control-flow language (such as C, C++, Python, and many more) and that is why it compiles down to the machine code level and executes on a machine based on the von Neumann architecture. The host program utilizes accelerator for BigData computations by sending data to and receiving results from accelerator.

Files are run through the MaxCompiler which first generates a Kernel graph which is a graph of pipelined arithmetic units. And then, using the third-party tools, MaxCompiler converts graph to a *.max* file. This file is linked with the *.c* code, which presents the executable code and then all that is left is to execute the code.

First, the Manager checks if the hardware is configured. If it is not, the DFE starts configuring on the basis of the *.max* file. When this process is completed, BigData streams start and the process begins.

After this, there is no need for reconfiguring hardware if the same code is being executed. But if the same part of the hardware is used for another application, it is necessary to reconfigure the hardware for the purposes of this new application.

Over the years Maxeler has developed a method on how to transform an algorithm to its fully accelerated implementation, utilizing the Maxeler's static dataflow approach. As we will do the same with the golden-section search algorithm, now we will explain the steps of the process that has helped us achieve the acceleration of this algorithm. The five steps of the Maxeler method are [21]

- make original application in a language that's representative of the control-flow paradigm—write the code for the implementation of the desired algorithm; this code will be used as a reference point against which the accelerated version would be compared
- determine code segments with a potential for acceleration—basically, it is needed to locate the loops inside the application since that is the code that can be accelerated due to the nature of dataflow, then these segments of code will from C to MaxJ which is a superset of Java, with dozens of new functionalities (built-in classes) added on the top of the classical Java
- debugging of designs at simulation time—generating a working hardware implementation is time-consuming, so initial functional testing should be done using simulation
- debugging designs in hardware—after we are satisfied with the simulation results, the next step is to proceed to hardware build
- optimizations—having a working design is not the only final goal; the other is maximal acceleration that can be achieved with additional optimization and better utilization of hardware resources.

Now that we have explained the steps which we need to undertake to accelerate our algorithm, in the next section we will shift our focus briefly and dedicate some paragraphs to the derivation of the algorithm.

3.4   Minimization or Maximization of Functions

So far we have talked about the optimization problem without defining it precisely, which was enough for the time being, but now we have reached the point in which it would be good to know what we are dealing with before we begin to explain the golden-section search algorithm.

Since this is a numerical algorithm, we will define the optimization problem in an appropriate way. Let's say that we are given a function $f : A \to \mathbb{R}$ from some set A to the set of real numbers. Our goal is to find an element x_0 in A such that

- $f(x_0) \geq f(x)$ for all x in A (maximization)
- $f(x_0) \leq f(x)$ for all x in A (minimization)

As said in the previous section, we have equalized the optimization problem with the minimization or maximization problem but since these two tasks are trivially related to each other (one's function f can be easily another's $-f$), in the continuation of this part of the chapter, we will only talk about minimization. But this isn't the only reason for our choice. The other is that conventional optimization problems are usually stated in terms of minimization.

Usually, A is some subset of the Euclidean space $\mathbb{R}^n$. If $n = 1$, f is the function of one independent variable. In other cases, f depends on more than one variable. Set A can be specified by a set of constraints or (in)equalities that its members have to satisfy. The domain of function f, the set A, is called the search space or the choice set, while the elements of A are called candidate solutions or feasible solutions.

The function f is called an objective function. A feasible solution that minimizes the objective function is called an optimal solution.

Unless both the objective function and the feasible region are convex, the optimal solution is not unique since there can be several local minima.

A local minimum x^* is defined as a point for which there exists some $\delta > 0$ such that for all x where

$$\|x - x^*\| \leq \delta,$$

the expression

$$f(x^*) \leq f(x)$$

holds. Simply put, on some region around x^* all of the function values are greater than or equal to the value at that point.

While a local minimum is at least as good as any nearby points, a global minimum is at least as good as every feasible point. In a convex problem, if there is a local minimum that is interior (not on the edge of the set of feasible points), it is also the global minimum, but a nonconvex problem may have more than one local minimum, not all of which need be global minima.

We will only consider functions and feasible regions that are convex since the nonconvex problems go out of the scope of our chapter. The branch of applied mathematics that is concerned with the development of deterministic algorithms that are capable of guaranteeing convergence in finite time to the actual optimal solution of a nonconvex problem is called global optimization.

Practical optimization problems usually do not require the evaluation of the function over its whole domain, neither they are concerned with finding the global minima. In majority of the cases, we are interested in locating and calculating the minimal value of the objective function in one dimension in the given interval.

As said before, there are many techniques that can be used to solve this minimization problem. All of them have their pros and cons, but if we are given a unimodal continuous function over an interval and we want to find its minimum without using derivatives, one of the efficient methods is the golden-section search. The importance of the unimodality of the objective function lies in the fact that a unimodal function contains only one minimum or maximum on the interval.

3.4.1 Unimodality

Since the golden-section search method is used to find the extrema of a unimodal function, before the explanation of the method itself, we will define unimodality.

In relevant literature, there are several definitions of a unimodal function. The reason for this is that the definition depends on whether the function is supposed to have a unique minimum or a unique maximum (we will consider minima).

Kowalik and Osborne [22] say that f in unimodal on $[a, b]$ if f has only one stationary value on $[a, b]$. This definition has two disadvantages. First, it is meaningless unless f is differentiable on $[a, b]$, but we would like to say that $|x|$ is unimodal on $[-1, 1]$. Second, functions which have inflexion points with a horizontal tangent are prohibited, but we would like to say that $f(x) = x^6 - 3x^4 + 3x^2$ is unimodal on $[-2, 2]$ (here $f' \pm (1) = f''(\pm 1) = 0$).

Wilde [23] gives another definition that does not assume differentiability, or even continuity. However, his definition has its flaws as well. Since Wilde considers maxima rather than minima and we are interested in finding the minima of a given function, some of his inequalities have been reversed. The following definition after the modification defines unimodal function as following: function f is unimodal on $[a, b]$ if, for all $x_1, x_2 \in [a, b]$,

$$x_1 < x_2 \Rightarrow ((x_2 < x^* \Rightarrow f(x_1) > f(x_2)) \wedge (x_1 > x^* \Rightarrow f(x_1) < f(x_2))),$$

where x^* is a point in which f attains its least value in $[a, b]$. The disadvantage of this definition lays in the existence of the point x^*. To check if a function f is unimodal, we need to verify that f satisfies the condition of the definition. We need to know the point x^* (and such a point must exist).

Since it has been proven difficult to find the point x^*, Wilde's definition is hard to apply to functions. Thus we give another definition that is nearly equivalent to Wilde's.

Brent [24] defines unimodality as follows: function f is unimodal on $[a, b]$ if, for all $x_0, x_1, x_2 \in [a, b]$,

$$(x_0 < x_1 \wedge x_1 < x_2) \Rightarrow ((f(x_0) \leq f(x_1) \Rightarrow f(x_1) < f(x_2)) \wedge$$
$$(f(x_1) > f(x_2) \Rightarrow f(x_0) > f(x_1))).$$

Two possible configurations of the points x_0, x_1, x_2 and x^* that illustrate Wilde and Brent's definitions are shown in Fig. 3.3.

Fig. 3.3 Unimodal functions

Theorem 3.1 *If a point x^* at which f attains its minimum in $[a, b]$ exists, then Wilde and Brent's definitions are equivalent.*

For those who are curious, the proof of the theorem can be found in [24].

We will address another theorem that gives a simple characterization of unimodality. There is no assumption that f is continuous. Since a strictly monotonic function may have stationary points, the theorem shows that Kowalik and Osbonre's definition is essentially different from the other two definitions given, even if f is continuously differentiable.

Theorem 3.2 *Function f is unimodal on $[a, b]$ (according to Brent's definition) if and only if, for some (unique) $\mu \in [a, b]$, either f is strictly monotonic decreasing on $[a, \mu)$ and strictly monotonic increasing on $[\mu, b]$, or f is strictly monotonic decreasing on $[a, \mu]$ and strictly monotonic increasing on $(\mu, b]$.*

Proof of the theorem is omitted since this is not a mathematical textbook. However, we will give a few significant corollaries that help with finding the minimum of the function f using the golden-section search method.

Corollaries of the previous theorem

- If f is unimodal on $[a, b]$, then f attains its least value at most once on $[a, b]$. (If f attains its least value, then it must attain it at the point μ by Theorem 3.2.)
- If f is unimodal and continuous on $[a, b]$, then f attains its least value exactly once on $[a, b]$.
- If $f \in C^1[a, b]$ then f is unimodal iff, for some $\mu \in [a, b]$, $f' < 0$ almost everywhere on $[a, \mu]$ and $f' > 0$ almost everywhere on $[\mu, b]$. (Note that f' may vanish at a finite number of points.)

From the previous theorems and corollaries we can conclude that if f is unimodal on $[a, b]$, then the minimum of f (or, if the minimum is not attained, the point μ given by Theorem 3.2) can be located to any desired accuracy by the golden-section search method. However, we should ensure that the coordinates of the points at which f is evaluated are computed in a numerically stable way.

The method which will be explained in detail in the next section is based on the fact that the interval known to contain μ can be reduced in size as much as it desired. That way it is possible to find the point in which the function reaches its minimum.

3.5 Golden-Section Search

The golden-section search is a technique used for finding the minimal or maximal value of a strictly unimodal function by successively narrowing down the range of values inside which the extremum is known to exist. The name of the method comes from the fact that the algorithm maintains the function values for triples of points whose distances form a golden ratio. The algorithm is the limit of Fibonacci search for a large number of function evaluations. Fibonacci search and golden-section search were discovered by Kiefer [25].

Despite the fact that the golden-section search can it used to find any extrema of the objective function, to illustrate it, we will only consider the case of minimizing the function.

After we derive the method, we will show and explain the implementation of the technique in two different programming paradigms—control-flow and dataflow.

3.5.1 Derivation of the Method

Let f be a function over the interval $[a, c]$. We assume that $f(x)$ is both continuous and unimodal over $[a, c]$. Unimodality of the function f means that $f(x)$ has only one minimum in $[a, c]$.

The conditions above guarantee that the minimum of function f in the interval $[a, c]$ exists. But these conditions also remind us of the bisection method and we will apply a similar idea: narrow the interval that contains the minimum comparing function values.

The bisection method is an iterative method for finding the root of the given function. It is known that a root of a function f is bracketed in the interval (a, b) if $f(a)$ and $f(b)$ have opposite signs. The algorithm for narrowing the range of values in which the root is known to exist consists of the following steps:

1. isolate the root to the interval (a, b)
2. evaluate the function at the intermediate point x
3. split the interval into two—(a, x) and (x, b)
4. take as a new interval the one that contains the root, i.e. interval (a, x) if $f(a)$ and $f(x)$ have opposing signs, or interval (x, b) if $f(x)$ and $f(b)$ have opposing signs
5. return to the first step of the algorithm and repeat the process until the root is found within the given precision.

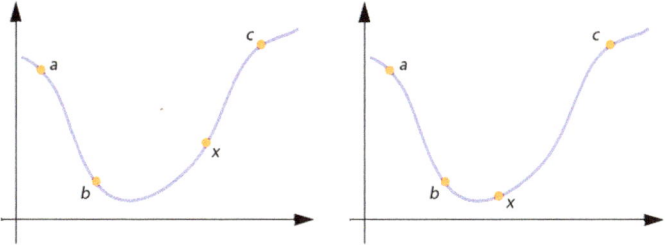

Fig. 3.4 Left: Bracketing triplet (a, b, x), Right: Bracketing triplet (b, x, c)

In a similar fashion, a minimum of a function can be bracketed. It is known that a minimum is bracketed when there's a triplet of points (a, b, c) so that the following holds:

$$a < b < c$$
$$f(b) < f(a)$$
$$f(b) < f(c)$$

Following the idea of the bisection method, it is required to choose a new point x, either between a and b or between b and c. Let us assume that the new point is in the subinterval (b, c). After the evaluation $f(x)$, two cases can occur

- if $f(b) < f(x)$, then the new triplet is (a, b, x) (Fig. 3.4, Left)
- if $f(b) > f(x)$, then the new triplet is (b, x, c) (Fig. 3.4, Right)

In both cases, the middle point of the new triplet is the abscissa whose ordinate is the best minimum achieved so far. This iteration of bracketing the minimum is continued until the required accuracy is reached. In other words, until the interval (a, c) is small enough.

At this point, an important question is being posed and that is how small is small enough.

If the minimum is at $x = b$, we might naively think that it can be bracketed in as small a range as $(1 - \varepsilon)b < b < (1 + \varepsilon)b$, where ε is the floating point precision of the computer the algorithm is being executed. But this is not correct.

The shape of the function $f(x)$ near the point b is described by Taylor's theorem

$$f(x) \approx f(b) + \frac{1}{2} f''(b)(x - b)^2$$

The second term on the right-hand size is negligible compared with the first term (it will be a factor ε smaller and it will act just like zero when added to it) when

$$\frac{1}{2} f''(b)(x - b)^2 < |f(b)|\varepsilon$$

The previous inequality can be transformed into

$$\frac{|x - b|}{|b|} < \sqrt{\varepsilon}\sqrt{\frac{2|f(b)|}{b^2 f''(b)}}$$

Assuming that the term under square root is the order of unity (for most functions, it is), we get the result

$$\frac{|x - b|}{|b|} < \sqrt{\varepsilon} \tag{3.1}$$

which shows us that the (relative) accuracy that we can reach is only $\sqrt{\varepsilon}$.

The minimum-finding routine presented in the next section will require an argument tol that a user running the program should supply. At the end of the execution, the routine will return an abscissa whose fractional precision is about $\pm tol$. Unless there's a better estimate for the right hand side of the Eq. (3.1) that the user knows of, tol should be set equal to (or not much less than) $\sqrt{\varepsilon}$ where ε is the floating point precision of the computer [14]. We have shown that smaller values will gain the user nothing.

Now that we have defined the accuracy of the method, what is left is to decide on a strategy for choosing the new point x if we're given the bracketing triplet (a, b, c).

Let us denote with ω the relative position of b at the interval (a, c)

$$\omega = \frac{b - a}{c - a} \qquad\qquad 1 - \omega = \frac{c - b}{c - a}$$

Suppose that the next trial point x is an additional fraction z beyond b,

$$z = \frac{x - b}{c - a}$$

Now the length of the next interval is (in relative units) either $z + \omega$ or $1 - \omega$. If we want to avoid or, at least, minimize the worst case possibility, then the optimal choice is to set these two equal. By doing this, we gain

$$z + \omega = 1 - \omega \Rightarrow z = 1 - 2\omega$$

One can see that the new point x is symmetrical with the point b in the original interval

$$|b - a| = |x - c|$$

This implies that the point x lies in the larger of the two subintervals. But where exactly is the point x? On a related note, where did the value ω come from? Presumably from the previous iteration of applying the same strategy, which means that if z was chosen to be optimal, then so was ω before it. This means that the distance of x from b compared to interval (b, c) is equal to the distance of b from a compared to interval (a, c), in other words,

$$\frac{x - b}{c - b} = \frac{b - a}{c - a}$$

or in relative units

$$\frac{z}{1 - \omega} = \frac{\omega}{1}$$

Combining this with out optimal choice for point z, we get the following system of equations:

$$z = 1 - 2\omega$$

$$\frac{z}{1 - \omega} = \omega$$

This gives us the quadratic equation

$$\omega^2 - 3\omega + 1 = 0$$

which yields

$$\omega = \frac{3 - \sqrt{5}}{2} \approx 0.38197$$

Thus, the optimum triplet of points (a, b, c) is such that its middle point b is a fractional distance 0.38197 from one end and 0.61803 from the other end. These fractions are those of the so-called *golden section* whose aesthetic properties were acknowledged in the ancient Greece. This is the reason why this optimal method of function minimization is called the golden-section search.

After deriving the method, we can see that the algorithm is the following:

- at each stage, we are given the bracketing triplet of points (a, b, c) from the previous iteration
- choose the new point x such that it is at the relative distance of 0.38197 from b to the direction of the larger subinterval (either (a, b) or (b, c))
- the new triplet is the one that has the lower value at the midpoint
- continue this process until the given accuracy is achieved.

At each iteration step, the interval decreases by a factor 0.618034, i.e., the minimum is bracketed to an interval just 0.618034 times the size of the preceding interval.

The convergence of this method is linear, meaning that sufficiently near to the minimum $x*$ the errors $\varepsilon_k = x_k - x*$ in the position of the minimum x_k at two consecutive iterations are related as $|\varepsilon_{k+1}| \leq C|\varepsilon_k|$, where C is a constant that depends on the function being minimized.

In the following section, we will present the implementation of the golden-section search algorithm in the programming language C which is one of the languages prevalent in the control-flow paradigm.

3.5.2 The control-flow Implementation

Code 3.1 presents one control-flow implementation of the golden-section search algorithm in programming language C. In this implementation, given a function f and the interval (a, b) in which the minimum of the function f is located, this routine performs a golden-section search for the minimum, isolating it to the fractional precision of about tol. The abscissa of the minimum is returned as value calculated as $(a + b)/2$. The rest of the code is self-explanatory.

```c
float golden(float a, float b, float (*f)(float) , float tol){

    int i;
    float c, d, fc, fd;
    float golden_ratio = (sqrt(5) - 1)/2;

    c = b - golden_ratio * (b - a);
    d = a + golden_ratio * (b - a);

    while(abs(a-b) > tol){

        fc = (*f)(c);
        fd = (*f)(d);

        if(fc < fd){
            b = d;
            d = c;
            c = b - golden_ratio * (b - a);
        }
        else{
            a = c;
            c = d;
            d = a + golden_ratio * (b - a);
        }
    }

    return (a+b)/2;
}
```

Code 3.1 control-flow implementation of the golden-section search

This function is called from the *main*() function of the implementation.

In the next section, we will provide the dataflow implementation of the golden-section search algorithm.

3.5.3 The DataFlow Implementation

As seen in the control-flow implementation of the golden-section search algorithm, it is required to calculate the value of the given function almost countless of times. Since functions can be diverse, their evaluation time varies. This can make the testing of the algorithm troublesome, but there are ways to improve it.

To simplify the testing, we have chosen our test functions to be quadratic functions. This can be done without loss of generality because quadratic functions are unimodal and other, more complex functions can be reduced to or approximated with quadratic functions. The other reason for doing this is to avoid the localization of the minimum, which is out of the scope of this paper.

More about the testing can be found in Sect. 3.6 and now we will proceed with the dataflow implementation of the algorithm.

To know what type of code to write, we need to know what a dataflow application requires. A Maxeler dataflow supercomputer consists of DataFlow Engines (DFEs) and CPUs. That is why a dataflow application consists mostly of CPU code, with small pieces of the source code, and large amounts of data, running on dataflow engines.

Just like with the control-flow implementation, the golden-section search routine is not a part of the *main*() function of the application, but a procedure of its own that is called within the *main*() function.

As a part of the CPU code, we define the function *golden*() which presents the control-flow implementation of the golden section code. Code 3.3 shows this. Note that this function does not evaluate the objective function, but rather calls the function *f_val* that does it instead. The implementation of the function *f_val* is described in Code 3.2.

The function *golden*() requires five arguments—two represent the end points a and b of the interval in which the minimum is bracketed, and the remaining three arguments are the coefficients of the objective function.

```
float f_val(float x, float a, float b, float c){

    //evaluates a function f
    float f_value;
    f_value = a*(x*x)+ b*x +c;

    return f_value;
}
```

Code 3.2 Evaluation of the quadratic function with the coefficients a, b and c

```
float golden(float c1, float c2, float c3, float a, float b){

    //golden section search CPU\index{Central Processing Unit
        (CPU)}

    int i;
    float c, d, fc, fd;
    float golden_ratio = (sqrt(5) - 1) / 2;

    c = b - golden_ratio * (b - a);
    d = a + golden_ratio * (b - a);

    for(i = 0; i < 50; i++){

        fc = f(c, c1, c2, c3);
        fd = f(d, c1, c2, c3);

        if(fc < fd){
            b = d;
            d = c;
            c = b - golden_ratio * (b - a);
        }
        else{
            a = c;
            c = d;
            d = a + golden_ratio * (b - a);
        }
    }
    return (b+a)/2;
}
```

Code 3.3 Changed implementation of the golden-section search

It can be noticed that the *while* loop from the control-flow implementation has been replaced with the *for* loop in the dataflow implementation. The reason behind this decision lies in the fact that loop unrolling plays a major role in the dataflow implementation and it is much easier to unwind a *for* loop compared to the *while* loop. We have replaced the *while* loop with the *for* loop with the maximum of 50 iterations which was the estimated number of iterations for achieving the greatest precision when it evaluating the minima of the functions from our test example. For different test examples, a different number of iteration can be required. With one small modification of the function presented in Code 3.4, the number of iterations can be any number that the end user of the application desires. This modification is nothing but adding another argument to the function which would represented the maximum number of iterations and then in the *for* loop replacing the number 50 with this argument.

After inputting the number of functions we want to run the algorithm for, inside the *main*() function, we generate this many quadratic functions. This number is represented with variable n. Once that is done, all that is left is to call the function *golden*() and measure the time it takes for it to execute. This is shown in Code 3.4.

```
printf("CPU works\n");
clock_gettime(CLOCK_MONOTONIC, &start_t1);

for(j = 0; j < n; j++)
    xCPU[j] = golden(coefa[j], coefb[j], coefc[j], a, b);

clock_gettime(CLOCK_MONOTONIC, &end_t1);
```

Code 3.4 Time measurement as well as the execution of the golden-section search

Functions that are used for time measurement are explained in Sect. 3.6.2. If we run this code segment for different values of n, we will get the estimated times it takes for the control-flow implementation to process this many functions.

Inside the CPU code, we pass a call to Kernel where the code that will be projected in space, i.e., on a 2D card, is located. The name of the project is *MemStream*, therefore we ought to call the function of the same name. This function requires six arguments just unlike the *golden*() function. In the call of *MemStream*() shown in Code 3.5, n presents the number of objective function, a and b are the end points of the interval in which the minimum is located, while the remaining three arguments present arrays of coefficients of objective functions. The last argument which is x_dfe is an array in which we will save the abscissa values of the minima found.

The segment of code shown in Code 3.5 presents the part that that we have just described with the added calls to the functions that measure the execution time of the Kernel.

```
printf("Kernel works\n");

float *x_dfe = (float *) malloc(sizeof(float)*n);

clock_gettime(CLOCK_MONOTONIC, &start_t2);

MemStream(n, a, b, coefa, coefb, coefc, x_dfe);

clock_gettime(CLOCK_MONOTONIC, &end_t2);
```

Code 3.5 Time measurement for Kernel

Since the code that runs on the dataflow engine is written in MaxJ which is a subset of the programming language Java, we use a Java library to describe it. To execute actions on the DFE, which include sending data streams and sets of parameters to the DFE, we need call the Simple Live CPU (SLiC) API functions. To be able to use these functions, we need to include the following headers:

```
#include "Maxfiles.h"
#include "MaxSLiCInterface.h"
```

Code 3.6 Headers that are used in the implementation

To create dataflow implementations (*.max* files), we write MaxJ code and then execute this code to generate the *.max* file. This file can then be linked and called via the SLiC interface that we mentioned before. A *.max* file generated by MaxCompiler for Maxeler DFEs consists of Kernel and Manager files.

The Manager is responsible for connecting the CPU code and Kernels. It defines data streams between the host program and the accelerator as well as the type of variables in those streams. The code we write for the Manager is used to create the Kernel which attributes can be modified to fit our requirements. Code 3.7 presents an example of a Manager file.

```
public class MemStreamManager extends CustomManager{

    public MemStreamManager(EngineParameters arg0){
        super(arg0);
    }

    private static final String s_kernelName = "MemStreamKernel";

    public static void main(String[] args) {

        EngineParameters params = new EngineParameters(args);

        Manager manager = new Manager(params);

        Kernel kernel = new
            MemStreamKernel(manager.makeKernelParameters());

        manager.setKernel(kernel);

        manager.setIO(IOType.ALL_CPU);

        manager.createSLiCinterface();

        manager.build();
    }
}
```

Code 3.7 Manager code

But the Manager also describes the data flow choreography between Kernels, the DFE's memory and various available interconnects depending on the particular dataflow machine. By decoupling computation and communication, and using a flow model for off-chip I/O to the CPU, DFE interconnects and memory, Managers allow us to achieve high utilization of available resources such as arithmetic components

Type	Size (items)	Host write speed	Chip area cost
Scalar input/output	1	Slow	Low
Mapped memory (ROM / RAM)	Up to a few thousand	Slow	Moderate
Stream input/output	Thousands to billions	Fast	Highest

Fig. 3.5 Chip characteristics

and memory bandwidth. Maximum performance in a Maxeler solution is achieved through a combination of deep-pipelining and exploiting both inter- and intra-Kernel parallelism. The high I/O-bandwidth required by such parallelism is supported by flexible high-performance memory controllers and a highly parallel memory system [19].

Generally speaking, there are many different ways to transfer data to and from the accelerator. Data can be transferred as scalar or stream values. Depending on that, the time it takes to transfer it between the host and the accelerator can be different. Figure 3.5 shows the different types of chips that can be used for this as well as their type, size, host write speed and chip area cost.

Now that we have explained what Managers are, let's us do the same with Kernels.

Kernels are graphs of pipelined arithmetic units which can be logical or arithmetical. Without loops in the dataflow graph, data simply flows from inputs to outputs. As long as there is a lot more data than there are stages in the pipeline, the execution of the computation is extremely efficient. With loops in the dataflow graph, dataflows in a physical loop inside the DFE, in addition to flowing from inputs to outputs.

Each arithmetic operation requires a certain amount of time to compute the result. During this period the accelerator is idle, which does not work in favour of the main goal and that is achieving significant speedup compared to the implementation of the same algorithm in a different paradigm. How long it will take for the arithmetic unit to execute the operation and get the result depends on the number representation. It often happens that if a custom number representation is used, that the computing time will decrease [26]. That is the reason why in out proposed solution, we define a new floating-point type which representation is shown in Code 3.8.

```
private static final DFEType ioType = dfeFloat(8,24);
```

Code 3.8 Custom number representation

As we mentioned before, the host program is based on the control-flow paradigm and as such it does not allow a fixed-point number representation. This means that all variables are floating point. When variables arrive to the accelerator, they are converted to fixed-point type that we have defined above. With this their representation changed and every operation performs computation with fixed-point numbers. Once the computation is done, the result is converted to the floating-point type before it is transferred to the host program.

While this approach drastically improves execution time of the algorithm, it also slightly reduces the accuracy of the result. But time savings are enough compensation of the lost of accuracy.

```
DFEVar a = io.scalarInput("a", ioType);

DFEVar b = io.scalarInput("b", ioType);

DFEVar a1 = io.input("a1", ioType);

DFEVar a2 = io.input("a2", ioType);

DFEVar a3 = io.input("a3", ioType);

DFEVar c, d, fc, fd, rez, temp;

DFEVar golden_ratio = constant.var(0.618034);
```

Code 3.9 Receiving data from host program

In the case of the golden-section search algorithm, the dataflow implementation is not drastically different from the control-flow implementation. The *while* loop is replaced with a *for* loop with a definite number of iterations. Each of these iterations can be unrolled, i.e., definite number of times replicated on the 2D card. Code 3.10 shows the way the algorithm is executed after data is transferred from the host program.

```
c = b - golden_ratio * (b - a);
d = a + golden_ratio * (b - a);

for (int i = 1; i <= max; i++){
    fc = a1 * c * c + a2 * c + a3;
    fd = a1 * d * d + a2 * d + a3;

    a = fc < fd ? a : c;
    b = fc < fd ? d : b;

    temp = fc < fd ? c : d;

    d = fc < fd ? temp : a + golden_ratio * (b-a);
    c = fc < fd ? b- golden_ratio * (b-a) : temp;
}

io.output("r", ((b+a) / 2), ioType);
```

Code 3.10 Essence of the dataflow implementation

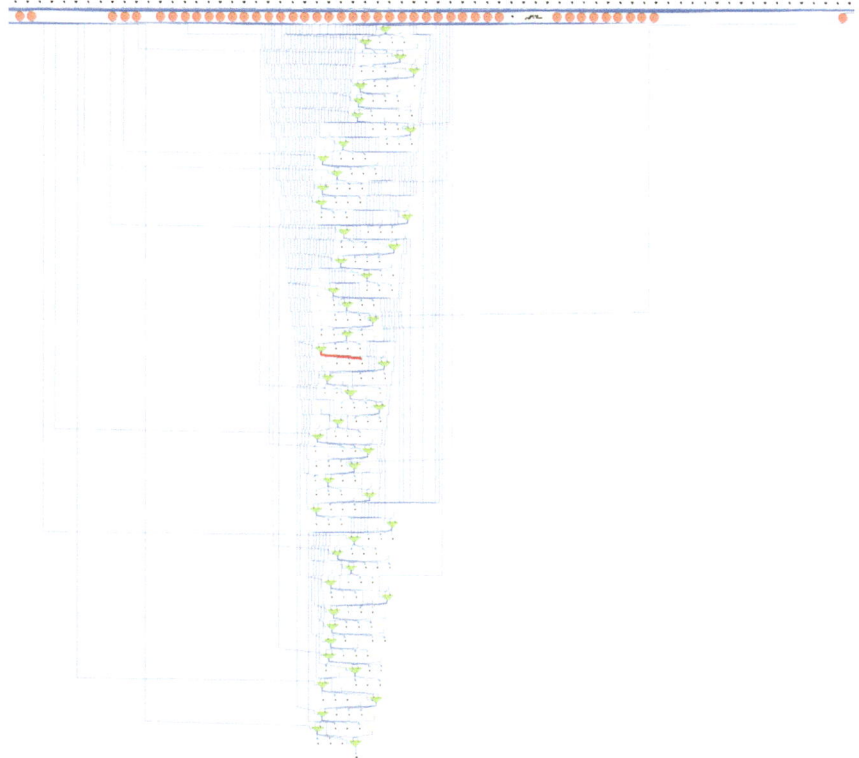

Fig. 3.6 Execution graph of golden-section search algorithm

At the end of this section, we would like to present the execution graph of the dataflow implementation of the golden-section search algorithm that can be seen on Fig. 3.6. Due to the big number of iterations, the graph is large and almost unreadable. But from what can be seen, each node presents one simple arithmetical or logical operation. Connections between nodes stand for a flow of data between functional units. After data is transferred to the accelerator, it flows through plenty of functional units, up to the end, computing the result which is then transferred back to the host program.

Not to be fooled into thinking that only numeric algorithms can be implemented using the dataflow paradigm, we would like to mention that many of the other algorithms that have been implemented and accelerated this way can be found on Maxeler AppGallery website [27, 28].

The goal of the dataflow implementation is to achieve better performance than the implementations using the control-flow microprocessors. In the next section, we will see if we managed to achieve just that. We will evaluate the performance of the control-flow and the dataflow implementations that we have proposed.

3.6   Performance Evaluation

Technology is developing fast. This growth is followed by the changes in computer architecture. The computer components are becoming more and more powerful with every passing day. But these improvements are sometimes unable to follow the speed in which the amount of data is multiplied and collected. That is why the complexity and the execution time of algorithms continue to play a crucial role in choosing the right algorithm for processing the given amount of data.

As we have mentioned before, dataflow paradigm presents an alternative to the traditional control-flow paradigm. Not only that, but dataflow paradigm has been proven to be more suitable for analyzing BigData.

Despite this, it is still essential to compare these two paradigms. One of the important performance indicators is speed and in the next couple of sections we will compare the results we have gotten when we tested the implementations of the golden-section search algorithm in these two paradigms.

At the beginning of this section, we will explain the way that we measured various indicators and then we will compare the results.

3.6.1   MAX4 Card Usage Evaluation

To solve BigData problems, the dataflow paradigm replies on hardware accelerators, which in turn rely on FPGA chips. In the case of dataflow, programs are written to configurate the hardware. When implementing the golden-section search algorithm, we wrote those programs too in the form of the Kernel program.

The application written this way is partly executed on the host computer and partly on the accelerator. CPU nodes can utilize as many DFEs as are required for a particular application and release DFEs for use by other nodes when they are not running computation, this way ensuring that all cluster resources are optimally balanced at runtime.

For our implementation of the algorithm, we have used the MAX4 Card which is contained in the MPC-X2000 node.

In Fig. 3.7 is shown the usage evaluation of the MAX4 Card. Let it be noted that the percentage use of block memory and logic utilization are high while the usage of DSP blocks (which are used for digital signal processing) is low.

3.6.2   Execution Time

To evaluate the execution time of the algorithm, we used standard functions from the header $time.h$ and variables of the $structtimespec$ type. The following part of the CPU code shows that we start measuring the time before calling the function

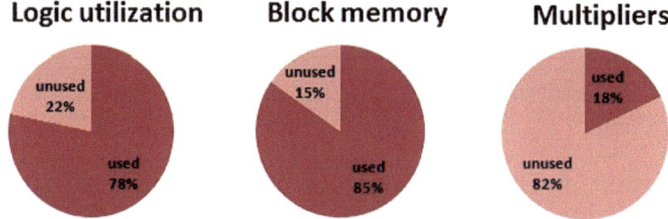

Fig. 3.7 Usage evaluation of MAX4 card

golden() which finds the points in which functions from the given array reach their minimum. We stop measuring the time once all iterations of the *for* loop are done.

```
clock_gettime(CLOCK_MONOTONIC, &start_t1);

for(j = 0; j < n; j++)
    xCPU[j] = golden(coefa[j], coefb[j], coefc[j], a, b);

clock_gettime(CLOCK_MONOTONIC, &end_t1);
```

Code 3.11 Part of the CPU code which is responsible for measuring the execution time

To measure the time we use the *clock_gettime*() function that is a part of the *time.h* header. The declaration of this function is shown in Code 3.12. The *clock_gettime*() function gets the current time of the clock specified by *clock_id* (in our case, we use the predefined $CLOCK\_MONOTONIC$ which represents the absolute elapsed wall-clock time since some arbitrary, fixed point in the past), and puts it into the buffer pointed to by tp.

```
#include <time.h>
int clock_gettime(clockid_t clock_id,
                  struct timespec *tp );
```

Code 3.12 Declaration of the function *clock_gettime*

Since we call the function *clock_gettime*() two times, we get two points in time. The first one marks the beginning of the loop execution and the other one marks its end. All that is left to do is to calculate the time that has passed between these two points. The function shown in Code 3.13 does just that and as a result returns elapsed time.

```
double get_elapsed_time(struct timespec *before, struct timespec
    *after){

    double deltat_s = after->tv_sec - before->tv_sec;
    double deltat_ns = after->tv_nsec - before->tv_nsec;
```

```
  return deltat_s + deltat_ns * 1e-9;
}
```

Code 3.13 Calculating Elapsed Time

3.6.3 Test Examples

Without loss of generality, we have decided use test our algorithm in a set of quadratic functions whose minimum is located in the interval (0, 1000). One of the reasons for doing this is because quadratic functions are unimodal and many other functions can be reduced or approximated with quadratic functions. Our goal is to generate functions that are just like that.

A univariate (single-variable) quadratic function has the form

$$f(x) = ax^2 + bx + c$$

where $a, b, c \in \mathbb{R}$ and $a \neq 0$.

The vertex of a quadratic equation is the minimum or maximum point of the equation. In our case, we are looking for the minimum. An easy way to calculate the vertex α is

$$\alpha = \frac{2a}{b}$$

Since we want our minimum to be bracketed in the interval (0, 1000), then the inequations $0 < \alpha < 1000$ can help us derive the conditions for the coefficients of the quadratic function. Note that the coefficient c isn't required for calculating the vertex. Therefore c can be a random number.

If we choose the coefficient a in a similar fashion, then the coefficient b will have to satisfy the condition $0 < \frac{2a}{b} < 1000$.

We generate the coefficients of test functions as shown in Code 3.14.

```
void fgenerator(float *a, float *b, float *c){

    *a = abs(rand() % 1000) + 1;
    *b = -abs((2 * rand() * (float)(*a) + 1) % 1000);
    *c = rand() % 1000;
}
```

Code 3.14 Generating the coefficients of test functions

The function $fgenerator()$ generates the coefficients of quadratic functions whose minimum can be found in the interval (0, 100) while the function $farraygen()$ (shown in Code 3.15) for the argument n which represents the required number of functions, generates arrays of coefficients for n quadratic functions. These arrays are used to testing the implementation of the algorithm in two different paradigms.

```
void farraygen(float* ar1, float * ar2, float *ar3, int n){

    int i;
    for(i = 0; i < n; i++)
        fgenerator(ar1 + i, ar2 + i, ar3 + i);
}
```

Code 3.15   Generating arrays of coefficients

3.6.4   Test Results and Comparison with control-flow paradigm

The testing was done in the Maxeler offices in Belgrade. In Fig. 3.8 is shown execution time of golden-section search algorithm using test examples of different size, starting from just function whose minimum needed to be found and going as far as sixty millions.

As was expected, for test examples where the number of functions is relatively small, the control-flow implementation achieves better performance but not by much. The reason for this lies in the fact that dataflow paradigm is suitable for solving BigData problems. If the dataset on which we test the implementation is small, the

Fig. 3.8 Execution speed of the golden-section search algorithm

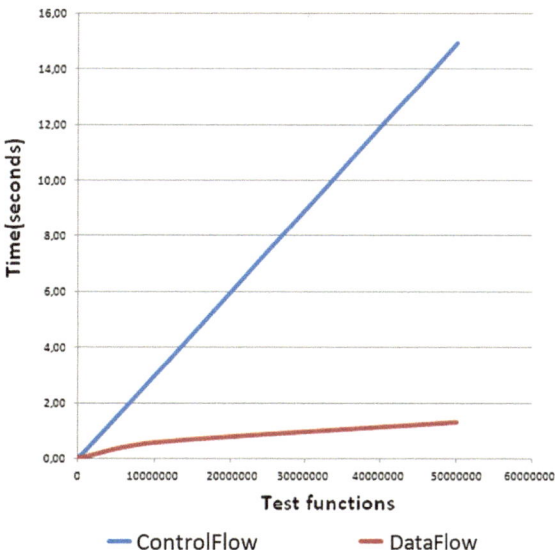

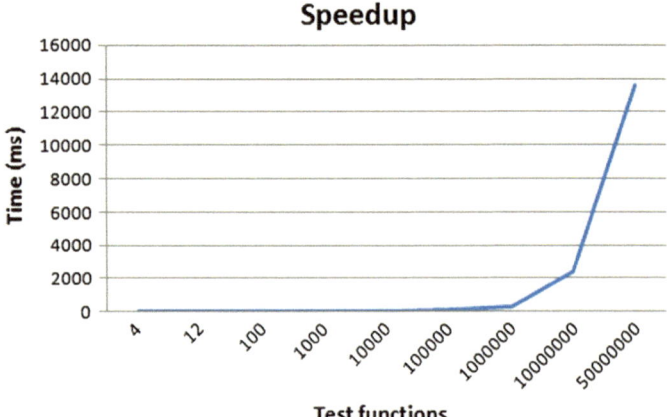

Fig. 3.9 Achieved speedup

dataflow implementation cannot show advantages which provides FPGA hardware. The speedup is achieved with increasing number of functions in the test example. The first time when the acceleration is noticeable is when the number of test functions reaches one million.

Using the MAX4 Isca @ 200 MHz card over the Intel(R) Core(TM) i5-3350P CPU @ 3.10 GHz microprocessor, the dataflow implementation achieves acceleration about 11.5 times, which can be seen in Fig. 3.9.

While doing the testing, we have noticed that storing the data into memory has significantly prolonged the execution time of the control-flow implementation. This was not the case with the dataflow implementation due to the fact that data is being streamed.

3.6.5 Cluster Testing

Besides the control-flow and dataflow computers, BigData can be processed with clusters. A cluster is a set of connected computers that work together. In many respects, they can be viewed as a single system. The computers that are connected this way are called nods and each of them is set to perform the same task that is being controlled and scheduled by software. Clusters have proven to be of great use in many areas of science and economy, and they present some of the fastest supercomputers in the world [29].

To compare the execution time between clusters and dataflow accelerator card, we have tested our test functions on a cluster with Intel(R) Core(TM) i7-6700K CPU @ 4.00 GHz microprocessor. The results show that running a control-flow program on a cluster is faster than running it on a regular computer, but when we compare the execution time of programs on clusters with the time we achieved with the MAX4

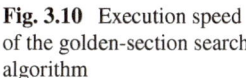

Fig. 3.10 Execution speed of the golden-section search algorithm

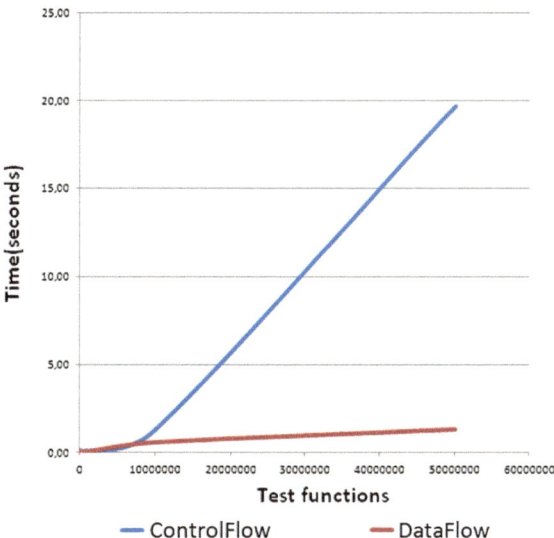

Isca @ 200 MHz card. Figure 3.10 shows execution time gotten when using clusters (control-flow on the graph) and the Maxeler card (dataflow on the graph). Clusters are faster on a smaller number of functions, which was to be expected, while the dataflow implementation is proven to be faster on a large scale, that is when the number of function is bigger than a million functions.

The dataflow implementation achieves acceleration about 11.24 times compared to the clusters. This speedup is smaller than the speedup achieved when we compare the execution times on the regular and dataflow computer. This is because the execution time on clusters is shorter compared to the regular computer of the Von Neumann architecture.

To conclude, it is safe to say that when working with large amounts of data, or in our case, with huge numbers of function, that the dataflow implementation is a much better choice than control-flow implementation. The achieved speedup can testify that.

3.7 Conclusion

By reading this chapter you might have noticed that the focus shifted many times. It balanced somewhere on the line between mathematics and computer science. The reason for this is the nature of the presented algorithm. In its essence, the golden-section search algorithm is a numerical method which was used way before digital computers became a prevalent tool for conducting researchers.

This is the reason why the authors have decided to derive the algorithm instead of just describing and explain in detail which might have not been needed the mathematical background of the method. We sincerely hope that reading those sections of the chapter was not tedious. Creativity in this work follows the path referred to as Specialization in [30].

After that we have shown the implementation of the algorithm using two different paradigms—the traditional control-flow and the emerging dataflow paradigm. The purpose of doing this was not to show that the algorithm can be written in a different programming paradigms, but rather to see if it can be accelerated.

In this chapter the implementation of the golden-section search algorithm using dataflow accelerator provided by Maxeler technology is presented. As seen in Sect. 3.6, this implementation achieves better performance compared to the control-flow implementation.

The results of testing show that the dataflow implementation achieves remarkable speedup not only compared on the control-flow implementation run on a regular computer but also compared to the same implementation run on a cluster. In the first case, the achieved speedup is about 11.5 times, while in the second case it is 11.24 times. The reason for slightly worse result when it comes to comparing the execution times of dataflow computer and a cluster is due to the fact that a cluster is faster than a regular computer. It is worth noting that this acceleration is achieved when the amount of data, in our case, the number of test functions, is big enough to be considered BigData.

Since the golden-section search algorithm is quite straightforward with only one loop in which the minimum of the objective function is calculated, the dataflow implementation was not too different from the control-flow implementation. All we needed to do was transform the *while* loop from Code 3.1 to the *for* loop with the definite number of iterations in Code 3.3. Later, this loop transformed into the dataflow implementation.

While doing the performance evaluation, we have noticed that a big problem for the control-flow implementation was the transfer of data to and from memory. This significantly increased the overall execution time. With the dataflow implementation, we did not have this problem. This is because the date is being streamed.

At the end of this paper, all that remains to be said is that we hope that our work can serve as a motivation for implementations of other numeric algorithms using the dataflow implementation.

However, there might be cases where the dataflow implementation will not be drastically better in terms of speed, but there is a high probability that this will not happen if the amount of data that needs to be processed is big enough to be considered BigData.

Acknowledgements The authors would like to thank Professor Milutinovic for inviting us to contribute to this book and for his encouragement throughout this project, as well as Milos Kotlar for providing guidance during the process of writing this work. This research was supported by Maxeler Technologies, Serbia, Belgrade. We want to thank our families, colleagues who provided insight and expertise that greatly assisted the research.

References

1. Lyman P, Varian H (2004) How much information 2003? Tech Rep
2. Fisher D, DeLine R, Czerwinski M, Drucker S (2012) Interactions with big data analytics. Interactions 19(3):50–59
3. Hurson A, Lee B (1993) Issues in dataflow computing. Adv Comput 37:285–333
4. Kos A, Toma S, Salom J, Trifunovic N, Valero M, Milutinovic V (2015) New benchmarking methodology and programming model for big data processing. Int J Distrib Sens Netw 11:1–7
5. Milutinovic V (1996) The best method for presentation of research results. IEEE TCCA Newsl 1–6
6. Jakus G, Milutinovic V, Omerovic S, Tomazic S (2013) Concepts. Concepts, ontologies, and knowledge representation. Springer, Berlin, pp 5–27
7. Furht B, Milutinovic V (1987) A survey of microprocessor architectures for memory management. Computer 20:48–67
8. Milutinovic V (1996) Surviving the design of a 200 MHz RISC microprocessor: lessons learned. IEEE Computer Society, Washington
9. Milutinovic V (1989) High-level language computer architecture. Computer Science Press Inc., Washington
10. Milutinovic V, Stenstrom P (1999) Special issue on distributed shared memory systems. Proc IEEE 87:399–404
11. Trifunovic N, Milutinovic V, Salom J, Kos A (2015) Paradigm shift in big data supercomputing: dataflow vs. controlflow. J Big Data 2:4
12. Gill PE, Murray W, Wright MH (1981) Practical optimization. Academic Press Inc. (London) Limited, London
13. Antoniou A, Lu WS (2007) Practical optimization: algorithms and engineering applications. Springer, Berlin
14. Press WH, Teukolsky SA, Vetterling WT, Flannery BP (2007) Numerical recipes 3rd edition: the art of scientific computing. Cambridge University Press, Cambridge
15. Patron RSF, Botez RM, Labour D (2012) Vertical profile optimization for the flight management system CMA-9000 using the golden section search method. In: IECON 2012-38th annual conference on IEEE industrial electronics society. IEEE, pp 5482–5488
16. Tsai CH, Kolibal J, Li M (2010) The golden section search algorithm for finding a good shape parameter for meshless collocation methods. Eng Anal Bound Elem 34(8):738–746
17. Nazareth L, Tseng P (2002) Gilding the lily: a variant of the Nelder-Mead algorithm based on golden-section search. Computational optimization and applications. Springer, Berlin
18. Dennis JB, Misunas DP (1974) A preliminary architecture for a basic data-flow processor. In: Newsletter ACM SIGARCH Computer Architecture News Homepage, vol 3(4)
19. Maxeler (2013) Multiscale dataflow programming
20. Milutinovic V, Salom J, Trifunovic N, Giorgi R (2015) Guide to dataflow supercomputing. Springer, Berlin
21. Maxeler (2016) Maxeler DFE debugging and optimization tutorial
22. Kowalik J, Osborne MR (1968) Methods for unconstrained optimization problems. American Elsevier Publishing Company, New York
23. Wilde DJ (1964) Optimum seeking methods. Prentice Hall, Upper Saddle River
24. Brent RP (1973) Algorithms for minimization without derivatives. Prentice Hall, Upper Saddle River
25. Kiefer J (1953) Sequential minimax search for a maximum. Proc Am Math Soc 4(3):502–506
26. Fu H, Osborne W, Clapp B, Pell O (2008) Accelerating seismic computations on FPGAs from the perspective of number representations. Rome
27. http://appgallery.maxeler.com/
28. Trifunovic N, Milutinovic V, Korolija N, Gaydadjiev G (2016) An AppGallery for dataflow computing. J Big Data

29. Bader DA, Pennington R (2001) Cluster computing: applications. Int J High Perform Comput 15(2):181–185
30. Blagojevic V et al (2016) A systematic approach to generation of new ideas for PhD research in computing. Advances in computers, vol 104. Elsevier, Amsterdam, pp 1–19

Chapter 4
Matrix-Based Algorithms for DataFlow Computer Architecture: An Overview and Comparison

Jurij Mihelič and Uroš Čibej

Abstract The focus of this chapter is on algorithms for dataflow computer architecture, which use matrices and vectors as the underlying data structure. DataFlow algorithms usually run as an optimized part of a classical control-flow procedure, which includes reading, writing, and other manipulation of data in such a way that it is easily processed by the dataflow part. Such data handling is usually referred to as data choreography and, in order to obtain an efficient algorithm, it must be appropriately examined. Our main goal is to describe various data choreographies that may be used when streaming matrices to a dataflow engine. Additionally, we have performed an experimental evaluation of these choreographies while using them with algorithms for various matrix-based computational problems. Finally, we have also given an overview of how several problems have been solved from graph theory using matrix multiplication as the basis.

4.1 Introduction

Most modern day computational technology has been dominated by the von Neumann architecture, which is so omnipresent that even many computer engineers are unaware of its alternatives. One of the reasons for this is simply because of the standard curriculum in computer science presents it as the only way computers are constructed. DataFlow architecture is one such alternative, which was once viewed as a competitor to the well-known von Neumann architecture, but has never managed to become a mainstream technology.

In the decades of its domination, users of von Neumann architecture have periodically been spoilt by its exponential improvements, which is most likely the main reason why any alternatives have not managed to keep pace. These exponential speedups were mainly due to an increase in clock speeds, which was the easiest

J. Mihelič (✉) · U. Čibej
Faculty of Computer and Information Science, University of Ljubljana,
Večna pot 113, Ljubljana, Slovenia
e-mail: jurij.mihelic@fri.uni-lj.si

© Springer Nature Switzerland AG 2019
V. Milutinovic and M. Kotlar (eds.), *Exploring the DataFlow
Supercomputing Paradigm*, Computer Communications and Networks,
https://doi.org/10.1007/978-3-030-13803-5_4

technique to increase the number of operations per unit of time. However, due to physical limitations, von Neumann architecture is no longer managing to provide exponential improvements.

More precisely, it is not possible to use the simple technique of increasing clock speed. This is due to the many engineering problems that occur at the manufacturing sizes that modern chips are being made. One of the biggest problems is heat removal, since thermal losses increase drastically with higher clock frequencies. So, instead of increasing clock frequency, chip manufacturers have tried another approach: increasing the number of CPUs on a chip, i.e., making multicore processors.

This increases the possibilities of parallelization, however, the responsibility for speedups is now entirely on the programmers (and/or compilers) who have to exploit multiple cores as efficiently as possible. This has proven to be a difficult task, so new hardware solutions have been sought. The two most successful have been, so-called, general-purpose graphical processing units (GPGPUs), and reconfigurable hardware based on field-programmable gate arrays (FPGAs) [6]. This chapter focuses on implementing dataflow algorithms on FPGA, more precisely on the Maxeler architecture [32], which uses FPGA as its enabling technology.

DataFlow architecture enables a programmer to more easily exploit parallelisms in an algorithm, however, it is definitely not a silver bullet. Naïve approaches to programming this architecture result in disappointing performance. In order to achieve the best possible performance, both the problem and the algorithm have to be suitable for the architecture. In addition, the programmer has to be aware of the many complex parts and finely tune all the parameters.

The main goal of this chapter is to present the many challenges a programmer is faced with when fine-tuning a particular algorithm for the given computer. In this exploration, we strived to follow the algorithm engineering methodology described in [2, 24, 25] in order to systematically and impartially evaluate the presented approaches.

The problems that we will deal with are basic linear algebra problems: multiplication of a matrix with a vector, and the multiplication of two matrices. We chose these two problems because they are practically very important and because, even though they appear to be very simple, there is a surprising amount of options for their implementation. The basic algorithms that we explore follow directly from the definition of these two operations, so a $\Theta(n^2)$ algorithm for multiplying a matrix with a vector and $\Theta(n^3)$ for matrix multiplication.

Another important approach to matrix multiplication is Strassen's algorithm [38], which is based on the divide and conquer algorithm design paradigm [10]. In practice, for smaller matrices, this algorithm (and many other asymptotically more efficient) must switch to basic multiplication, in order to maintain its efficiency. Thus, optimizing the basic multiplication has a potentially wider effect on many other practical multiplication approaches.

This chapter is structured as follows. In Sect. 4.2, we present the Maxeler dataflow computer architecture and its programming model, which we also used in our practical and experimental work. We also list several features and properties of problems suitable for the dataflow implementation as well as techniques, which may be used for acceleration of the algorithms. The focus of Sect. 4.3 is on the multiplication of a matrix with a vector. We discuss many options for the data choreography and experimentally evaluate them. In Sect. 4.4, we present the problem of matrix multiplication together with several algorithmic improvements and tuning techniques. We also add a presentation of experimental comparison and evaluation of these techniques. In Sect. 4.5 we explore the problems which can be solved via the matrix multiplication, such as problems in graphs. Finally, Sect. 4.6 concludes the chapter.

4.2 DataFlow Computation Paradigm

DataFlow paradigm is an abstract concept of computation, where the availability of data is guiding the execution of operations and not vice versa, as is the case with control-flow computers. A program for such a computer can be viewed as a directed graph, where nodes represent operations and edges are the flow of the data into these operations. Many different concrete approaches and implementations to executing such dataflow graphs are possible [37].

One direction for dataflow computers are token-based dataflow computers. The dataflow graph is decomposed into a set of tokens, and the computer has to check the availability of data for each token and fire the execution of tokens if all the data is available. These are usually designed to be general-purpose computers, two of the most prominent examples of such computers include the Manchester dataflow computer [15] and the MIT tagged-token dataflow architecture [1].

The second direction is based on, so-called, spacial computation [3]. Since the emergence of FPGAs, there have been various experiments into exploiting them by compiling algorithms straight into hardware. A dataflow graph is a very suitable candidate for the realization on an FPGA. By embedding a dataflow graph directly into hardware, operations are executed in parallel naturally, without any overhead for matching and firing the tokens. One of the more successful approaches of this type is the Maxeler architecture [21]. It has proven to be a very successful approach in many practical applications [12, 22, 39].

In this section, we will describe this architecture in more detail. In order to fully exploit the potential of such architecture, we also identify specific features which make a problem amenable to speedups. We also present a set of the most common acceleration techniques used. This is a very useful practical toolbox of methods, which one can attempt to use when engineering a specific algorithm.

4.2.1 Maxeler Architecture

Maxeler architecture is a hybrid architecture which utilizes both a classical central processing unit (CPU) as well as a so-called dataflow engine (DFE). These two units have the following functionality.

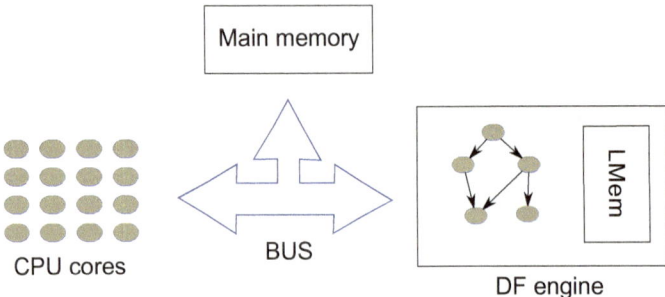

Fig. 4.1 A schematic overview of Maxeler architecture. The guiding program is run on the CPU cores that communicate through a bus with the DFE which is used to accelerate certain parts of the program

- A control-flow unit is responsible for the setup and supervision of the dataflow engine. Depending on the problem it does some (pre/post)processing of the data as well as running part of the algorithm for solving a specific problem.
- A DFE—a unit where the dataflow part of the computation is executed. This unit comprises an FPGA chip with some additional memory and logic. The dataflow on this FPGA is constructed using a set of Maxeler tools, which are responsible for compilation as well as the orchestration of the execution of a particular dataflow.

The two units communicate through input and output ports, all the communication being initiated by the CPU. In a nutshell, the execution of a program on this platform can be summarized as follows (See Fig. 4.1):

1. the CPU creates a set of data streams (in its main memory),
2. it pushes them through to the DFE, and
3. it receives the transformed data back to its main memory.

These three steps can be executed several times, depending on the application at hand. However, due to a relatively slow bus between the CPU and the DFE, the data transfer to/from the DFE is time-consuming. In order to overcome this bottleneck, the DFE has a substantial amount of local memory. Some intermediate results can thus be saved on the DFE, reducing the transfer times between the two computational units. The memory on the DFE is of two types. The first type of memory is Large Memory (LMem), which is a typical random access memory. The second type is Fast Memory (FMem), which is implemented on the FPGA. For a programmer these two types of memory differ in two important aspects. The first difference is the speed, LMem having a bandwidth in the order of GB/s, whereas the FMem has a bandwidth in the order of TB/s. The second difference is the size, FMem being very limited to typically a few megabytes, compared to several gigabytes for LMem. These differences guide the development of applications that are searching for a compromise between the speed and capacity of the available memory. Roughly speaking, FMem is used for temporary results, while main data is usually held in LMem.

In several of the following sections, we also present the results of experimentally evaluating various implementations of dataflow algorithms. In performing the corresponding experiments, we employed a computer, containing an Intel i7-6700K processor, with 8 MB and 64 GB of cache and main memory, respectively. The computer was equipped with a Maxeler dataflow unit, namely Vectis MAX3424A, available as a PCI-express extension card [31]. The unit is based on a Xilinx Virtex 6 SXT475 FPGA chip offering the following modules: 297600 lookup tables (LUT), 2×297600 flip-flops (FF), 2016 digital signal processors (DSP), and 2128 cells of 18 kbit block RAM (BRAM).

4.2.2   Suitable Problems for DataFlow Implementation

As is the case for any attempt at computational acceleration, there is no silver bullet for all the problems. The first task is to identify the characteristics of problems, those that are good candidates for Maxeler architecture. To be able to achieve an acceleration with a dataflow computer, the problem should exhibit one or several of the following features [17, 28, 36].

Amount of data: the characteristic of the Maxeler dataflow computer is the acceleration of data movement. In order to achieve significant speedups, the applications need to deal with a large amount of data. Such applications have recently been denoted as big data. The main speedups are gained when data is being streamed through a dataflow graph, i.e., the longer the graph, the greater the speedups. In addition, when running a program on a Maxeler computer, an initial delay is to be expected due to many factors, from the operating system schedulers, to the initial setup of all the data. Larger instances of problems are thus required to compensate for all such delays.

Data reuse: the most time-consuming part of the architecture is the transfer of data between the control-flow part and the dataflow part of a computer. So, the ideal situation is when the data can be moved to the DFE only once and then all the data manipulation can be done on the DFE.

Loop structure: the core of the algorithm should be constructed of simple loops that take the vast majority of time. These loops can be laid out in space, thus drastically increasing the number of operations performed in parallel.

Data independence: the patterns of data dependency must be relatively simple. The simplest possibility being an *embarrassingly parallel* problem where every data unit is completely independent from all the rest. Maxeler architecture can handle more complex data dependencies using a more complex topology of a dataflow graph, however, there are limits to this complexity because such a graph has to be physically laid out on the chip, and this is sometimes practically not possible.

Tolerance to initial delay: the algorithm needs to tolerate the initial delay before the dataflow engine starts actually processing the data. The data needs to travel through the entire chip with a frequency significantly lower than the frequency of a typical CPU. Currently, the ratio of frequencies is about 10. When this initial delay is overcome the results can be obtained with frequencies that can surpass the CPU frequency (because of the large parallelism that can be achieved inside the dataflow engine).

4.2.3   Acceleration Techniques

As previously mentioned, the speedups in dataflow implementations of algorithms are not trivially achievable. Previously we outlined some characteristics that the problem itself must have in order to be amenable for this approach. However, even when there is a perfectly suitable problem, we still need to address many specifics of the architecture to get a maximally fast implementation. (For a discussion on benchmarking see also [19], and for examples of successful dataflow algorithms see, e.g., [7, 14, 18, 30, 34].) Of course, this is true for any program on any platform, however, the differences on the Maxeler architecture are even more significant. The following list presents several acceleration techniques [17, 28, 36].

Data choreography: the planning of data movement is the most crucial factor. The DFE does not use prefetching in order to predict which data will be used next in the computation. Programmers need to ensure that the data is available for the computing element at the moment when it is required. Failing to do so can result in bottlenecks and thus severe slowdowns.

Efficient pipelining: programs can be written as very different configurations (i.e., directed graphs), which conceptually represent the same algorithm. However, slight variations, e.g., switch of the order of two (independent) operations, can result in drastic changes in performance. For beginners in this architecture this can be a frustrating experience, since users have been spoiled by modern compilers which usually offer a higher abstraction, eliminating the need for programmers to know the fine hardware details of the underlying platform.

Data precision: classical architectures have fixed size data types, e.g., a 32-bit integer or a 64-bit float. However, the Maxeler architecture is very flexible in defining data types and the data precision can be precisely specified for almost any operation. In many applications, it is possible to reduce the precision of the computation, while maintaining the correctness of the implementation. The result of the precision reduction is also the reduction of the transferred data, and the number of logical elements on the FPGA chip for the program.

Different algorithms: the mathematical analysis of an algorithm usually targets the RAM model, which models the actual performance on a von Neumann computer. In a dataflow world, such an analysis can give a completely false impression of the ranking of algorithms. When searching for the fastest algorithm for a Maxeler computer, a much wider set of algorithms should be considered, i.e., even algorithms that are deemed useless on a von Neumann computer [18, 34]. Each algorithm should have a rigorous empirical evaluation in order to extract the features that make them the most suitable in the given context.

In this chapter, we will explore these acceleration techniques to give a more detailed insight and a resource to learn how to program such computers more efficiently.

4.2.4   DataFlow Programming

In order for a programmer to devise a complete dataflow-based program, three components need to be written [21, 28, 39]:

CPU code: It is typically written in the C programming language. It controls the execution and uses the DFE as a processing unit by calling suitable functions exposed by a Maxeler compiler.

Set of kernels: Each kernel implements a certain functionality and is roughly an equivalent of a function abstraction. It has a set of input streams and a set of output streams attached.

Manager: The manager is the component that connects the data streams from the CPU to the recipient kernels and vice versa. It establishes connections between the kernels and the LMem as well as interconnecting the kernels. The manager also constructs the interfaces with which the CPU code interacts with the DFE.

The manager and the kernels are written in a domain specific language called MaxJ. This language is a superset of the Java programming language, with a few extensions which are more suitable for easier creation of the dataflow programs (Fig. 4.1).

The compiler transforms the description of the kernels into a dataflow graph which is then physically laid out on the FPGA chip by the compiler backend. The backend is typically very computationally intensive, since there are many structural constraints to be taken into account. A schematic overview of the programming components can be seen in Fig. 4.2.

4.3   Multiplication of a Matrix and Vector

In this section, we focus on dataflow implementations of the algorithms for the multiplication of a matrix with a vector, which is a fundamental operation in linear algebra. These algorithms are a basis of many scientific disciplines, such as applied mathematics, physics, and engineering. Often the dimension of input matrices and vectors are large, hence further improvements of such algorithms are always desirable.

In what follows, we show three different techniques for multiplication of a matrix and vector, which are suitable for dataflow architecture. Our categorization is mainly

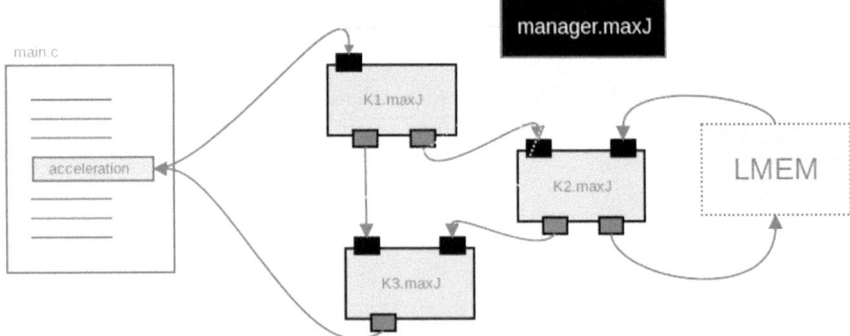

Fig. 4.2 The main program is defined in the .c file, where usually a loop is substituted with a call to the DFE. The kernels (K1, K2, K3) are specified in the MaxJ language, a superset of Java, and are reusable pieces of code that are unaware of the specific context in which they are being used. The third component of the program is the manager, also written in MaxJ. The manager specifies flow of data between the main program and the kernels

focused on the way the data is choreographed, i.e., how the matrix and vector are represented as streams: in particular, the matrix is examined by row, column, and stripped streaming. Moreover, for each technique, we also present several variations of parallelizing data access. Our approach to categorizing techniques is a systematic one and may be considered as an instance of the Mendeleyevization method [4], where the idea is to also explore categories with no examples in existing taxonomy; we draw the ideas from [26] for the purpose of presentation.

Of course, many other choreographies are possible: in general, one can process the elements of the input matrix in an arbitrary order while ensuring they are multiplied with the correct vector elements and accumulated into correct elements of the output matrix. However, not all choreographies are efficient in the exploitation of the dataflow as well as control-flow architecture, and hence may be deemed impractical. Nevertheless, we believe that our choice of choreographies represent the most efficient and practical solutions.

Furthermore, these techniques may also serve as an introductory example of dataflow programming. Other similar examples may be found in [21]. It should also be noted that in all three techniques, we only work with dense matrices, i.e., the majority of the elements are nonzero. Working with sparse matrices introduces many different techniques to succinctly represent such matrices, which are not, however, considered in this chapter.

Finally, it should also be noted that the problem we deal with in this case is not considered to be perfectly suitable for the dataflow architecture (e.g., a small loop body with few operations and low data reuse). Thus, its exploration is also interesting from a point of view of current and future possibilities and prospects.

4.3.1 Computer Representation

The term computer representation of vectors and matrices refers to their description in computer memory. Since in dense vectors and matrices most of the elements are nonzero, it is reasonable to use a computer representation which lists all their elements in a predefined order. A vector is usually represented with a sequence (i.e., an array) of elements. Often vectors are augmented with additional data, e.g., the length of the vector.

Two main representations exist for the matrices. The first is a *row-major* order, where elements are listed consecutively as they are ordered in rows from top to the bottom as well as from left to the right within each row, and, the second is *column-major* order, where elements are listed consecutively as they are ordered in columns from left to the right as well as from top to the bottom within each row. Hence, the element $a_{i,j}$ of the matrix stored as an array in computer memory can be obtained, in a row-major representation by taking the $(i \cdot n + j)$th element (i.e., the index of the element starting from 0) from the array, while in column-major representation it can be obtained by taking the $(i + j \cdot n)$th element. Let us denote with A the array to store the matrix; thus, we have

$$a_{i,j} = A[i \cdot n + j]$$

in row-major representation, and

$$a_{i,j} = A[i + j \cdot n]$$

in column-major representation.

Conversion of a matrix between the two representations is made straightforward by using the transposition operation. See Fig. 4.3 for details. However, if one wishes to perform the operation in-place, i.e., to reuse the storage space, transposition may become a much more elaborate task for non-square matrices.

Let A denote a matrix, then A^T denotes a transposition of A. If A is given in a row-major representation then A^T is its column-major representation and vice versa. In the following text, we assume the row-major order if not noted otherwise.

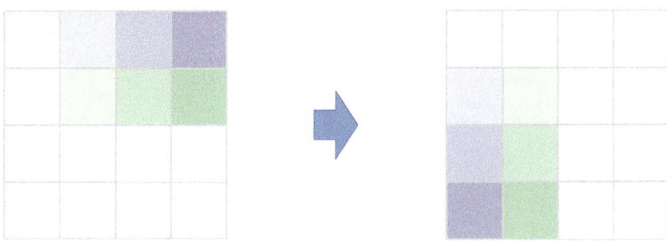

Fig. 4.3 Matrix transposition: the ith row becomes the ith column

4.3.2   Problem Definition

Let us begin with a more formal definition of the problem. Consider a matrix $A = [a_{i,j}]$ of dimensions $m \times n$, whereby m is the number of rows and n is the number of columns. Let $B = [b_i]$ be a vector of dimension n, where n is the number of elements. One can think of a vector as a matrix of dimension $n \times 1$, often called a column vector.

The result of the multiplication AB is a column vector $C = [c_i]$ of dimension m, where the ith element of the resulting vector C is a dot product of the ith row of the matrix A and the vector B, i.e.,

$$c_i = \sum_{j=1}^{n} a_{i,j} b_j, \tag{4.1}$$

where $1 \le i \le m$. In tabular form, multiplication of the matrix A and vector B is represented as follows:

$$\begin{bmatrix} a_{11} & a_{12} & \dots & a_{1n} \\ a_{21} & a_{22} & \dots & a_{2n} \\ \vdots & \vdots & \ddots & \vdots \\ a_{m1} & a_{m2} & \dots & a_{mn} \end{bmatrix} \begin{bmatrix} b_1 \\ b_2 \\ \vdots \\ b_n \end{bmatrix} = \begin{bmatrix} a_{11}b_1 + a_{12}b_2 + \dots + a_{1n}b_n \\ a_{21}b_1 + a_{22}b_2 + \dots + a_{2n}b_n \\ \vdots \\ a_{m1}b_1 + a_{m2}b_2 + \dots + a_{mn}b_n \end{bmatrix}.$$

It should be noted that in the computation of C, each element of matrix A is used only once while each element of vector B is used m-times. Since the elements of the resulting vector C do not depend on each other, at least some parallelization should be, in general, possible.

The definition can be translated into the algorithm shown in Fig. 4.4. The first loop iterates through all of the rows of matrix A while the second loop calculates (by accumulating it in c_i, which should be initialized to zero unless the previous content is added), the dot product as specified by Eq. (4.1). It should be noted that the order of loops can be arbitrarily permuted while still obtaining the correct algorithm.

Let us now count the primitive operations, i.e., scalar multiplications and additions, required to perform the matrix and vector multiplication: There are $mn = \Theta(mn)$ of both. Thus, the total number of elementary operations is

$$T_{vec}(m, n) = 2mn = \Theta(mn).$$

Fig. 4.4 The algorithm for multiplying a matrix and a vector

Input: a matrix A and a vector B
Output: the product vector $C = AB$
for $i = 1$ **to** m **do**
    for $j = 1$ **to** n **do**
        $c_i \leftarrow c_i + a_{i,j}b_j$

Hereinafter $m = n$ (one can also think of n as the larger dimension) is considered, and n is used as a notation for a square matrix of dimensions $n \times n$.

4.3.3   Rowwise Matrix Access

4.3.3.1   Basic Computation

First, let us explore the most straightforward approach to multiplication of a matrix and vector. In main memory, matrices are often stored in a row-major order, i.e., elements are stored sequentially by rows from left to right, starting in the top-left corner and ending in the bottom-right corner.

Considering a user's point of view, it is easiest to stream the matrix elements as stored in main memory, which gives the data choreography as depicted in Fig. 4.5. Observe that the matrix is streamed once as stored in main memory, while the vector is streamed once and used multiple times, i.e., once for each multiplication (dot product) with the corresponding row of the matrix. When processing the first row, the vector elements are read from the input and stored into fast memory of the DFE unit, while subsequent iterations read the elements from fast memory.

Such choreography of data results in the dataflow kernel computing the following function

$$vec_{row}(A, x) \mapsto Ax,$$

i.e., a function which takes a matrix A and a vector x and returns their product Ax. This kernel is obviously suitable for direct use in many algorithms.

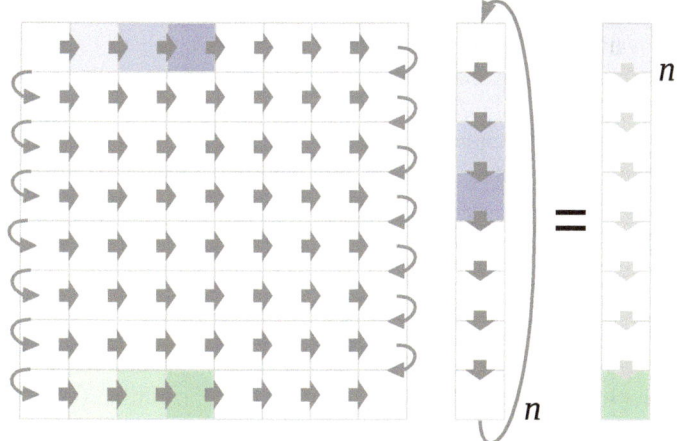

Fig. 4.5 Basic rowwise data choreography for multiplication of a matrix and vector

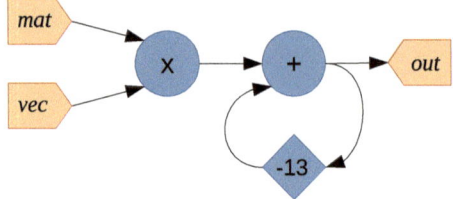

Fig. 4.6 A loop in the dataflow graph representing a calculation of a dot product

A disadvantage of such choreography is that it creates a loop in the resulting dataflow graph, hence causing an immediate slowdown of the whole kernel. In particular, each operation of addition (in the dot product calculation) has to wait for its result to become available and to be then reused again as an input. In Fig. 4.6 we give a simplified representation of the dataflow graph which corresponds to the calculation of a vector dot product.

In our hardware configuration the delay caused by the loop was 13 clock ticks. Notice that there are n additions (including the first adding to the initial zero) for each row of the matrix.

4.3.3.2   Replication of Stream Computation

The replication of a single stream computation (also called a pipe) within a kernel is a common design pattern for creating dataflow engine implementations. Using p pipes has the potential for a p-fold increase in the performance of a dataflow engine. However, other factors, such as maximum PCI-Express bus bandwidth may also affect the performance. The technique is in some sense similar to using several threads in control-flow processors.

Such replication of computation may sometimes mitigate an undesirable slowdown caused by a loop in the dataflow graph. Here, we present two options for such parallelization depending on the direction that the replication is performed, i.e., rowwise or columnwise. Both options are shown in Fig. 4.7a, b, respectively.

In the former, p elements in the corresponding row of the matrix are processed at a time and multiplied with corresponding p elements of the vector. The result is then added into the accumulator (containing partial sums) of length p. At the end of the row, these accumulated partial sums are summed together to obtain the final result, which is then output.

The easiest method for totaling the accumulated values inside the dataflow engine is to use the `TreeReduce.reduce` (or equivalent) function from the Maxeler MaxPower library [23], which constructs a binary tree of additions (other operations are also possible), and, thus, calculates the sum in $\Theta(\lg p)$ tree levels. See also Fig. 4.8) for an example of the reduce operation on the stream of eight elements. Another option is to output all accumulated values (i.e., p per each vector element) and do the summation in the control-flow part.

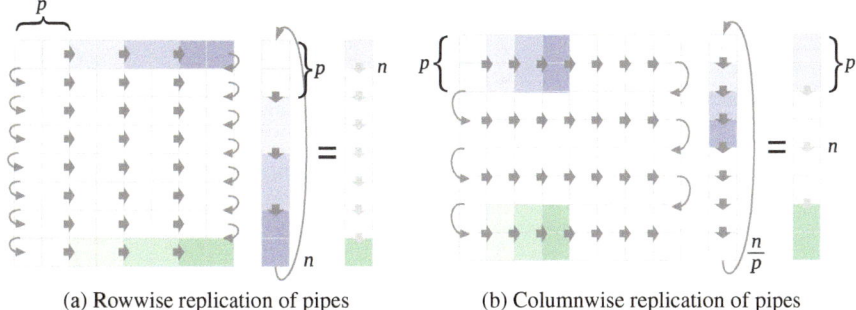

(a) Rowwise replication of pipes   (b) Columnwise replication of pipes

Fig. 4.7 Replication of stream computation for rowwise data choreography for multiplication of a matrix and vector

Fig. 4.8 The usage of the reduce operation to construct a binary tree of arithmetic operations

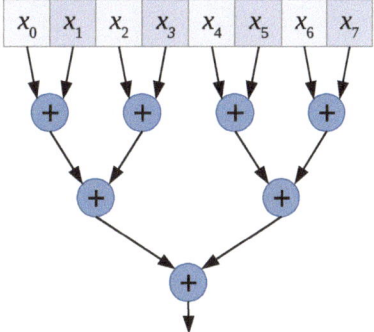

In the latter, p elements in the corresponding column of the matrix are taken and multiplied with the corresponding element of the vector; the result is again added to the accumulator, which is output at the end of a row. Hence, p elements of the output vector are produced at the same time. This technique needs no additional summation code, while a disadvantage is that the input matrix must be transformed to put the elements in the correct order. See also the stripped matrix techniques in Sect. 4.3.5 on how to transform the matrix into stripes of width p.

4.3.3.3 Performance Comparison

Now, let us look at experimental comparison of the described access techniques. In particular, we compare the running time and resource consumption of algorithms where we vary the number of dataflow pipes. The basic (no pipes) version of the algorithm is denoted with Row, and the vectorized versions with 2, 4, 8, 16, 32 are denoted with RowRowP2, RowRowP4, RowRowP8, RowRowP16, RowRowP32 and RowColP2, RowColP4, RowColP8, RowColP16, RowColP32 for rowwise and columnwise vectorization, respectively. The experiments were conducted on square input matrices and vectors of dimensions from 1024 to 16384 in steps of 1024.

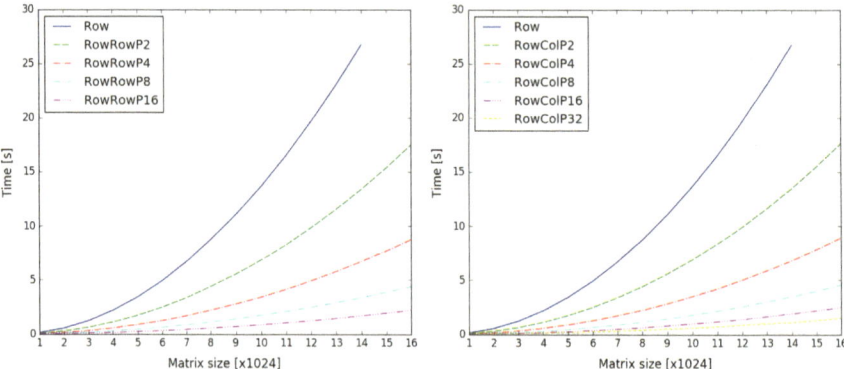

Fig. 4.9 Running-time comparison of rowwise matrix access according to the number of pipes

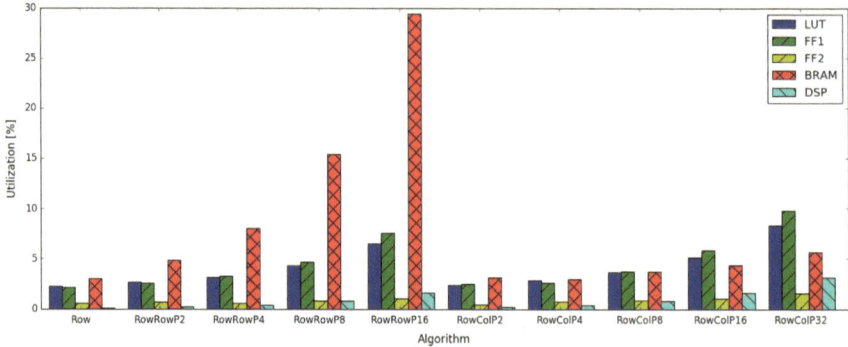

Fig. 4.10 Resource-consumption comparison of the rowwise matrix access techniques

For a graphical comparison of the running times see Fig. 4.9. Observe that, RowRowP32 is missing from the plot because we were unable to compile it due to substantial FPGA resource consumption. We also see that the algorithms using the same number of pipes perform very similarly irrelevantly to the rowwise or columnwise parallelization. Evidently, the larger the number of pipes, the faster the running time and better acceleration is achieved. In particular, for the larger matrices ($\geq$ 2084) the average acceleration over the basic version approximately equals the number of pipes: an exception here is RowColP32, which could not achieve 32-fold acceleration, but rather only about 25-fold.

Unfortunately, pipe-based parallelization is not infinitely scalable mainly for the following reasons: limited throughput and/or limited FPGA resources. See Fig. 4.10 for a comparison of the resource consumption. Here, using rowwise pipes, RowRowP16 is the maximum which can be compiled while using columnwise pipes there are still plenty of resources available even with the RowColP32 version. Thus, for the former, resources are the bottleneck (one may consider moving the reduce operation from the dataflow kernel to the control-flow code) while for the latter it is the throughput.

4.3.4 Columnwise Matrix Access

4.3.4.1 Basic Computation

As seen in the previous section, a loop in the dataflow graph, caused by a dependence of consecutive operations, results in a computation slowdown proportional to the size of the loop. Additionally, mitigation of this problem with large number of pipes may not be possible for the limited FPGA resources available.

Below, we present an alternative data choreography whereby the matrix is processed columnwise, i.e., the elements are accessed sequentially from top to bottom by columns, starting in the top-left corner and ending in the bottom-right corner. For a graphical representation of this data choreography see Fig. 4.11.

Initially, the elements of the first column of the matrix are multiplied with the first element of the input vector; the result is then added to the accumulator, which is initialized to zero. Thereafter, all the other columns are processed in the left to right manner. Observe that, in contrast to the technique from the previous section, the sequential results (products) are not dependent, since they are part of different dot products; hence, no additional explicit loop is needed.

Nevertheless, it is important to note that there is still a loop in the resulting dataflow graph. The reason for this is that, when a particular column is processed, the accumulated values from the previous iteration over the preceding column are needed. Fortunately, these accumulated values are already available when needed

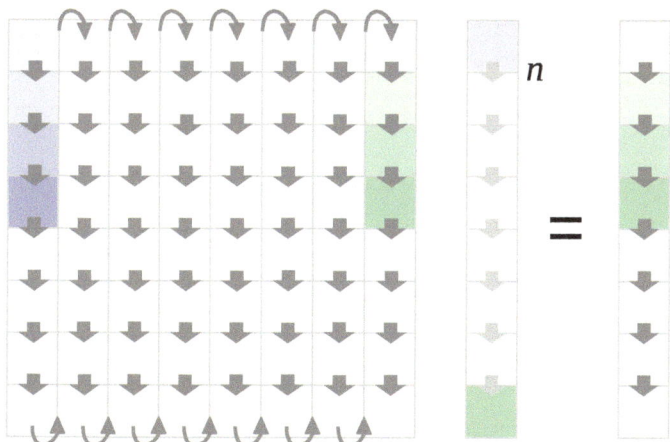

Fig. 4.11 Basic columnwise data choreography for multiplication of a matrix and vector

since the latency of the addition operation is in all practical cases much lower than the loop length.

To summarize, the loop length in this technique corresponds to the size of the vertical dimension of the input matrix. The loop is also *dynamic* because it depends on the input data, while the loop described in the previous section is *static* because it depends only on the dataflow computation.

Since matrices are usually stored in row-major order in a computer's main memory, the input matrix A must be transposed (see also Fig. 4.3) before it is fed to the kernel. The function the kernel computes is

$$vec_{col}(A, x) \mapsto A^T x.$$

4.3.4.2 Replication of Stream Computation

Below we explore two straightforward possibilities for additional parallelization using pipes. Figure 4.12 also shows respective data choreographies. The first choreography is based on the pipe parallelism in rows, i.e., to iterate through columns of the matrix, which have a width of p; thus, taking p elements at a time from a row of the matrix as well as p elements from the input vector and element-wise multiplying them together. Again, a dynamic loop is used to temporarily store the products, which are ultimately summed together to obtain the final result. Since the technique iterates through the whole column before proceeding to the next, the input vector is read only once; moreover, it is read in blocks of p elements with a waiting period of n ticks between the blocks to compensate for the column processing time. Finally, the matrix has to be preprocessed so the elements are available for the dataflow streaming in the order as shown in Fig. 4.12a.

The second choreography (see Fig. 4.12b) uses columnwise pipes, i.e., it takes several elements at a time from the current column, which are multiplied with the

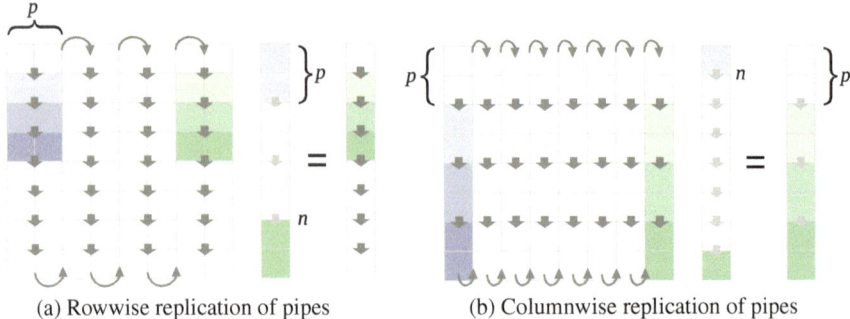

(a) Rowwise replication of pipes   (b) Columnwise replication of pipes

Fig. 4.12 Replication of a stream computation for columnwise data choreography for multiplication of a matrix and vector

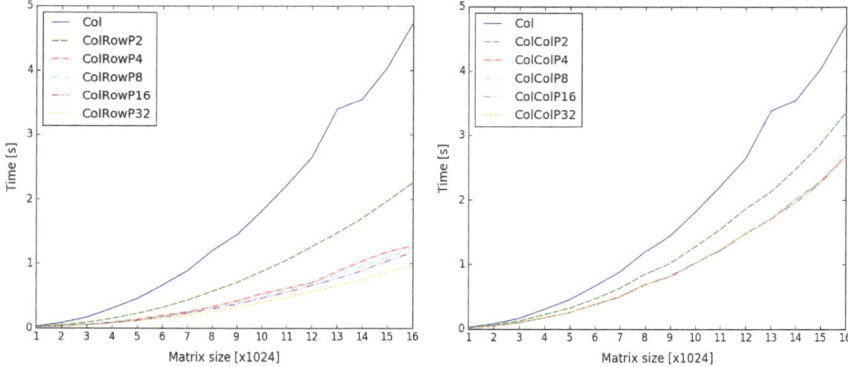

Fig. 4.13 Running-time comparison of rowwise matrix access according to the number of pipes

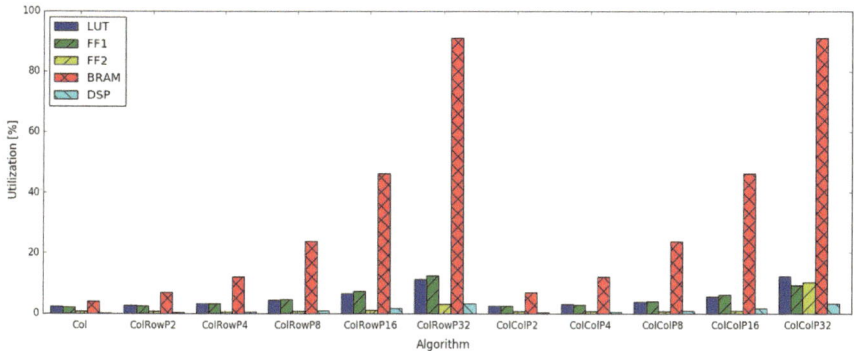

Fig. 4.14 Resource-consumption comparison of columnwise matrix access techniques

corresponding element in the vector. The products are then streamed to a dynamic loop to be available for summation when the next column is to be processed.

A disadvantage of columnwise techniques is the requirement of a dynamic loop whose length corresponds to the size of the matrix. Hence, a dataflow kernel reserving enough memory resources must be compiled in order to support matrices of larger sizes.

4.3.4.3 Performance Comparison

In Figs. 4.13 and 4.14, a comparison of running times and resource consumption can be found for the columnwise access techniques. Observe, that here, rowwise parallelization performs better than columnwise. In terms of resource consumption, the two variations are very similar, the bottleneck is reached in BRAM consumption at 32 streams.

4.3.5 Stripped Matrix Access

4.3.5.1 Basic Computation

In the previous two sections, we presented techniques including some of the draw-backs of using them. The first (rowwise) choreography, presented in Sect. 4.3.3, results in a static loop to overcome the latency of the addition operation, while the second (columnwise) choreography, presented in Sect. 4.3.4, results in a long dynamic loop intended to store intermediate results. Thus, the former technique introduces a multifold kernel slowdown, and the latter consumes significant amount of FPGA resources.

Below, we present another technique for data choreography, which mitigates both of the above-described issues. The main idea here is to combine both rowwise and columnwise techniques by decomposing a matrix into horizontal *stripes* (rowwise approach), which are furthermore processed in a vertical fashion (columnwise) as described in Sect. 4.3.4. See Fig. 4.15 for a graphical representation of the choreog-raphy.

Denote with s the stripe width, i.e., the number of elements in a column. Since the stripe length is n, we have $s \cdot n$ elements in total for each stripe. There are n/s stripes and for every stripe the whole vector must be streamed. Fast memory of the DFE unit is used for the re-streaming of the vector

The algorithm for transforming the given matrix A with rowwise stripes of width s into a stream of elements is shown in Fig. 4.16. Here the notation A^S is used to denote the stripped version of a matrix S. Observe that columnwise stripes are also possible and actually necessary in the choreography described in the previous section, where a rowwise replication of pipes with columnwise matrix access is used (see Fig. 4.12a).

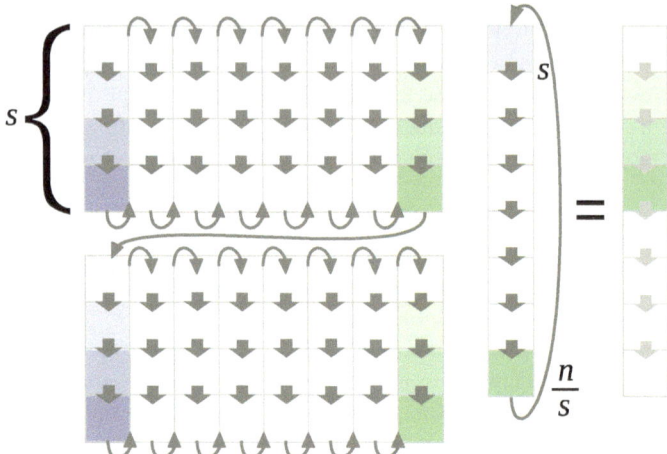

Fig. 4.15 Basic stripped data choreography for multiplication of a matrix and vector

Input: a matrix A, stripe width s
Output: the stripped matrix $B = A^S$
for $i = 1$ **to** n **step** s **do**
 for $j = 1$ **to** n **do**
  for $k = i$ **to** $i + s$ **do**
   $B[l++] \leftarrow A_{k,j}$

Fig. 4.16 The algorithm for transforming the matrix with rowwise stripes into a stream of elements

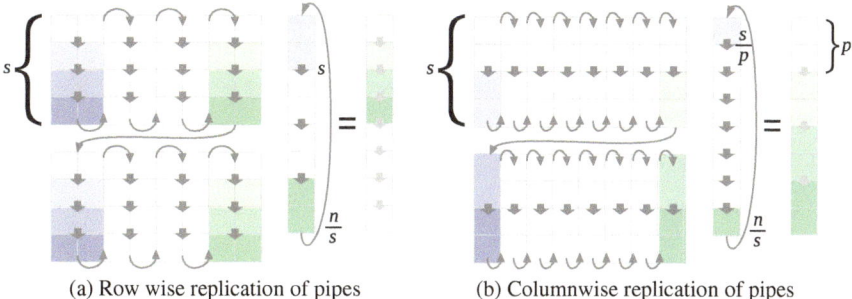

(a) Row wise replication of pipes   (b) Columnwise replication of pipes

Fig. 4.17 Replication of stream computation for stripped data choreography for multiplication of a matrix and vector

4.3.5.2   Replication of Stream Computation

Notice that processing of each stripe may be additionally parallelized using a pipe-based mechanism. See Fig. 4.17 for the two common techniques, both of which are described in Sect. 4.3.4. Here, each stripe is obviously parallelized separately. Additionally, observe that in rowwise parallelization $s \leq n$ and $p \leq n$ while in columnwise parallelization $p \leq s \leq n$ must hold.

4.3.5.3   Performance Comparison

In order to compare stripe-based approaches to data choreography, it is first necessary to inspect if the stripe size has any influence on the performance. As expected, the running time is approximately the same, unhampered by the stripe size, and a similar picture is also obtained when considering resource consumption. In particular, the running time for the largest matrix size is about 3.5 s while the resource consumption is below 3.6% in all five categories. We omit the corresponding comparison charts for all possible stripe sizes.

Now we focus on the pipe-based parallelization for which the charts of running times and resource consumption are shown in Figs. 4.18 and 4.19, respectively. Here we selected the stripe size of 128 and created several version of pipe-based parallelized algorithms. In particular, for 2, 4, and 8 pipes; for larger number of pipes the algorithms could not be compiled or they were no better in terms of performance. The

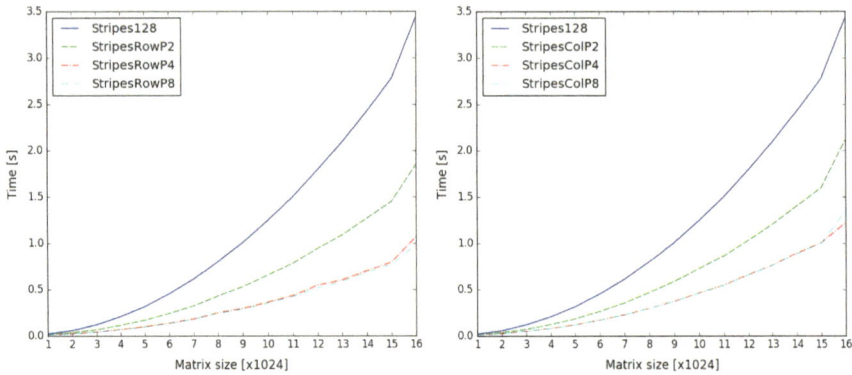

Fig. 4.18 Running-time comparison of stripped matrix access according to the number of pipes

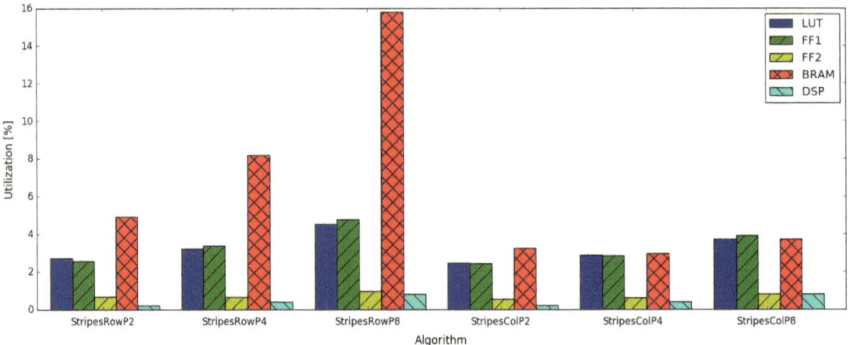

Fig. 4.19 Resource-consumption comparison of stripe-based matrix access techniques

latter can also be observed in the charts where 4- and 8-pipe versions exhibit a very similar performance; thus, the dataflow engine throughput is maximized. Finally, notice that rowwise parallelization produces slightly faster algorithms but are also more resource hungry.

4.3.6 Multiplying a Matrix with a Set of Vectors

4.3.6.1 Storing the Matrix in the DFE Unit

In Sect. 4.2.2 we stated several properties of the problems having a potential for an acceleration using dataflow architecture. Unfortunately, the problem of multiplying a matrix with a vector is not one which strongly exhibits such potential. The main reasons for this may be sought in the low reuse of data, reasonable sizes of inputs as well as too little computation per main loop iteration. Nevertheless, the techniques

Fig. 4.20 Enlarging a matrix from a dimension $n \times m$ to $n' \times n'$, where $n' \geq \max(m, n)$, and padding it with zeros

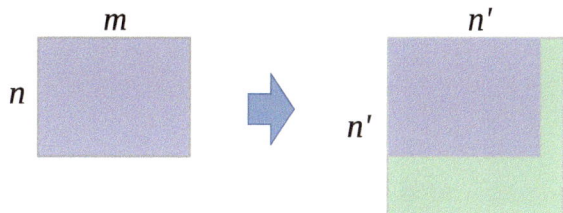

presented above in previous sections may have a potential future, when dataflow technology will catch up modern processors in terms of operational speed.

However, with a slight modification of the underlying problem, even today the techniques have great potential for use in practice. The extension of the problem presented uses many vectors, instead of one, to be multiplied with a given matrix. This way the data reuse criterion as well as massive data are now both more strongly satisfied. Observe that, now the matrix has been reused, not just the vector.

Some changes are needed to the implementations in order to store the matrix inside the DFE unit. Since fast memory may be too small to store matrices of a reasonable size, we chose DFE's large memory for this purpose. We base our implementation on the most advanced technique above, i.e., stripped matrix access. To do this, only the control-flow supervisor and the dataflow manager needs to be changed while the multiplication kernel remains unchanged. Observe also that, large memory transfers data in blocks of 384 bytes (may depend on the actual dataflow architecture). In our case, since one element of the matrix consumes 4 bytes, there is a constraint that a matrix dimension must be divisible by 96. If this is not the case, we resolve the issued by padding (in the control-flow part the matrix with zeros as is shown in Fig. 4.20. The same technique may also be used to obtain a square matrix.

4.3.6.2 Performance Comparison

In order to experimentally demonstrate the feasibility of the proposed approach, we performed the following experiment to compare control-flow and dataflow-based implementations. We base our dataflow implementation on the stripped matrix representation and use of 48 pipes with rowwise parallelization. The only adaptations needed are in the manager and control-flow part of the algorithm. In particular, we need to store the matrix into the large memory of the DFE before feeding it as a stream to the kernel.

We performed the experiments with matrices of dimensions 3072×3072 and the set of vectors of corresponding size, where the size of the set varies from $1 \cdot 1024$ to $20 \cdot 1024$. The results are shown in Fig. 4.21. The left y-axis gives the running time in seconds of the control-flow algorithm (denoted with cpu), and dataflow algorithm (denoted with dfe), while the right y-axis gives the acceleration (denoted with accel) achieved by using the dataflow approach. Observe that the acceleration is almost four times over the control-flow algorithm. Even though the dataflow cannot (yet) out-

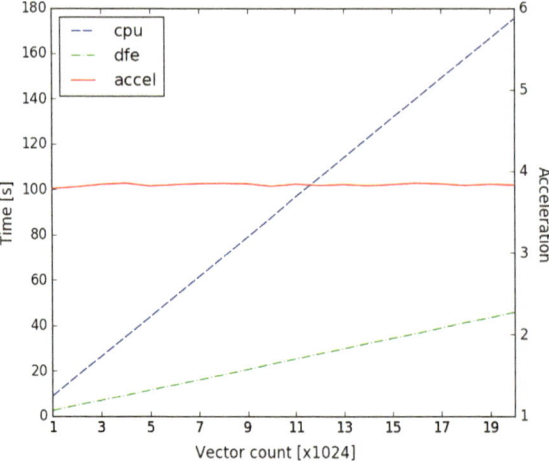

Fig. 4.21 Running-time comparison of multiplication of a matrix with a set of vectors

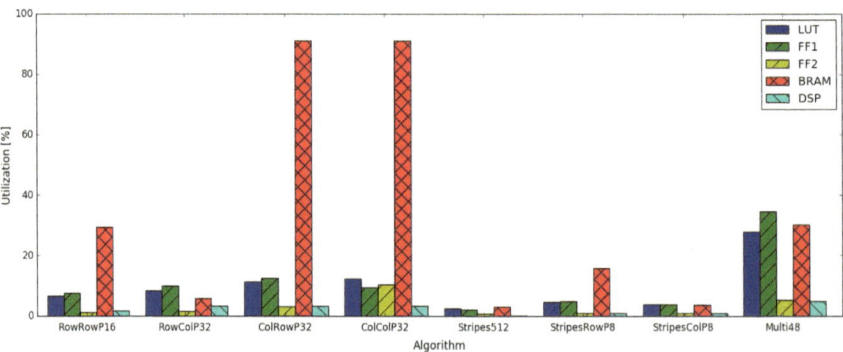

Fig. 4.22 Resource-consumption comparison of dataflow algorithms for multiplication of a matrix with a set of vectors

perform the control-flow when multiplying a matrix with one vector, the experiment clearly show that with many vectors a considerable speedup can be achieved. This holds true even for moderate matrix sizes and number of vectors.

In Fig. 4.22, we also give a resource consumption of the algorithm, which includes the best algorithms from each category. Observe that both algorithms using the row-wise choreography are economical in terms of resource consumption while both algorithms (only ColRowP32 shown) using the columnwise choreography are not. The former is also slow while the latter is much faster. The stripe-based choreography manages to combine the best properties from both worlds.

Table 4.1 Qualitative comparison of the techniques for multiplying a matrix with a vector

Algorithm	Matrix		Vector		DataFlow graph		
	Format	Minsize	Repeat	Wait	Loop	Ticks	Reduce
Row	Original	1	n	1	Static	$n^2 l$	No
RowRow	Original	p	n	1	Static	$n^2 l$	No
RowCol	Row-stripes	p	n/p	1	Static	$n^2 l$	Yes
Col	A^T	l	1	n	Dynamic	n^2	No
ColRow	Column-stripes	l	1	n	Dynamic	n^2	Yes
ColCol	A^T	pl	1	n/p	Dynamic	n^2	No
Stripes	Row-stripes	-	n/s	s	Dynamic	n^2	No
StripesRow	Row-stripes+pipes	-	n/s	s	Dynamic	n^2	Yes
StripesCol	Row-stripes	-	n/s	s/p	Dynamic	n^2	No

4.3.7 Discussion

The problem of multiplying a matrix with a vector is a very important one. Thus, it is worth researching and designing new algorithms as well as exploring new computer architectures. In the above sections, we presented several approaches to solving the problem with a dataflow architecture. We have show that in the specific domain of multiplying a matrix with a large set of vectors, it is worth considering a computation with a dataflow computer. However, in the simplest version with only one vector, the control-flow approach still dominates.

Table 4.1 shows a brief qualitative comparison of the techniques of multiplying a matrix with a vector. The first column of the table contains the short name of the approach. The second and the third columns describe the requirements for the input matrix and vector, respectively. Here, the sub-column format gives the input form required from the input matrix, while minsize denotes the minimum size for the matrix, the sub-column repeat gives the number of times the vector needs to be read while wait shows the wait time in ticks between two consecutive reads of the vector. The final column gives the properties of the dataflow graph: loop gives the type of loop in the graph, ticks the total number of ticks needed to perform a full multiplication, and reduce tells whether the additional operation of summation of accumulated values is also necessary.

To summarize, stripe-based choreography is probably the most usable approach in most cases. Nevertheless, rowwise choreography,despite its poor running-time performance, exhibits low resource consumption. On the other hand, columnwise choreography has good running times while also having large resource requirements.

4.4 Multiplication of Matrices

In this section, we focus on dataflow implementations of the algorithms for the multiplication of matrices. As is the case with matrix and vector multiplications, the problem is fundamental to linear algebra. It finds applications in computer graphics, machine vision, robotics, image manipulation, and signal processing.

Here, we mostly present various data choreographies and compare their performance on the dataflow architecture. We base our algorithms on those from the previous section on matrix and vector multiplication. Before delving into details, we describe the problem more formally, followed by various approaches to solving the problem, and, finally, we present results of the experimental performance evaluation, followed by a brief discussion.

4.4.1 Problem Definition

We start with a formal definition of the problem. Consider a matrix $A = [a_{i,j}]$ of dimension $m \times l$, where m is the number of rows and l is the number of columns, and a matrix $B = [b_{i,j}]$ of dimension $l \times n$, i.e., a matrix with l rows and n columns, i.e.,

$$
A = \begin{bmatrix} a_{1,1} & a_{1,2} & \cdots & a_{1,l} \\ a_{1,1} & a_{1,2} & \cdots & a_{1,l} \\ \vdots & \vdots & \ddots & \vdots \\ a_{m,1} & a_{m,2} & \cdots & a_{m,l} \end{bmatrix} \text{ and } B = \begin{bmatrix} b_{1,1} & b_{1,2} & \cdots & b_{1,n} \\ b_{1,1} & b_{1,2} & \cdots & b_{1,n} \\ \vdots & \vdots & \ddots & \vdots \\ b_{l,1} & b_{l,2} & \cdots & b_{l,n} \end{bmatrix}.
$$

The result of multiplication $A \times B$ is a matrix $C = [c_{i,j}]$ of dimension $m \times n$, the elements of which are

$$
c_{i,j} = \sum_{k=1}^{l} a_{i,k} \cdot b_{k,j}, \tag{4.2}
$$

where $1 \leq i \leq m$ and $1 \leq j \leq n$. As is the case in the matrix and vector multiplication, here the elements of C are also dot products of vectors, i.e., $c_{i,j}$ is a dot product of a ith row of A and jth column of B.

The definition can be translated into a naïve algorithm. See Fig. 4.23 for details. The first two loops iterate through all elements of the resulting matrix C in a row-major order, while the innermost loop calculates (by accumulating it in $c_{i,j}$, which should be initialized to zero) the dot product as specified by Eq. (4.2). Observe that the order of loops can be arbitrarily permuted while still obtaining the correct algorithm.

Let us determine the number of scalar operations needed to compute the matrix product. There are mn elements of C to calculate, each requiring l multiplications and l additions, giving the total count of elementary operations as

Fig. 4.23 The naïve matrix
multiplication algorithm

Input: a matrix A and a matrix B
Output: the product matrix $C = AB$
for $i = 1$ **to** m **do**
  for $j = 1$ **to** n **do**
    for $k = 1$ **to** l **do**
      $c_{i,j} \leftarrow c_{i,j} + a_{i,k} * b_{k,j}$

$$T_{mat}(m, n, l) = 2mnl = \Theta(mnl).$$

From now on $m = n = l$ are considered and the notation n is used for the dimension of the square matrix $n \times n$. Observe that, the above analysis of the naïve algorithm (i.e., by definition) for matrix multiplication gives computational complexity of $O(n^3)$.

4.4.2 Algorithmic Improvements

There are several algorithms that improve on the above bound for the naïve multiplication algorithm. One of them is the well-known Strassen's algorithm [10, 38] achieving complexity of $O(n^{\lg_2 7}) = O(2.808)$. It is based on the divide and conquer algorithm design method, where the product of two (block matrices of dimension 2×2 is calculated using only seven multiplications (of submatrices), which results in the above-stated complexity. Strassen's algorithm is usually considered to be practical since it is possible to devise an implementation which outperforms the naïve algorithm. To do this several algorithm engineering [24, 29] techniques must be used such as stopping the recursion with the naïve algorithm, use of cache-friendly loops, etc.

Unfortunately, almost all theoretically better algorithms are currently considered impractical due to their bad empirical performance and/or the complexity of their implementation. We give a list of several matrix multiplication algorithms in Table 4.2 together with their (approximate) year of development and a constant α such that the algorithm's complexity is $O(n^{\alpha})$. The horizontal line shows the divide between practical and theoretical algorithms.

4.4.3 Algorithm Tuning

Several computer-architecture-dependent techniques exist to speed up the above algorithms [24, 27], memory management being one of the most important issues [13]. One of the most common is to store matrices A and B in the computer main memory in such a way that the processor cache memory is efficiently used. In particular, the matrix A is stored in a row-major order, as usual, and B in a column-major

Table 4.2 Matrix multiplication algorithms and their computational complexities

Algorithm	Year	α	References
Naïve		3	[10]
Divide and conquer		3	[10, 35]
Strassen	1969	2.808	[10, 38]
Coopersmith-Winograd	1990	2.376	[9]
Stothers	2010	2.3736	[11]
Williams	2011	2.3728642	[40]
Le Gall	2014	2.3728639	[20]

order. Here, we assume sequential access of the elements [16]. Matrix B is thus transposed before it is used in a multiplication algorithm. Such techniques may give speedups of several orders of magnitude as cache memory may be up to 1000 times faster than main memory [5].

Another technique is to divide matrices into submatrices, as stated above in the divide and conquer paradigm, to obtain the so called *block* matrix multiplication. We can also use this technique in dataflow computation as presented in details in Sect. 4.4.5. it is also possible to also perform only a few divide steps and, thereafter, proceed with a naïve algorithm. See also [35] for an overview and experimental comparison of several matrix multiplication tuning techniques. In general, these techniques mostly exploit computer cache behavior to achieve performance improvements. A detailed cache analysis has already been done [8] and the cache-aware algorithm for idealized fully associative cache incurs $\Theta(\frac{n^3}{b\sqrt{M}})$ cache misses, where M is the number of cache lines and b is the number of bytes in each line.

A completely different approach is to exploit the parallelism possibilities of a given computer architecture. Generic algorithms exist for common shared-memory multiprocessors, which, again are based on the aforementioned divide and conquer design paradigm, where each of the eight multiplications of submatrices is performed in a separate thread of execution as is also the case for the four additions which are also necessary. The maximum possible speedup of the parallel algorithm is $\Theta(n^3/\log^2 n)$ however, unfortunately, is not practical due to the inherent cost of communication. More practical parallel algorithms achieve a speedup of $\Theta(n^2)$ [33].

Since in this chapter we are focused on the dataflow architecture, which has its own particularities, we use naïve algorithm as our starting point and explore other tuning techniques such as matrix transposition and block multiplication.

4.4.4 Naïve Matrix Multiplication

Let us first describe matrix multiplication algorithms which are based on the observation that columns (respectively rows) of a matrix of dimension $n \times n$ are basically vectors of dimension n. Furthermore, to multiply two matrices, one matrix can be decomposed into vectors, which are then multiplied with the other matrix. Here, we describe the following two possibilities to multiply matrices A and B:

- Decompose matrix A into rows A_i and then multiply every row A_i with matrix B to obtain rows of the product matrix $A \times B$.
- Decompose matrix B into columns B_j and then multiply every column B_j with matrix A to obtain columns of the product matrix $A \times B$.

In what follows we focus on the latter option while exploitation of the former may be carried out in a similar manner.

What we described above is basically a reduction of matrix multiplication to the problem of multiplying a matrix and a vector. Hence, the algorithms presented in the previous section can be completely reused for a new purpose. The only adaptation needed is primarily a minor modification of the control-flow program, which now takes two matrices as an input and produces one matrix as an output. Nevertheless, we have given three different data choreographies in the previous section. Namely, rowwise, columnwise, and stripped access are possible, which results in three different algorithms. We call these algorithms naïve, since they are straightforward to obtain and they also do not explore data reuse possibilities, which are necessary for adept dataflow performance. Let us briefly describe specifics of each.

4.4.4.1   Rowwise Matrix Access

First, let us take a look at the matrix multiplication algorithm which reuses the matrix-vector multiplication described in Sect. 4.3.3. Thus, matrix A is accessed in a rowwise order. See also Fig. 4.24 for a graphical representation of data choreography during the algorithm. Matrix B is decomposed into n columns, where each column is stored in the dataflow engine fast memory and repeatedly multiplied with matrix A, which must, hence, be fed n-times into the matrix-vector multiplication kernel.

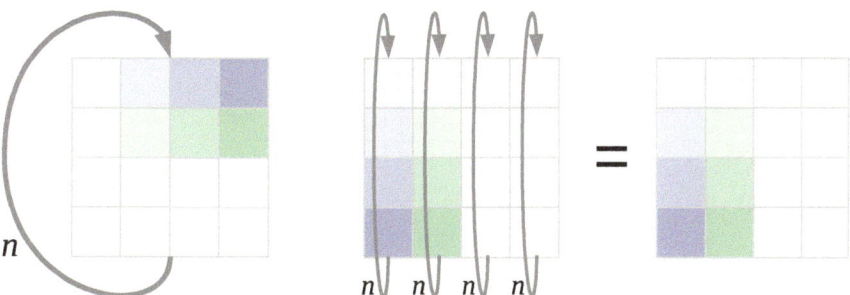

Fig. 4.24 Data choreography in naïve algorithms for matrix multiplication based on rowwise matrix access

Since matrices are stored in a row-major order in main memory, the decomposition of a matrix into rows is straightforward, however, decomposition into columns requires transposition. Consequently, the function such an algorithm computes (if none of the input matrices are transposed) is

$$mat_{row}(A, B) \mapsto (AB^T)^T = BA^T,$$

where we used well-known algebraic equalities, i.e., $(AB)^T = B^T A^T$ and $(A^T)^T = A$.
   Now, to obtain $C = AB$, there are two options:

- To transpose both the input matrix B as well as the output matrix. Indeed,

$$mat_{row}(A, B^T) = (AB)^T = C^T.$$

- To transpose B and also swap the inputs A and B. Thus,

$$mat_{row}(B^T, A) = AB = C.$$

Obviously, the latter option requires one transposition less than the former.

4.4.4.2 Columnwise Matrix Access

Here we reuse the matrix-vector multiplication algorithm that is described in Sect. 4.3.4. Matrix A is accessed in a columnwise order. See also Fig. 4.25 for a graphical representation of data choreography which is used in the algorithm. Again, matrix B is decomposed into n columns, where each column is stored in the dataflow engine fast memory and repeatedly multiplied with matrix A. However, the main difference is that the columnwise matrix-vector dataflow kernel requires the matrix A to

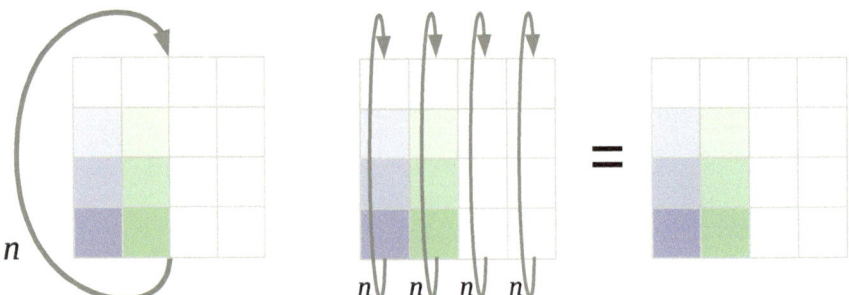

Fig. 4.25 Data choreography in naïve algorithms for matrix multiplication based on columnwise matrix access

be transposed before it is streamed; one can also consider that the matrix is processed by columns.

Since the decomposition into columns requires transposition, the function such an algorithm computes (if none of the input matrices are transposed) is

$$mat_{row}(A, B) \mapsto (A^T B^T)^T = BA.$$

Similarly to that above, to obtain $C = AB$, there are two options:

- To transpose both input matrices A and B as well as the output matrix. Thus,

$$mat_{row}(A^T, B^T) = (AB)^T = C^T.$$

- To swap the inputs A and B without any transpositions. Thus,

$$mat_{row}(B, A) = AB = C.$$

Obviously, the second option seems much more inviting as it requires three transpositions less than the first.

4.4.4.3 Stripped Matrix Access

For completeness, let us also give a few words about the reuse of the matrix-vector multiplication algorithm that is described in Sect. 4.3.5. Here, matrix A is decomposed into stripes which are streamed in a columnwise order. The graphical representation of the algorithm is similar to that presented in Figs. 4.24 and 4.25, except that matrix A is as seen in Fig. 4.15.

Since the decomposition into columns requires transposition, the function such an algorithm computes (if none of the input matrices are transposed) is

$$mat_{row}(A, B) \mapsto ((A^{-S})^T B^T)^T = BA^{-S}.$$

Similar to the above, to obtain $C = AB$, there are two options:

- To stripe-transform the matrix A and transpose B as well as the output matrix. Indeed,
$$mat_{row}(A^S, B^T) = (AB)^T = C^T.$$

- To swap inputs A and B while also stripe-transforming the matrix B. Indeed,

$$mat_{row}(B^S, A) = AB.$$

4.4.4.4 Usage of DFE's Large Memory

All the algorithms described in Sect. 4.4.4 stream matrix A to the dataflow engine several times, while matrix B is streamed only once and its rows/columns are stored in the dataflow engine fast memory. Since streaming from main memory to the dataflow engine is considered to be slow, to speed up the algorithms we can store the matrix A in the dataflow engine large memory. Most of the implementation remains the same, except a change in the dataflow manager code, where the input to the kernel is configured to be read from the DFE's large memory. Similarly, the control-flow code is extended with a few lines to store the matrix A into large memory before the kernel is used.

A drawback of using the large memory is that data is streamed in bursts of specific size (depending on the model), so the matrix must be padded with zeros to conform to the large memory constraints.

4.4.5 Block Matrix Multiplication

Here we discuss another popular technique for performing matrix multiplication, which is appropriate for large matrices and does not consume too many resources. In particular, each matrix is decomposed into many submatrices of a predefined size, and only these smaller matrices, called blocks, are multiplied with the kernel. The main goal for a programmer is thus to produce an implementation which efficiently multiplies two blocks. The technique is in many ways similar to the divide and conquer method for algorithm design [10].

4.4.5.1 Block Matrices

A block matrix is one that is interpreted as being composed of (nonoverlapping) submatrices called *blocks*. It can also be visualized as a collection of vertical and horizontal lines that partition the matrix into a collection of smaller submatrices. In general, blocks may be of different sizes, thus, there are many possible ways to interpret a particular matrix as a block matrix. In what follows we consider all blocks to be of the same size and the size is denoted with b.

Consider matrices $A = [a_{i,j}]$ of dimension $m \times l$ and $B = [b_{i,j}]$ of dimension $l \times n$. Now, a block matrix A with q row partitions and s column partitions, and a block matrix B with s row partitions and p column partitions are

$$
A = \begin{bmatrix} A_{1,1} & A_{1,2} & \cdots & A_{1,s} \\ A_{1,1} & A_{1,2} & \cdots & A_{1,s} \\ \vdots & \vdots & \ddots & \vdots \\ A_{q,1} & A_{q,2} & \cdots & A_{q,s} \end{bmatrix} \text{ and } B = \begin{bmatrix} B_{1,1} & B_{1,2} & \cdots & B_{1,p} \\ B_{1,1} & B_{1,2} & \cdots & B_{1,p} \\ \vdots & \vdots & \ddots & \vdots \\ B_{s,1} & B_{s,2} & \cdots & B_{s,p} \end{bmatrix},
$$

respectively. The matrix product $C = AB$ can be formed blockwise, yielding a $m \times n$ matrix C with q row partitions and r column partitions, where

$$C_{i,j} = \sum_{k=1}^{p} A_{i,k} \cdot B_{k,j},$$

and $1 \leq i \leq q$ and $1 \leq j \leq r$. Block product can only be calculated if blocks of matrices A and B are compatible, i.e., when the number of columns of block $A_{i,k}$ equals the number of rows of block $B_{k,j}$, for each $1 \leq i \leq q$, $1 \leq j \leq r$, and $1 \leq k \leq p$. In what follows we consider $m = n = l$ and $p = q = r$ as well as that p divides n: such blocks are always compatible.

4.4.5.2   Block Multiplication

Let us first explain how two blocks (often called tiles) are multiplied. Of course one could use some of the naïve approaches presented above. However, since we know that blocks are small compared to the whole matrix we may explore a different approach of parallelizing the multiplication.

The block multiplication algorithm is shown as MaxJ code in Fig. 4.26. In the proposed approach one row of the first block is multiplied with one column of the second block at the same time. Hence, the total tick count of the kernel equals to the number of the elements in the block. In order to do this, we replicate the stream

```
1  CounterChain cc = control.count.makeCounterChain();
2  DFEVar i = cc.addCounter(blockSize, 1);
3  DFEVar j = cc.addCounter(blockSize, 1);
4
5  DFEVar matA = io.input("matA", valType);
6  DFEVar matB = io.input("matB", valType);
7
8  DFEVar[] summands = new DFEVar[blockSize];
9  for (int k = 0; k < blockSize; k++) {
10     DFEVar a = Reductions.streamHold(stream.offset(matA, +k), j === 0);
11
12     Memory<DFEVar> bBuf = mem.alloc(valType, blockSize);
13     bBuf.write(j, matB, i === k);
14     DFEVar b = stream.offset(bBuf.read(j), + blockSize * blockSize);
15
16     summands[k] = a * b;
17  }
18
19  DFEVar matC = FloatingPointMultiAdder.add(summands);
20  io.output("matC", matC, valType);
```

Fig. 4.26 Source code in MaxJ programming language for multiplication of blocks

calculation using the static loop of length equal to the block size (one dimension, denoted with blockSize).

First, the counter chain is initialized (lines 2 and 3) with two counters i and j, where the latter is nested. Next we define (lines 5 and 6) address addr corresponding to j. Both inputs (block elements) are read (lines 8 and 9) using the streams aIn and bIn; here, valType denotes the type of elements. The main static loop calculates the products of elements in the corresponding row and column into the array summands defined in line 11. To obtain dot products these elements are then summed (line 22, function reduce) and the sum is the output (line 22).

Now let us explain the stream replication using the static loop (lines 11 to 19). In line 13, a "future" offset of k elements into the stream aIn is used and put into variable a. Instead of a, one can imagine, due to the static replication, a set of a's, each a_k having different lookup into the stream. Additionally, a stream hold construct with $j===0$ condition is used causing the value of a to remain the same while a particular row is processed, effectively causing the a_k containing the kth element of the row.

The second block is reused in the computation and is thus stored in the fast memory of the dataflow engine: each row resides in its own buffer bBuf (line 15) while kth buffer is filled in when ith row of the second block is read in (line 16). Similarly to the above, we need a set of b's containing the jth column of the second block, which can straightforwardly be obtained by reading the jth element, i.e., from address *addr* in the corresponding bBuf (line 17). Additionally, a future stream offset of full block size is used to ensure that the buffers are already filled in. Finally, the corresponding a's and b's are multiplied and the result is stored in the array summands.

4.4.5.3   Discussion

Several improvements of the above technique are possible. For example, double buffering may be used to improve the performance of fast memory reads and writes: for details see "Dense Matrix Multiplication" project on Maxeler's AppGalery [22].

Observe also, that the control-flow part of the algorithm must be carefully implemented, in order to stream the input blocks in the correct order and also to properly process the resulting block, which must be added to the previous blocks. Additionally, some blocks may also be stored in the dataflow engine large memory.

4.4.6   Performance Comparison

In this section, we compare several of the above matrix multiplication techniques. In general, the block multiplication technique should produce the fastest algorithms which, at the same, are also resource friendly. However, our experimental goal is not so much to identify the superior algorithm in terms of performance, but to determine

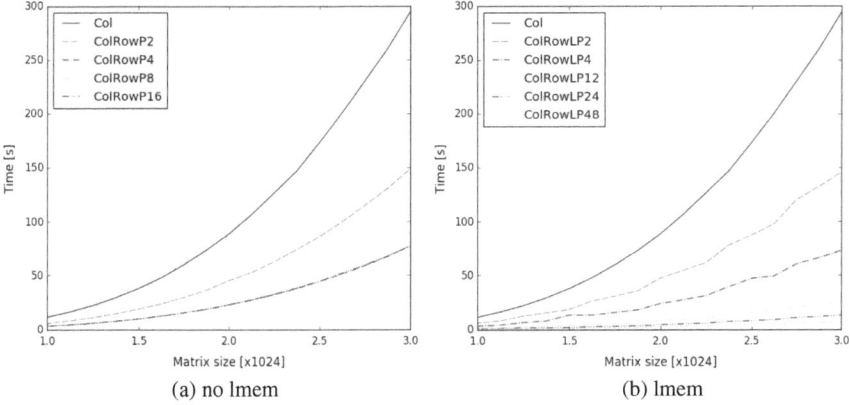

Fig. 4.27 Running-time comparison of columnwise matrix access according to the number of pipes

how different data choreography techniques compare to each other as well as how various dataflow techniques influence performance.

Let us first focus on the naïve approaches, where rowwise, columnwise, or stripped matrix choreography is used for repeated multiplication of a matrix with a vector. Due to its ineffectiveness (as described and shown by experiments in Sect. 4.3.3), the first has been skipped and therefore only the last two have been compared. In our first set of experiments we are interested in how the number of pipes influence the running time (see Fig. 4.27a) as well as the effect of the use of large memory with a various number of pipes (see Fig. 4.27b). In the former approach as many as 16 pipes are used, however, 4 or more pipes do not exhibit any significant improvement because the throughput of the available PCIe bus is already maximized. In the latter, as many as 48 pipes are used, where each increase in the number of pipes causes a notable improvement in performance. Here, the 48 pipes case was the highest that we were able to compile.

Similar results were obtained for the stripped matrix access. See Fig. 4.28 for running-time charts. Observe that the same levels of parallelism can be exploited as is the case with the columnwise technique. We were also able to compile the algorithm with 96 pipes, however, the experiment shows that 48 pipes already achieve the maximum throughput.

Now we focus on the block multiplication. As can be observed in Fig. 4.29a this group of algorithms was much better. In particular, when the matrix size is 3072×3072, one of the slowest block-based algorithms, i.e., Block128, the running time is about 6 s, while the running time of the fastest stripped-based algorithm StripesRowLP96 is about 7 s. In order to demonstrate the scalability and practicality of the algorithm (when used with larger matrices) we also present a graph of the running-time performance up to matrices of dimension 10240×10240. Observe that the larger the block size the better the performance. See also Fig. 4.29b for another similar comparison, where the ordinate axis uses logarithmic scale.

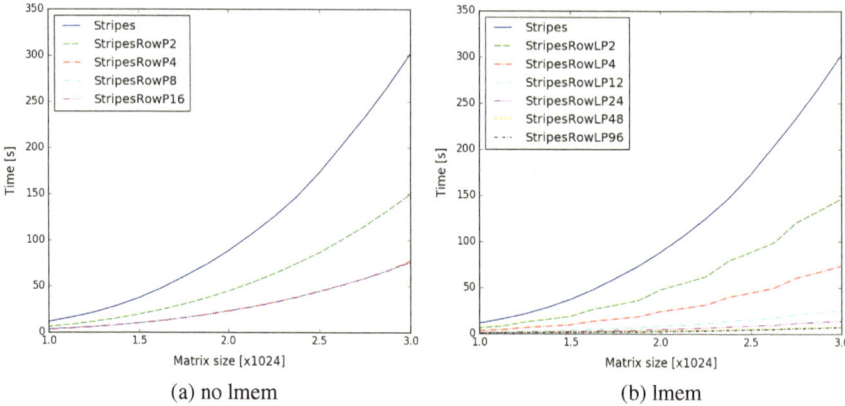

Fig. 4.28 Running-time comparison of stripped matrix access depending on the number of pipes

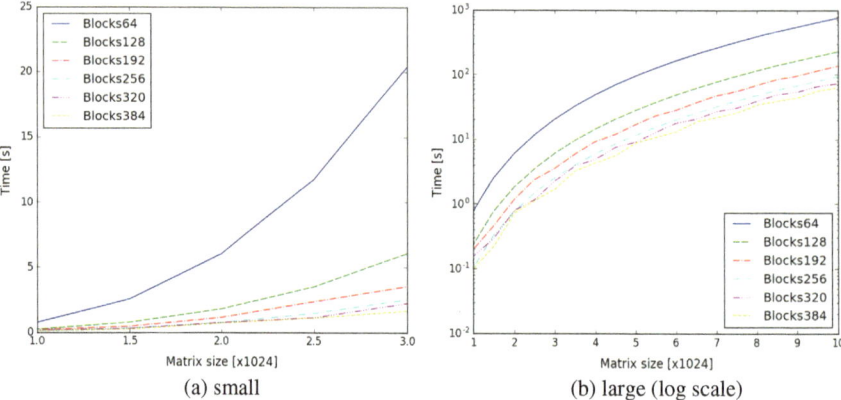

Fig. 4.29 Running-time comparison of block-based matrix access depending on the number of pipes

Now, we give a brief comparison of a representative technique from each of the different choreography techniques. Let us also include the control-flow implementation of the algorithm to give a better overview on the comparison. Note that the control-flow implementation is not highly optimized; however, we employed the classic technique of transposing the second matrix before multiplication, in order to get a better performance of cache memory due to a decrease of cache misses.

All the dataflow algorithms (selected for this comparison) are more efficient in running time performance than the control-flow algorithm. The performance of column-wise and stripped-based techniques is very similar (when the same number of pipes is used) while block multiplication with a block size of at least 128×128 outperforms all the other algorithms.

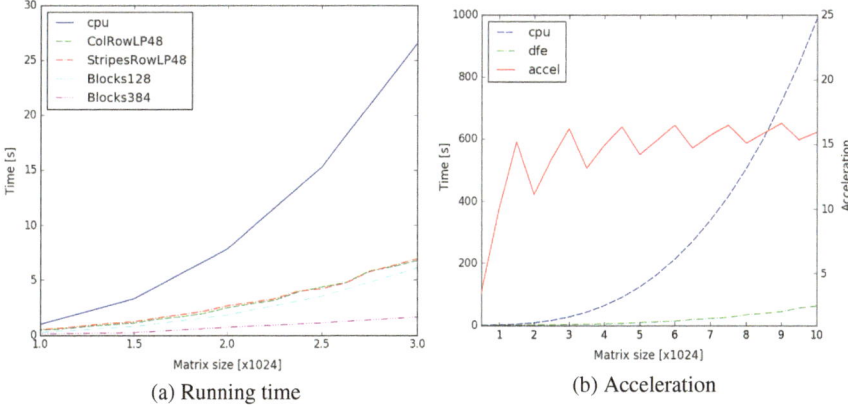

Fig. 4.30 Running-time comparison of various matrix multiplication algorithms depending on the number of pipes

Fig. 4.31 Resource-consumption comparison of block-based matrix multiplication algorithms

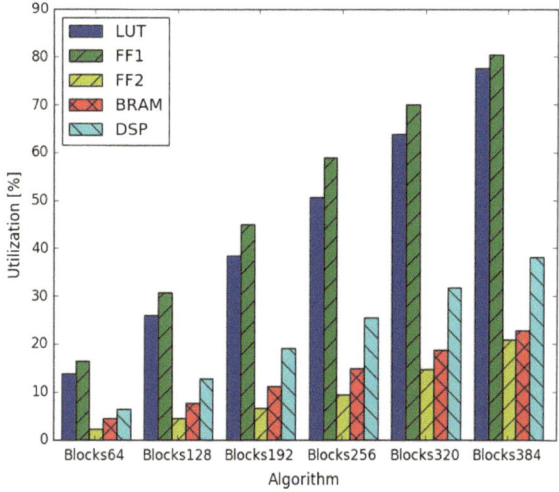

To conclude this comparison of the efficiency of various approaches, let us discuss the 1 acceleration, which can be obtained with the dataflow-based algorithm over the control-flow only implementation. See Fig. 4.30b for the plot of the running time (left y-axis) and acceleration (right y-axis). Observe that the acceleration achieved is about 15.

Finally, let us also present the resource consumption of the block multiplication algorithms. See Fig. 4.31. Observe that the consumption increased linearly with the block size. Here, the largest block size of 384×384 consumes already about 80% of the FF1 resources, thus not leaving much space for any other dataflow kernels.

4.5 Extending Matrix Algorithms

4.5.1 Matrix Exponentiation

Matrix exponentiation is a common mathematical extension of matrix multiplication. Given a matrix A, the kth power of A is defined as $A^k = A \cdot A \cdots A$, where multiplication is used $k - 1$ times. It can also be defined recursively as $A^k = A^{k-1} \cdot A$. Here, $A^1 = A$ and/or $A^0 = 1$, where 1 is the identity matrix, i.e., all elements are zero except the elements on the diagonal which are 1.

Naturally, these definitions lead to a straightforward algorithm for matrix exponentiation, where $O(k)$ matrix multiplications need to be performed. However, one can do a lot better.

A well-known technique for speeding up the exponentiation is matrix squaring: first, the square (A^2) is calculated, then a square of the square, i.e., $A^4 = (A^2)^2$, and so on, until A^k is obtained. Observe that one must be careful if k cannot be directly obtained by squaring. In particular, the algorithm is based on the following equation:

$$A^k = \begin{cases} A \cdot (A^2)^{\frac{k-1}{2}} & \text{if } k \text{ is odd} \\ (A^2)^{\frac{k}{2}} & \text{if } k \text{ is even.} \end{cases}$$

Hence, the matrix is first squared and the exponent is halved, where for odd exponents an additional multiplication is necessary. Observe also that the squaring trick would not work if the matrix multiplication had not been associative.

The number of matrix multiplications in this algorithm is $\sim 2 \log_2 k$, i.e., in each step there are at most two multiplications, and thus the asymptotic complexity (for the number of multiplications) is $O(\log k)$. Obviously, the total complexity of the matrix exponentiation is $O(n^3 \log k)$, where the dimension of the matrix A is $n \times n$ and k is the exponent .

In order to perform matrix exponentiation using the dataflow engine, only one of the above techniques is required, i.e., by repeated multiplication or by squaring, implementing them in the control-float part while calling the dataflow kernel from matrix multiplication, as described in Sect. 4.4.

4.5.2 Counting Walks in a Graph

Now let us focus on a completely different problem, where, given a graph $G = (V, E)$ with a set of vertices V and a set of edges $E \subseteq V \times V$, it is necessary to determine the number of different walks between any two vertices of the graph G. A *walk* of length l is a sequence of l edges or $l + 1$ vertices, i.e., $(v_1, v_2), (v_2, v_3), \ldots, (v_l, v_{l+1})$ or $v_1, v_2, \ldots, v_{l+1}$, respectively, where the source vertex of the then succeeding edge is the destination vertex of the preceding one. Observe that a walk allows for repeating of edges and vertices, as opposed to *trails* and *paths*, respectively.

Let A be the adjacency matrix of dimension $n \times n$ of a graph G with $n = |V|$ vertices, i.e., $m_{i,j} = 1$ if and only if $(i, j) \in E$; all the other elements are zero. Based on the formula $A^l = A^{l-1} \cdot A$ we observe that

$$a_{i,j}^l = \sum_{k=1}^{n} a_{i,k}^{l-1} a_{k,j}.$$

Here a^l denotes the element from the matrix A_l (not the exponentiation). Assume that A_{l-1} contains the number of walks of length $l - 1$. Then $a_{i,k}^{l-1} a_{k,j}$ is the number of walks of length $l - 1$ between the vertices i and k multiplied with the number of walks of length 1 (which can only be 0 or 1) between k and j. The vertex k thus has a role of intermediate vertex; now, going through all possibilities, we conclude that A_l contains the number of walks of length l. The base case of induction, i.e., that the adjacency matrix A contains the number of walks of length 1, is easy to see.

Hence, matrix exponentiation can be simply exploited to calculate the number of walks of length l in a given graph G. We explore this idea in more depth further in the next section.

4.5.3 Counting Triangles in a Graph

Here, we show how to count triangles in a given graph G using the algorithm for matrix multiplication. Let $\Delta(G)$ denote the number of triangles in graph G. A triangle in a graph is a cycle of length three, i.e., a walk of length three where the starting and terminating vertex are the same. Observe that in such a walk, no vertex can repeat. We can find such walks in the diagonal of matrix A^3, where A is the adjacency matrix of graph G. Notice that each triangle is counted six times as each of the three vertices can be a starting point of the cycle and there are two possible directions.

In order to calculate the total number of triangles in graph G, we need to sum the elements in the diagonal of the adjacency matrix A of graph G. Such a sum is often called the *trace* of matrix A and denoted with $\mathrm{tr}(A)$. Finally, the trace is divided by six to obtain the correct answer, i.e.,

$$\Delta(G) = \mathrm{tr}(A^3)/6.$$

Again, the dataflow-based algorithm can be most simply developed by using two matrix multiplications using the dataflow engine, and then performing the rest of the calculation in the control-flow part.

Table 4.3 Semirings and their use

R	$\oplus$	$\odot$	$\bar{0}$	$\bar{1}$	Problem
$\mathbb{R}$	$+$	$\cdot$	0	1	Real matrix multiplication
$\{0, 1\}$	$\vee$	$\wedge$	0	1	Vertex reachability
$\mathbb{R} \cup \{\infty\}$	min	$+$	∞	0	All-pairs shortest paths
$\mathbb{R} \cup \{-\infty\}$	max	$+$	$-\infty$	0	Critical paths
$\mathbb{R}^+ \cup \{\infty\}$	min	max	∞	0	Minimum spanning tree
$\mathbb{R}^+ \cup \{\infty\}$	max	min	0	∞	Widest path

4.5.4 Semiring Generalizations

Matrix multiplication can be generalized in the sense that the two operations needed, i.e., scalar addition and multiplication, can be replaced. In particular, many problems can be solved if the two operations conform to the *semiring* properties. A semiring is an algebraic structure $(R, \oplus, \odot)$, where R is a non-empty underlaying set of elements, and $\oplus$ and $\odot$ are two binary operations, called addition and multiplication, respectively, on the elements from R, with the following properties:

- $(R, \oplus)$ is a commutative monoid with the identity element $\bar{0}$,
- $(R, \odot)$ is a monoid with the identity element $\bar{1}$,
- left and right distributivity of multiplication over addition,
- the additive identity $\bar{0}$ annihilates R, i.e., $r \odot \bar{0} = \bar{0} \odot r = \bar{0}$ for every $r \in R$.

In Table 4.3 we give several examples of semirings together with the problem that they solve.

4.5.5 All-Pairs Shortest Paths

We demonstrate how easy it is to create a dataflow algorithm for one of the semiring-based problems on the well-known all-pairs shortest paths problem [10]. Given a graph $G = (V, E)$ with a cost matrix A, where $a_{i,j} \in \mathbb{R}$ gives a cost for each edge $(i, j) \in E$; if $(i, j) \notin E$ then $a_{i,j} = \infty$, the goal of the all-pairs shortest path problem is to compute the cost of a shortest path for each and every pair of vertices of G. Of particular interest is the matrix D, where $d_{i,j}$ is the cost of a shortest path form i to j.

There are several algorithms for solving this problem and the one based on the semiring theory is to replace $\oplus$ with min and $\odot$ with $+$, i.e., by using semiring $(\mathbb{R} \cup \{\infty\}, \min, +, \infty, 0)$. Now, assume that A^{l-1} contains the costs of shortest paths containing at most $l - 1$ edges (i.e., hops), then by $A^l = A^{l-1}A$ we have

$$A^l_{i,j} = \min_{k=1}^{n}\{a^{l-1}_{i,k} + a_{k,j}\}.$$

Observe that the minimum goes through all the possible intermediate vertices k and that $a_{i,k}^{l-1} + a_{k,j}$ is the length of path going through k. Thus, by selecting the minimum one it gives the shortest. The number of possible hops has been incremented at each step of the algorithm. Finally, by observing that any path in a graph with n vertices can have at most $n - 1$ hops, it is evident that A^{n-1} contains the costs of shortest paths. Observe also that $A^l = A^{n-1}$ for all $l \geq n$. Thus the algorithm is based on repeated squaring until A^l is obtained for some $l \geq n - 1$. Asymptotic complexity of this algorithm is $O(n^3 \log n)$.

Now let us focus on the dataflow implementation. Consider the matrix multiplication given in Fig. 4.26. Only two lines need to be changed:

- line 16 becomes `summands[k] = a + b;`, and
- line 19 becomes `DFEVar matC = TreeReduce.reduce(new Min(), summands);`.

Obviously, similar changes may be necessary in the control-flow part of the matrix multiplication algorithm as well as adaptation to do a repeated squaring.

4.6 Conclusions

In this chapter, we focused on dataflow algorithms which use matrices and vectors as their underlaying data structure. In particular, we comprehensively discussed matrix and vector multiplication as well as matrix and matrix multiplication. Our main goal was to compare various implementation techniques for dataflow architecture wherein the main focus has been on the data representation and handling.

An experimental evaluation of the techniques showed that multiplying a matrix with a set of vectors has the potential to be solved more efficiently with the dataflow computer. Furthermore, even greater potential is exhibited for the matrix multiplication problem. Finally, we showed how to easily transform the basic multiplication algorithms to solve several problems from the graph theory, such as the all-pairs shortest paths problem.

We also listed several of the semiring-based problems which can be solved similarly to that of dataflow architecture. These problems represent an interesting research direction, where an experimental evaluation would give much deeper insights into the dataflow architecture potential.

Acknowledgements We would like to thank Matej Žniderič who performed the initial work of implementing and experimentally evaluating most of the algorithms as well as Ivan Milanković who handled Maxeler's dataflow computers. Gratitude also goes to Veljko Milutinovic and Nemanja Trifunović for allowing us to use their infrastructure and for all the ideas and discussions we had.

References

1. Arvind, Nikhil RS (1990) Executing a program on the mit tagged-token dataflow architecture. IEEE Trans Comput 39(3):300–318
2. Bader DA, Moret BME, Sanders P (2002) Algorithm engineering for parallel computation. Springer, Berlin, pp 1–23
3. Becker T, Mencer O, Gaydadjiev G (2016) Spatial programming with OpenSPL. Springer International Publishing, Cham, pp 81–95
4. Blagojević V, Bojić D, Bojović M, Cvetanović M, Đorđević J, Đurđević D, Furlan B, Gajin S, Jovanović Z, Milićev D, Milutinović V, Nikolić B, Radivojević Z, Protić J, Stanisavljević Ž, Tartalja I, Tomašević M, Vuletić P (2016) A systematic approach to generation of new ideas for PhD research in computing. Advances in computers, vol 104. Elsevier, Amsterdam, pp 1–19
5. Bryant RE, O'Hallaron DR (2010) Computer systems: a programmer's perspective, 2nd edn. Addison-Wesley Publishing Company, USA
6. Chen D, Cong J, Pan P et al (2006) Fpga design automation: a survey. Found Trends® Electron Des Autom 1(3):195–330
7. Čibej U, Mihelič J (2017) Adaptation and evaluation of the simplex algorithm for a data-flow architecture. Advances in computers, vol 104. Elsevier, Amsterdam
8. Cohn H, Umans C (2003) A group-theoretic approach to fast matrix multiplication. In: Proceedings of the 44th annual IEEE symposium on foundations of computer science. IEEE computer society, pp 438–449
9. Coppersmith D, Winograd S (1990) Matrix multiplication via arithmetic progressions. J Symb Comput 9(3):251–280
10. Cormen TH, Leiserson CE, Rivest RL, Stein C (2009) Introduction to algorithms, 3rd edn. MIT Press, Cambridge
11. Davie AM, Stothers AJ (2013) Improved bound for complexity of matrix multiplication. Proc R Soc Edinb Sect Math 143(2):351–369
12. Flynn MJ, Mencer O, Milutinović V, Rakocević G, Stenstrom P, Trobec R, Valero M (2013) Moving from petaflops to petadata. Commun ACM 56(5):39–42
13. Furht B, Milutinovic V (1987) A survey of microprocessor architectures for memory management. Computer 20:48–67
14. Guo L, Thomas DB, Luk W (2014) Customisable architectures for the set covering problem. SIGARCH Comput Archit News 41(5):101–106
15. Gurd JR, Kirkham CC, Watson I (1985) The manchester prototype dataflow computer. Commun ACM 28(1):34–52
16. Jacobs A (2009) The pathologies of big data. Commun ACM 52 (ACM Queue 7(6), 2010)
17. Korolija N, Popovic J, Cvetanovic M, Bojovic M (2017) Dataflow-based parallelization of control-flow algorithms. In: Hurson AR, Milutinovic V (eds) Creativity in computing and dataflow supercomputing, vol 104. Advances in computers. Elsevier, Amsterdam, pp 73–124
18. Kos A, Ranković V, Tomažič S (2015) Sorting networks on maxeler dataflow supercomputing systems. Adv Comput 96:139–186
19. Kos A, Tomažič S, Salom J, Trifunović N, Valero M, Milutinović V (2015) New benchmarking methodology and programming model for big data processing. Int J Distrib Sens Netw 11(8)
20. Le Gall F (2014) Powers of tensors and fast matrix multiplication. In: Proceedings of the 39th international symposium on symbolic and algebraic computation. ACM, pp 296–303
21. Maxeler Technologies (2015) Multiscale dataflow programming
22. Maxeler Technologies. Maxeler AppGallery. http://appgallery.maxeler.com/. Accessed 17 Feb 2017
23. Maxeler Technologies. Maxpower standard library
24. McGeoch CC (2012) A guide to experimental algorithmics, 1st edn. Cambridge University Press, New York, NY, USA, p 474
25. Mihelič J, Čibej U (2017) Experimental algorithmics for the dataflow architecture: guidelines and issues. IPSI BgD Trans Adv Res 13(1):1–8

26. Milutinovic V (1996) The best method for presentation of research results. IEEE TCCA Newsl
27. Milutinovic V (1989) High-level language computer architecture. Computer Science Press Inc.,
    New York
28. Milutinović V, Salom J, Trifunović N, Giorgi R (2015) Guide to dataflow supercomputing:
    basic concepts, case studies, and a detailed example. Computer communications and networks.
    Springer International Publishing, Berlin
29. Müller-Hannemann M, Schirra S (2010) Algorithm engineering: bridging the gap between
    algorithm theory and practice. Lecture notes in computer science. Springer, Berlin, Heidelberg
30. Oriato D, Tilbury S, Marrocu M, Pusceddu G (2012) Acceleration of a meteorological limited
    area model with dataflow engines. In: 2012 symposium on application accelerators in high
    performance computing (SAAHPC). IEEE, pp 129–132
31. PCI-SIG: Peripheral component interconnect special interest group (2015). 354–356. http://
    pcisig.com/. Accessed 17 Feb 2017
32. Pell O, Averbukh V (2012) Maximum performance computing with dataflow engines. Com-
    put Sci Eng 14(4):98–103. http://home.etf.rs/~vm/os/vlsi/predavanja/Part%202%20Paper
    %201%20Jakob%20Salom%20Selected%20HPC%20Solutions%20Based%20on%20the
    %20Maxeler%20DataFlow%20Approach%20(2).pdf
33. Randall KH (1998) Cilk: efficient multithreaded computing. PhD thesis, Massachusetts Institute
    of Technology
34. Ranković V, Kos A, Milutinović V (2013) Bitonic merge sort implementation on the maxeler
    dataflow supercomputing system. IPSI BgD Trans Internet Res 9(2):5–10
35. Rozman M, Eleršič M (2019) Matrix multiplication: practical use of a strassen-like algorithm.
    IPSI BgD Trans Internet Res 15(1)
36. Salom J, Fujii HA (2013) Selected HPC solutions based on the Maxeler data-flow approach
37. Šilc J, Robič B, Ungerer T (2012) Processor architecture: from dataflow to superscalar and
    beyond. Springer, Berlin
38. Strassen V (1969) Gaussian elimination is not optimal. Numer Math 13(4):354–356
39. Trifunović N, Milutinović V, Salom J, Kos A (2015) Paradigm shift in big data supercomputing:
    dataflow vs. controlflow. J Big Data 2(1):1–9
40. Williams VV (2011) Breaking the Coppersmith-Winograd barrier

Chapter 5
Application of Maxeler DataFlow Supercomputing to Spherical Code Design

Ivan Stanojević, Mladen Kovačević and Vojin Šenk

Abstract An algorithm for spherical code design, based on the variable repulsion force method is presented. The iterative nature of the algorithm and the large number of operations it performs make it suitable for implementation on dataflow supercomputing devices. Gains in computation speed and power consumption of such an implementation are given. Achieved minimum distances and simulated error probabilities of obtained codes are presented.

5.1 Introduction

A spherical code is a set of N D-dimensional real vectors on the unit sphere. Two standard optimization problems are associated with spherical codes:

- given N and D, find a spherical code such that the minimum Euclidean distance between any two code vectors is maximized over the set of all such codes (packing problem);
- given N and D, find a spherical code such that the radius of equally sized balls centred at code vectors and whose union contains the unit sphere, is minimized over the set of all such codes (covering problem).

Here, only the first problem will be addressed. The design of spherical codes is both an interesting theoretical and practical problem. There are only a few cases in which the exact solution is known [1, 2, 4, 9, 11, 28, 29].

I. Stanojević (✉) · V. Šenk
Faculty of Engineering (a.k.a. Faculty of Technical Sciences), University of Novi Sad,
Novi Sad, Serbia
e-mail: cet_ivan@uns.ac.rs

V. Šenk
e-mail: vojin_senk@uns.ac.rs

M. Kovačević
Department of Electrical and Computer Engineering, National University of Singapore,
Singapore, Singapore
e-mail: mladen.kovacevic@nus.edu.sg

© Springer Nature Switzerland AG 2019
V. Milutinovic and M. Kotlar (eds.), *Exploring the DataFlow
Supercomputing Paradigm*, Computer Communications and Networks,
https://doi.org/10.1007/978-3-030-13803-5_5

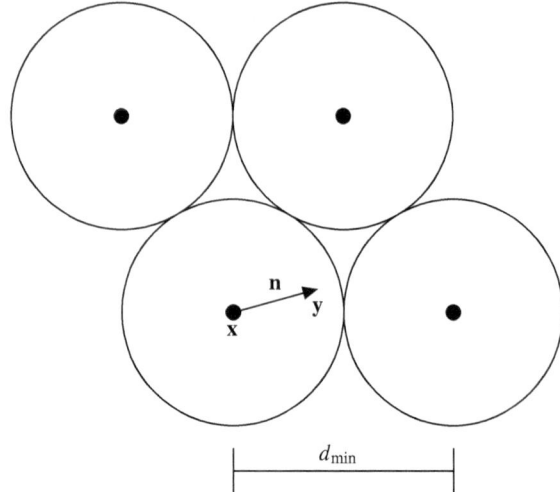

Fig. 5.1 Spherical code used as a channel code, $\mathbf{y} = \mathbf{x} + \mathbf{n}$ ($\mathbf{x}$ - transmitted vector, $\mathbf{n}$ - noise vector, $\mathbf{y}$ - received vector); the minimum noise amplitude, $|\mathbf{n}|$, which can produce an error is $d_{min}/2$

Different constructions of spherical codes can be used for source coding [17, 30] and channel coding [7, 14, 19, 36, 38], or both [15, 16], as illustrated in Fig. 5.1. By increasing the minimum distance of the code while keeping the number of vectors and their length fixed, noise resilience of the code is improved, for the same transmitted energy per symbol (vector) and the same code rate.

For other interesting properties of spherical codes, see also [3, 10] (universal optimality), [6] (stability), [12, 37] (rigidity) and [39] (exact description using algebraic numbers).

Dataflow supercomputing implemented on Maxeler systems [27] is a new computing paradigm, which allows more time and energy efficient implementations of algorithms that are inherently parallelizable. The main idea is translation of the code that performs an algorithm into a dataflow structure in which streams of data are exchanged between simple processing blocks of programmable logic. If the number of operations that can be performed at the same time is high, acceleration can be achieved by using space multiplexing in an FPGA circuit, for which the host machine only provides input data and collects its output.

There are several distinctive differences between dataflow acceleration and other competing technologies such as simultaneous execution of multiple threads on the multicore central processing units (CPUs) or many-core processing units (most well-known representatives of which are general-purpose graphics processing units or GPGPUs), together termed control-flow systems:

- In the dataflow approach, the most expensive part of the algorithm (in terms of the number of operations) is directly mapped into the corresponding dataflow structure, with a very deep pipeline (typically several thousand stages, limited only by the available resources of the FPGA), in which all the operations are executed every cycle. Unlike in control-flow systems, there is no "instruction execution" (fetching an instruction opcode from memory, decoding it, fetching the operands and storing the result to memory). Movement of data is minimized by feeding all the intermediate results directly into blocks which need them as inputs, thus eliminating memory as a possible bottleneck.

- Instead of minimizing latency, the aim of dataflow systems is maximizing data throughput. Scientific calculations most often require a large number of identical operations on input data and in such cases hardware resources are used efficiently in the deep pipeline, since except for the transient periods (initial filling of the pipeline and flushing it at the end) all the processing in it is done once every cycle, as its different stages act on different data.

- In dataflow systems, the programmer has complete control over the processing in the FPGA (which operation is done in which block and how the blocks are interconnected), including its exact timing. This is not the case in multicore/many-core systems, in which the runtime environment (the operating system scheduler in multicore systems, or the execution framework, such as OpenCL or CUDA, in many-core systems) assigns portions of code to be executed to different processor cores. Although the latter approach adds flexibility and simplifies programming, as it makes execution of the same code possible with different numbers of cores, it is usually not optimum, since the automatic assignment algorithm is unaware of the problem being solved and can introduce unpredictable delays due to data dependencies, which can be hand optimized in dataflow systems.

- Control-flow systems have a limited set of built-in number types, usually 8-, 16-, 32- or 64-bit integers/fixed-point numbers and 32- or 64-bit floating-point numbers. The programmer must use the "smallest" number type, large enough to hold values of a variable. This may use hardware inefficiently and increase the power consumption, as it may be possible to store values form the same range in a number type with a smaller number of bits. Since dataflow systems allow the processing structure to be defined at the level of logical units (e.g. flip-flops and logical functions), the minimum numbers of bits can be allocated for every variable, depending on the necessary value ranges. Maxeler library facilitates conversion between different number types in a design by built-in conversion substructures.

For some interesting applications of the dataflow supercomputing, see [18, 20, 21].

5.2 Optimization Methods

5.2.1 Direct Optimization

The optimization problem corresponding to optimal spherical code design can be formulated as follows: find D-dimensional vectors $\{\mathbf{r}_1, \mathbf{r}_2, \ldots, \mathbf{r}_N\}$ lying on the unit sphere, $|\mathbf{r}_i| = 1$, such that the cost function

$$U = - \min_{i \neq j} |\mathbf{r}_j - \mathbf{r}_i| \tag{5.1}$$

attains its minimum value. A usual procedure for finding the minimum of a function like this one is to create an initial set of vectors and iteratively adjust every one of them until a minimum is reached.

Although this optimization would produce a desired code, algorithms that perform it suffer from serious practical problems [31]. Since U only depends on the smallest distance between any two vectors, it locally depends only on the code vectors which have their nearest neighbours at the current minimum distance. If vectors are moved in the direction away from their nearest neighbours in every iteration, it is difficult to choose the length of those moves. Since the minimum distance should be increased at every move, when a vector has two or more neighbours at approximately the same minimum distance, the length of its move may be only very small, leading to extremely slow convergence.

Another problem is that U is not differentiable with respect to vector coordinates, so the gradient descent method or Newton's method cannot be applied directly to its minimization.

5.2.2 Variable Repulsion Force Method

A different approach, successfully used in the literature [9, 22–26], and in this work, is to choose the differentiable cost function (assuming no vectors overlap)

$$V = \sum_{i < j} \frac{C}{|\mathbf{r}_j - \mathbf{r}_i|^{\beta-2}}, \tag{5.2}$$

where $C > 0$ is an arbitrary constant and $\beta > 2$ is an adjustable parameter. V can be thought of as the potential of N particles repelling by the conservative central force (from the ith to the jth)

$$\mathbf{F}_{i \rightarrow j} \overset{\triangle}{=} - \nabla_{\mathbf{r}_j} \left(\frac{C}{|\mathbf{r}_j - \mathbf{r}_i|^{\beta-2}} \right) = C(\beta - 2) \frac{\mathbf{r}_j - \mathbf{r}_i}{|\mathbf{r}_j - \mathbf{r}_i|^{\beta}}. \tag{5.3}$$

As β is increased, the force other particles exert on one particle starts to be dominated by the force of its nearest neighbours. If $\beta \rightarrow \infty$, the force equilibrium (the stationary point of V) is attained at a position where any particle has two or more nearest neighbours at an equal distance which is as large as possible. In such a position, the force exerted on the jth particle only has the component normal to the surface of the unit sphere, i.e. collinear with $\mathbf{r}_j$.

Iterative minimization of V usually produces only a local minimum, and there is no guarantee that it is also a global minimum. In order to approach a global minimum as closely as possible, the procedure must be repeated multiple times with different initial vectors, which can be chosen at random. Since it is desirable that all randomly chosen points on the unit sphere be equally probable, i.e. that no directions be privileged, a simple two-step procedure is used [31]:

1. choose coordinates $g_{i,1}^*, \ldots, g_{i,D}^*$ as independent samples of a normalized Gaussian random variable ($\mathcal{N}(0, 1)$);
2. normalize the obtained vector,

$$\mathbf{r}_i = \frac{\mathbf{g}_i^*}{|\mathbf{g}_i^*|}. \tag{5.4}$$

For a fixed value of β, V can be minimized using the gradient descent method or Newton's method. Even though Newton's method has a faster rate of convergence, the gradient descent method is chosen here for the following reasons:

- The overall solution of the problem is not the minimum of V for a particular value of β, but for β as high as possible, so β must be gradually increased after every iteration. The number of iterations which produce the minimum for a fixed β is irrelevant, since after β is increased the minimization is restarted (with a better initial position).
- Gradient descent is much simpler and a single iteration is much faster.

The negative partial gradient of V with respect to the jth vector is

$$-\nabla_{\mathbf{r}_j} V = \sum_{i \in \{1,\ldots,N\}\setminus\{j\}} \mathbf{F}_{i \to j} \triangleq \mathbf{F}_{\to j}, \tag{5.5}$$

which is the total force exerted on the jth particle. In order to preserve the constraint that all vectors are on the unit sphere, their new values can be calculated by moving them along the directions of the corresponding total forces and normalizing them,

$$\mathbf{r}_j^* = \mathbf{r}_j + \alpha \mathbf{F}_{\to j}, \tag{5.6}$$

$$\mathbf{r}_j := \frac{\mathbf{r}_j^*}{|\mathbf{r}_j^*|}. \tag{5.7}$$

The constant $\alpha > 0$ determines the speed of the procedure and should be chosen so that the procedure is stable (vectors converge) and the error (the difference between the current and the final values) diminishes as fast as possible.

Although α could be optimized to achieve the previous goals, when β is high enough the iteration in (5.6) and (5.7) suffers from numerical difficulties in a finite precision implementation:

- if for some j, $|\mathbf{r}_j - \mathbf{r}_i| > 1$ for all $i \neq j$, $|\mathbf{r}_j - \mathbf{r}_i|^{-\beta}$ can be calculated as 0 in (5.3) due to numerical underflow and $\mathbf{F}_{\to j}$ can be calculated as 0 as well;
- if for some i and j, $|\mathbf{r}_j - \mathbf{r}_i| < 1$, the calculation of $|\mathbf{r}_j - \mathbf{r}_i|^{-\beta}$ in (5.3) can cause numerical overflow.

The likelihood of these conditions increases as β increases. In order to avoid these problems, a slight modification of the procedure will be used. Since every vector is moved along the direction of the corresponding force, that direction can be calculated by first scaling the force and then normalizing it. A convenient way to scale the force is

$$\mathbf{F}^*_{\to j} = \sum_{i \in \{1,\ldots,N\}\setminus\{j\}} \left(\frac{\mu_j}{|\mathbf{r}_j - \mathbf{r}_i|^2} \right)^{\beta/2} (\mathbf{r}_j - \mathbf{r}_i), \tag{5.8}$$

where

$$\mu_j = \min_{i \in \{1,\ldots,N\}\setminus\{j\}} |\mathbf{r}_j - \mathbf{r}_i|^2. \tag{5.9}$$

The value of the subexpression in the first parentheses in (5.8) is in $(0, 1]$ for all i, and is equal to 1 for at least one term, so neither underflow nor overflow can occur in the calculation of (5.8).

The normalized force

$$\mathbf{F}_{\to j} = \frac{\mathbf{F}^*_{\to j}}{|\mathbf{F}^*_{\to j}|} \tag{5.10}$$

is used in a modified version of (5.6),

$$\mathbf{r}^*_j = \mathbf{r}_j + \alpha_j \mathbf{F}_{\to j}, \tag{5.11}$$

where

$$\alpha_j = \frac{\mu_j}{2\beta}. \tag{5.12}$$

It is shown in the appendix that this value of the step size, α_j, ensures the stability of the process and fast convergence.

5.2.3 Force Loosening

In order to maximize the minimum distance, β should be as high as possible. If vectors are moved with a high β immediately from their random initial position, the convergence is slow and the obtained local minimum of V usually produces a lower minimum distance than if β is gradually increased (the force gradually loosened). Given $\mathbf{r}_1, \ldots, \mathbf{r}_N$ and β, the total move norm of one iteration,

$$S = \sum_j |\mathbf{r}'_j - \mathbf{r}_j|^2, \tag{5.13}$$

can be calculated ($\mathbf{r}'_1, \ldots, \mathbf{r}'_N$ are new vectors). It is observed in numerical results that for a continuous chain of iterations from a starting set of vectors and for fixed

β, S decreases asymptotically exponentially with the number of iterations. It is also observed that if β is increased by the same amount after each iteration (linearly), after a transient period S starts decreasing as $S \sim \beta^{-1}$. In order to reach high values of β quickly, yet without forcing the vectors into a position of numerical deadlock (which happens whenever β is increased too rapidly), the following strategy for increasing it is used:

1. Set $n := 1$ and $l := 2$.
2. Starting from $\mathbf{r}_1, \ldots, \mathbf{r}_N$ and using $\beta = 2^l$, calculate new vectors $\mathbf{r}'_1, \ldots, \mathbf{r}'_N$ and the corresponding total move norm S'.
3. Starting from $\mathbf{r}_1, \ldots, \mathbf{r}_N$ and using $\beta = 2^{l(1+1/n)}$, calculate new vectors $\mathbf{r}''_1, \ldots, \mathbf{r}''_N$ and the corresponding total move norm S''.
4. If $S' > S''$, set $\mathbf{r}_j := \mathbf{r}'_j$ for all j.
   Otherwise, set $\mathbf{r}_j := \mathbf{r}''_j$ for all j and set $l := l(1 + 1/n)$.
5. If more iterations are necessary, set $n := n + 1$ and go to step 2.

5.3 Implementation

The majority of operations in the iterative procedure is performed in the loop for calculating new vectors. This is the part in which a number of operations can be done at the same time and can thus benefit from hardware acceleration.

For convenience, an overview of steps performed for a single vector is given here:

1. $\mu_j = \min_{i \in \{1, \ldots, N\} \setminus \{j\}} |\mathbf{r}_j - \mathbf{r}_i|^2$.
2. $\mathbf{F}^*_{\to j} = \sum_{i \in \{1, \ldots, N\} \setminus \{j\}} \left(\frac{\mu_j}{|\mathbf{r}_j - \mathbf{r}_i|^2} \right)^{\beta/2} (\mathbf{r}_j - \mathbf{r}_i)$.
3. $\mathbf{F}_{\to j} = \frac{\mathbf{F}^*_{\to j}}{|\mathbf{F}^*_{\to j}|}$.
4. $\mathbf{r}^*_j = \mathbf{r}_j + \frac{\mu_j}{2\beta} \mathbf{F}_{\to j}$.
5. $\mathbf{r}_j := \frac{\mathbf{r}^*_j}{|\mathbf{r}^*_j|}$.

The new vector, calculated in the last step, can be stored:

- in the same memory location as the old one, which is more convenient for a software implementation, since only one block of memory is used;
- in the appropriate location of a different block of memory or another data stream, which is more convenient for a streamed hardware implementation, since it simplifies dependencies on input data.

Although the new values obtained in these ways differ, they can both be successfully used for finding good spherical codes.

5.3.1 Software Implementation

In the software implementation, vector coordinates are stored as floating-point numbers. Operations with them are performed by the floating-point unit (FPU) of the CPU and by the runtime library functions, which are highly optimized. Their data type is the standard IEEE 754 double precision type [40], with 53 bits of mantissa and 11 bits of exponent.

Since vector coordinates and vectors themselves are accessed sequentially in all the operations performed on them, the algorithm has good cache locality. Typically, for codes used in practice, values N and D are such that the total amount of iterated data is small enough to fit completely in the cache of modern processors.

5.3.2 Hardware Implementation

In the hardware implementation, vector coordinates are stored as signed fixed-point numbers with 3 integer bits and 44 fractional bits. Maxeler dataflow engines support highly configurable number formats and this choice is the result of the following limitations:

- Vector coordinates are in the interval $[-1, 1]$, their differences are in $[-2, 2]$ and their squares are in $[0, 4]$, so at least 3 integer bits are necessary.
- Some Maxeler library functions (e.g. for calculating square roots) have an upper bound on the total number of bits of fixed-point numbers they can operate on. Since other fixed-point types with a higher number of integer bits are used for storing intermediate results, a total of 47 bits is obtained as the highest value for which all the calculations can be performed using only library functions.

Although floating-point numbers can be used as well, fixed-point numbers occupy much less hardware resources.

Operations on vectors, such as addition (subtraction), multiplication (division) by a scalar or norm calculation are performed in parallel on all their coordinates, as shown in Figs. 5.2, 5.3 and 5.4. Although it may seem that these calculation structures yield their results instantaneously, it is not the case, since each of them introduces a pipeline delay of several cycles, which depends on used number types and low-level compiler decisions. This delay only increases the overall (long) pipeline delay, i.e. the time from the moment when the first input data block enters the dataflow structure until the first result block can be read at its output, which is added to the total calculation time only once (not once for every block of input data). If the number of cycles needed to perform all the calculations is much larger than the pipeline delay, that delay can be neglected, and it can be considered that all the operations in the structure are done in parallel, once every cycle.

In the algorithm above, μ_j, which is calculated in step 1, is used in steps 2–5. In the software implementation, two separate inner loops over i are necessary for steps

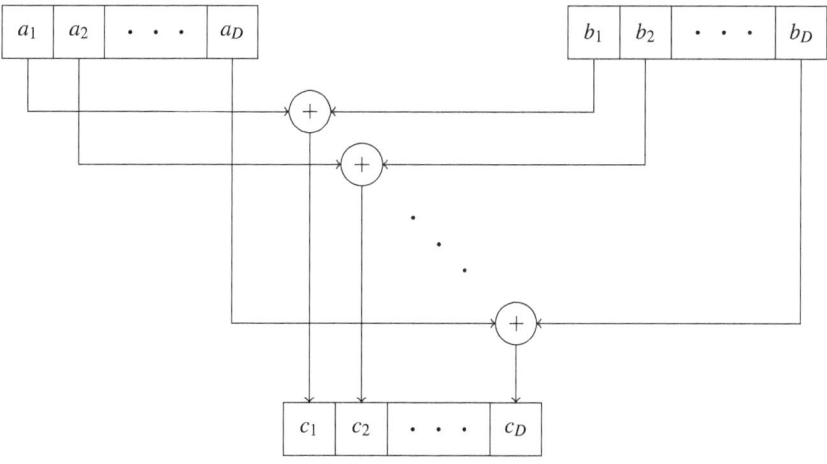

Fig. 5.2 Dataflow substructure for parallel addition of vectors

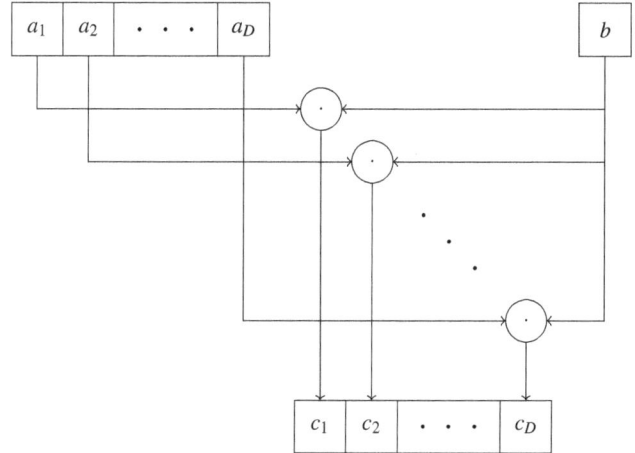

Fig. 5.3 Dataflow substructure for parallel multiplication of a vector and a scalar

1 and 2. Whereas, this is also the case in the hardware implementation, another level of parallelism is possible: while μ_j is being used in steps 2–5, μ_{j+1} can be calculated, as shown in Fig. 5.5. Since one inner loop (for calculating a single value of μ_j or $\mathbf{r}_j$) lasts N cycles, one complete iteration lasts $N(N + 1)$ cycles.

The maximum number of vectors, $N_{\max}$, and their maximum dimension, $D_{\max}$, must be hard coded in the dataflow structure. In order to provide a flexible solution, the actual number and dimension of vectors, N and D, are configured as runtime

Fig. 5.4 Dataflow
substructure for calculation
of the (squared) norm of a
vector

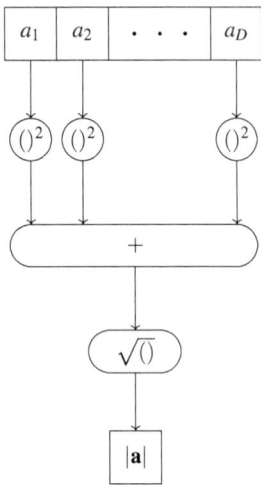

Fig. 5.5 Order of
calculation: dependent
values are in the same
columns, and independent
values calculated at the same
time are in the same rows

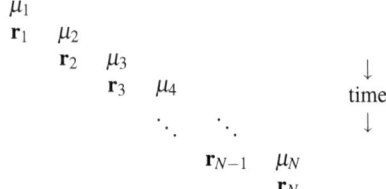

Table 5.1 MAX2 FPGA resource usage

	Used	Maximum	Percentage
LUTs	157716	207360	76.06
FFs	193082	207360	93.11
BRAMs	225	324	69.44
DSPs	141	192	73.44

parameters (scalar inputs). On a Maxeler MAX2[1] device, possible values are $N_{max} = 16384$ and $D_{max} = 6$. These values are obtained by fine tuning the design, since FPGA circuits have a limited number of programmable logic elements: look-up tables (LUTs), flip-flops (FFs), block RAM modules (BRAMs) and digital signal processing modules (DSPs). The actual resource usage is shown in Table 5.1.

The architecture of the system is shown in Fig. 5.6. The host application is responsible for creating the initial vectors, sending the current vectors in each iteration to the dataflow device and collecting their updated values. Vector coordinates are transferred serially from the host and back to the host. On the dataflow device, they are

---

[1]MAX2 is a PCI Express expansion card with a ×8×8 interface. It has two Xilinx Virtex-5 LX330T FPGAs operating at 100MHz, each connected to 6GB of DDR2 SDRAM.

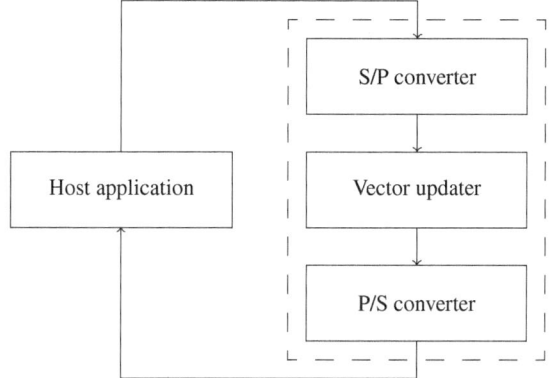

Fig. 5.6 Architecture of the system: blocks inside the dashed box are on the Maxeler MAX2 dataflow device

internally converted to the parallel form in the serial-to-parallel (S/P) converter and after the update back to the serial form in the parallel-to-serial (P/S) converter. Both conversions last ND cycles, but since outputs of each block are available before its execution ends, the total execution time of one iteration is less than $2ND + N(N + 1)$ cycles.

5.3.3 Performance

Measured execution times of a single iteration are shown in Fig. 5.7. The host CPU is Intel i7-3770K@3.5GHz and the software implementation is single-threaded, executing in 64-bit mode. The host operating system is Linux (distribution CentOS 6.4, kernel version 2.6.32-358.18.1). During the measurements, all motherboard functions which change the CPU clock frequency depending on the load are disabled.

For low values of N, the software implementation is considerably faster than the hardware one, since in the latter case, the majority of time is spent on the communication between the host and the dataflow device.

For $N \gtrsim 50$, the hardware implementation becomes faster than the software one despite all the overhead of communication and control the host needs to perform.

For $N \gtrsim 500$, the execution time of the hardware implementation starts to show similar asymptotic behaviour as that of the software one, since it is dominated by the actual calculations. Both procedures are essentially the same and quadratic in N.

It is interesting to note that the time of execution in software increases as the vector dimension, D, increases, but is not proportional to it. This can be explained by the time needed for other operations apart from vector operations. In hardware, the times for different values of D are practically the same (their curves in Fig. 5.7 almost

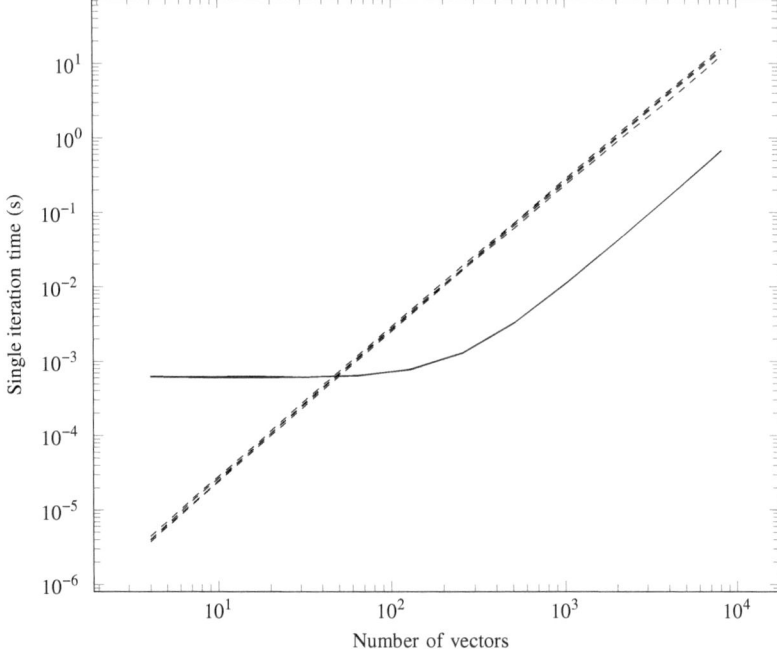

Fig. 5.7 Execution times, $N \in [4, 8192]$ and $D \in [3, 6]$: results for the software implementation are shown in dashed lines and for the hardware implementation in solid lines

Table 5.2 Host power consumption

Mode	Power (W)
Idle	68
Software execution	88
Hardware execution	74

overlap) since vector operations are done in parallel. Their only small differences are caused by different amounts of data exchanged between the host and the dataflow engine, but those are within measurement errors.

The asymptotic speed gains of hardware acceleration are approximately $(18 \div 24)\times$, for $D \in [3, 6]$.

Figures 5.8 and 5.9 show FPGA device usage and temperature obtained by the maxtop utility, with ambient temperature of $23\,°C$. FPGA device temperature in idle mode is $45.5\,°C$.

Power consumption of the host in different execution modes is shown in Table 5.2. The measurements are performed on the mains cable of the host, with an estimated relative error of 5%.

Although the majority of power is used for idle operation of the host, which includes power for its auxiliary devices such as hard disks and fans, additional con-

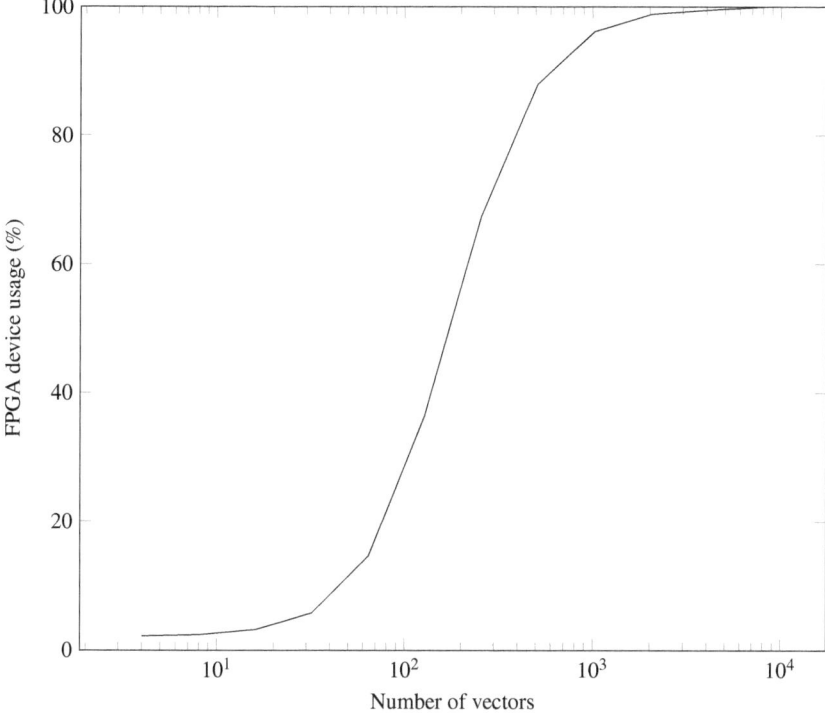

Fig. 5.8 FPGA device usage, $N \in [4, 8192]$ and $D = 6$

sumption for hardware execution is lower than for software execution, even when only a single thread is active.

5.4 Results

5.4.1 Minimum Distances

Minimum distances of spherical codes obtained by taking the best of 32 optimizer runs for different N and D are shown in Fig. 5.10. Even though these results are not the absolute optimum values, they are very close to them (the reader is referred to online tables compiled by Sloane [41] for a detailed list of the best known and other interesting spherical codes), and can serve as a starting point for calculating possible coding gains in systems which employ the constructed codes.

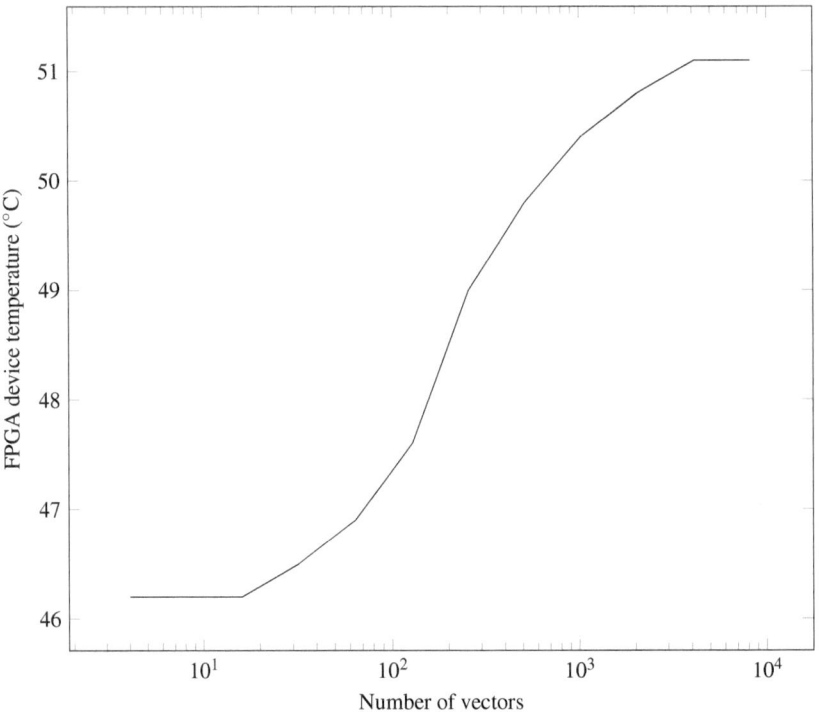

Fig. 5.9 FPGA device temperature, $N \in [4, 8192]$ and $D = 6$

5.4.2 Code Performance

Simulated frame error probabilities of some spherical codes are shown in Figs. 5.11, 5.12, 5.13 and 5.14. Optimum decoding is performed for the channel with additive white Gaussian noise, with signal/noise ratio

$$\text{SNR} = 10 \log \frac{1/D}{\sigma^2}, \tag{5.14}$$

where σ is the variance of noise per real dimension. Since all codewords are of unit energy, the energy per transmitted symbol is $1/D$.

In all figures, solid lines represent the simulated code performance and dashed lines represent the performance of hypercube codes with the same dimension and number of symbols, whose codewords are $\left[\pm 1/\sqrt{D} \quad \cdots \quad \pm 1/\sqrt{D} \right]$, which correspond to uncoded binary transmission. In the latter case, frame error probability can be calculated as

$$P_f = 1 - (1 - P_s)^D, \tag{5.15}$$

where

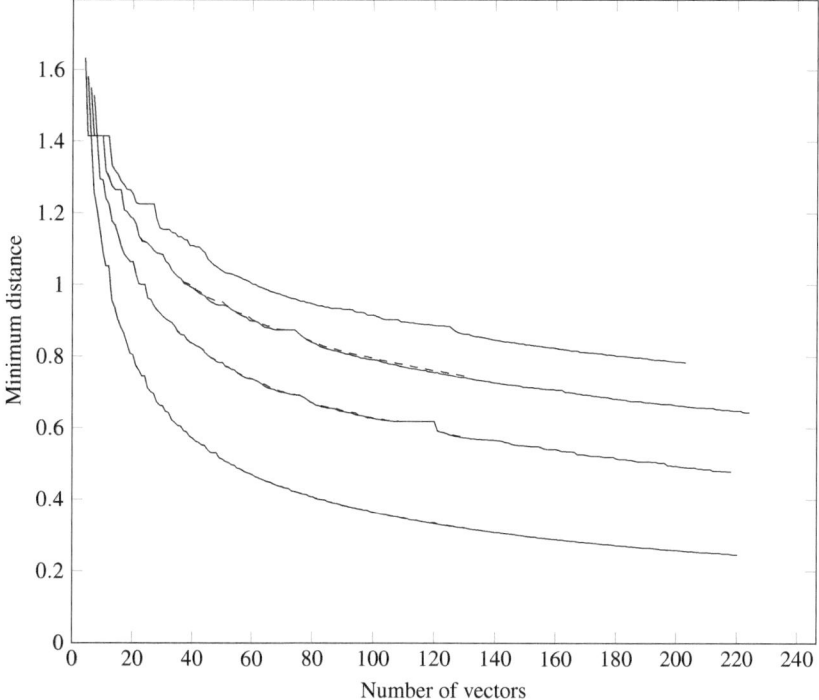

Fig. 5.10 Minimum distances of spherical codes: achieved minimum distances of obtained codes for $D = 3, 4, 5, 6$ (from bottom to top) are shown in solid lines and minimum distances of the best known spherical codes for $D = 3, 4, 5$ (from Sloane's tables) are shown in dashed lines

$$P_s = Q\left(\frac{1}{\sqrt{D}\sigma}\right) \tag{5.16}$$

is the symbol error probability, and

$$Q(x) = \frac{1}{\sqrt{2\pi}}\int_x^\infty e^{-\frac{t^2}{2}}dt. \tag{5.17}$$

For high SNRs, error probability is dominated by the code minimum distance, which determines the asymptotic coding gain of spherical codes

$$\Delta SNR = 20\log\frac{d_{min}}{2/\sqrt{D}}, \tag{5.18}$$

where d_{min} is the minimum distance of a spherical code, and $2/\sqrt{D}$ is the minimum distance of the corresponding hypercube code. Achieved coding gains for the simulated codes are shown in Table 5.3.

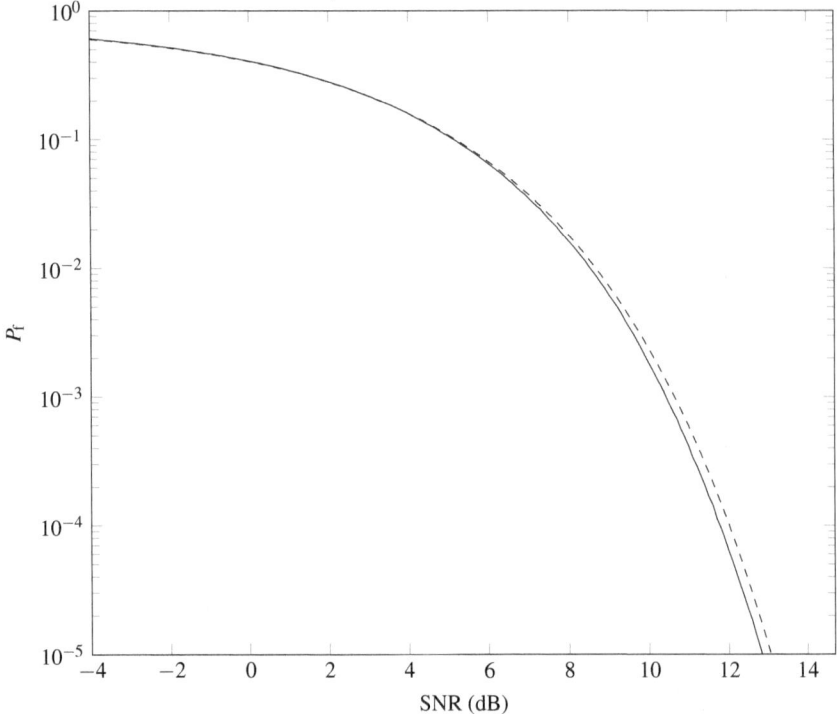

Fig. 5.11 Frame error probability, $D = 3$, $N = 8$

Table 5.3 Coding gains

D	N	$d_{\min}$	ΔSNR (dB)
3	8	1.216	0.45
4	16	1.107	0.88
5	32	1.056	1.45
6	64	0.989	1.66

5.5 Methodology Considerations

In [5], an interesting classification of different approaches to the generation of new ideas for PhD research in computing is given, without purporting that the list of 10 different methods covers all possible cases.

According to that list, we used the following methods:

- Extraparametrization
  Instead of solving the original problem of minimizing the "hard potential" cost function (5.1), a different "soft potential" cost function (5.2) is minimized, which depends on a new parameter, β. As explained in Sect. 5.2, this ensures the numerical

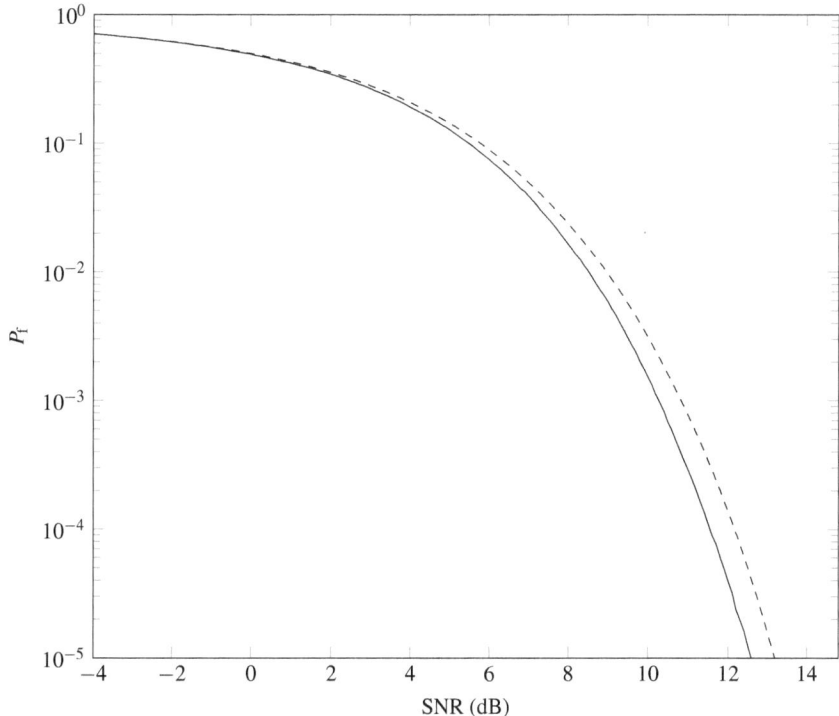

Fig. 5.12 Frame error probability, $D = 4$, $N = 16$

stability of the procedure and enables the use of general optimization methods, which require differentiable cost functions.

- Transgranularization
  Even though the solutions of the two problems are in general different for any finite value of β, by carefully increasing it, the sequence of vector positions obtained while solving the second problem can be made arbitrarily close to a local optimum of the first problem. The procedure for this, described in Sect. 5.2.3, is a novelty of the present method. Unlike in previous approaches, e.g. [22–26, 39], where β was kept fixed or increased in discrete steps after numerical convergence, continuous adaptation of β is equivalent to its fine-grained control and it results in a lower number of iterations and faster convergence.

- Revitalization
  Advances in high-performance computing and the development of the dataflow computing paradigm revitalized the problem, which was being solved by traditional methods during the last three decades. This broadened the range of parameters N and D for which solutions can be obtained and increased the precision of those solutions.

In general, there are numerous methodologies for generating ideas, to name a few: brainstorming [33], lateral thinking [13], mind mapping [8] and TRIZ [32, 34, 35]. The last one focuses more on techniques of problem-solving than on psychological

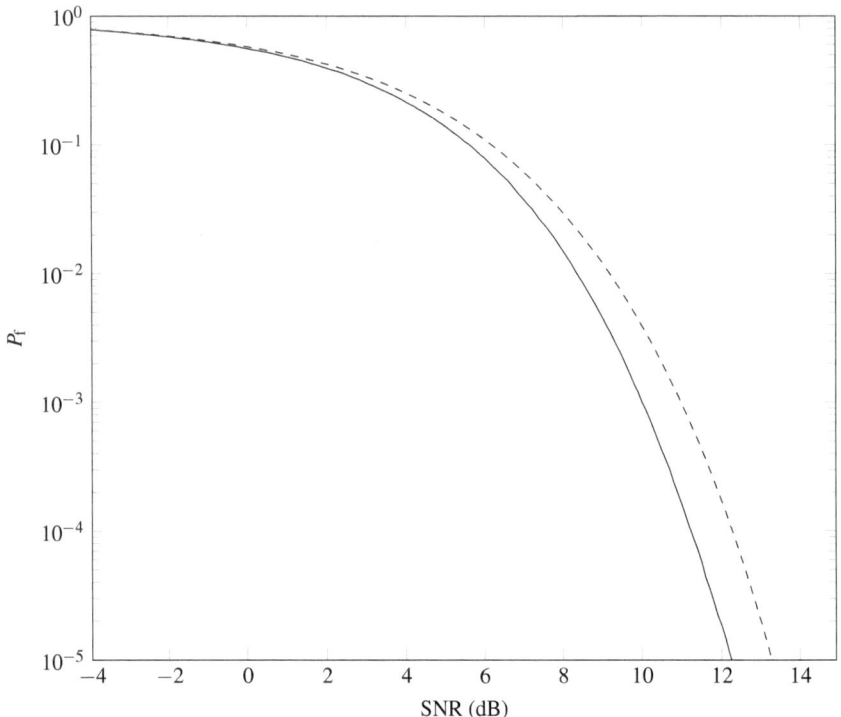

Fig. 5.13 Frame error probability, $D = 5$, $N = 32$

aids to achieve the intellectual freedom necessary to solve problems (the so-called "thinking outside the box"), so we will focus on it here. Although rather elaborate and complicated to understand and apply, it is based on millions of studied patents and it yielded a list of 40 principles from mechanical engineering that can, in a broader sense, be applied to all areas of engineering in order to yield efficient solutions to nonstandard problems.

From the list of 40 TRIZ principles, the following ones can be identified:

- Nonlinearity (standard name: Spheroidality-Curvature)
  The solutions to the problem are confined to a D-dimensional sphere, and its curvature enforces nonlinearities that are tackled by minimizing a nonlinear potential cost function.
- Dynamics
  The parameter of the cost function, β, is dynamically changed according to the estimated speed of convergence to the solution to the initial problem (Sect. 5.2.3).
- Partial, overdone or excessive action
  The optimization is not completed for any finite value of β. Rather, β is increased as soon as that is estimated to be beneficial for faster convergence to the final solution.

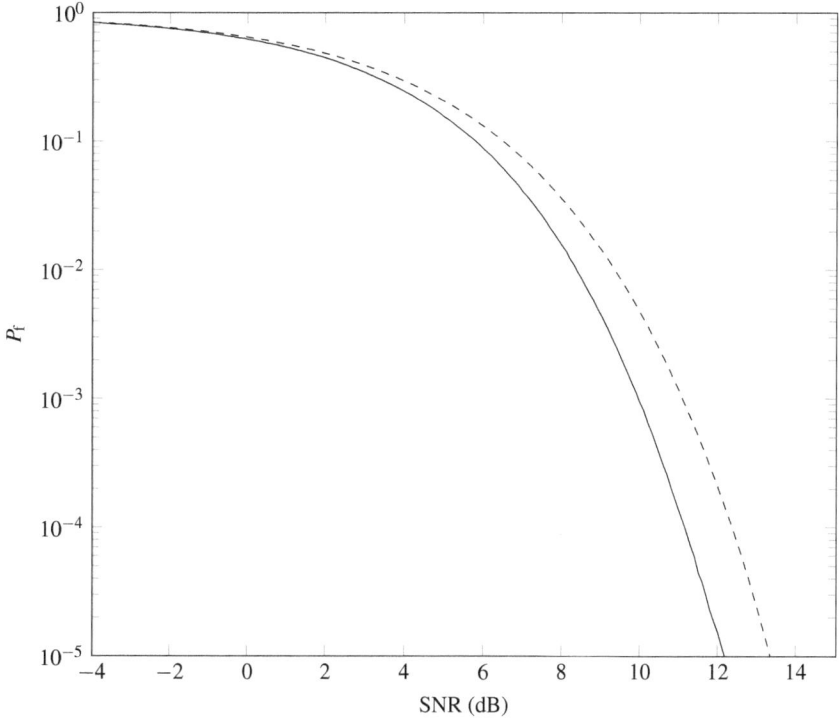

Fig. 5.14 Frame error probability, $D = 6$, $N = 64$

- Feedback
  The decision whether to increase β or not is made by calculating the move norm (5.13) of both options in every iteration.
- Parameter change
  The gradient descent step size (5.12) is calculated for every particle being moved in order to ensure convergence.
- Cheap short-living objects
  For given N and D, a number of different initial sets of vectors are randomly generated and the procedure is run for every one of them. The results are obtained as the optimum of all the runs.

Acknowledgements This work was supported by the Ministry of Education, Science and Technological Development of the Republic of Serbia, grant 451-03-00605/2012-16/198, project "Cloud Services for Applications with High Performance Requirements," and grant III44003, project "Conflagration Detection and Estimation of its Development by means of an Integrated System for Real Time Monitoring of Critical Parameters."

Appendix

We intend to justify our choice of the parameter α (see (5.12)), since it has been observed experimentally that:

- this choice ensures the stability and convergence of the algorithm,
- some other choices of α can lead to oscillatory behaviour, or very slow convergence.

Here, we wish to give a formal argument of convergence when (5.12) holds by proving the following claim: If a particle is in a neighbourhood of a stable balance position,[2] then the above algorithm will make it converge towards this position (all other particles being fixed). We will first analyze the case of infinitesimally small neighbourhoods. This analysis will enable us to obtain explicitly some relations needed for the finite neighbourhood case stated in Theorem 3.

Two-Dimensional Case

We start with the case $N = 3$, $D = 2$, i.e. 3 particles on the unit circle. This simple example illustrates the main idea and makes the general case (analyzed in the following subsection) easier to follow. Also, it enables one to explicitly obtain the condition (5.12) which ensures convergence of the algorithm.

Theorem 1 *Let $N = 3$, $D = 2$. If one of the particles is in an infinitesimally small neighbourhood of a stable balance position, then the above algorithm will make it converge towards this position.*

Proof Observe the configuration of points illustrated in Fig. 5.15:

$$
\begin{aligned}
\mathbf{m} &= \begin{bmatrix} 1 & 0 \end{bmatrix}, \\
\mathbf{r}_1 &= \begin{bmatrix} \cos\gamma & \sin\gamma \end{bmatrix}, \\
\mathbf{r}_2 &= \begin{bmatrix} \cos\gamma & -\sin\gamma \end{bmatrix}.
\end{aligned}
\tag{5.19}
$$

The particle $\mathbf{m}$ is in a stable balance position. The force exerted on this particle is (assuming for convenience $C = 1/(\beta - 2)$)

$$
\mathbf{F} = \sum_i \frac{\mathbf{m} - \mathbf{r}_i}{|\mathbf{m} - \mathbf{r}_i|^\beta} = \sum_i \frac{\mathbf{p}_i}{|\mathbf{p}_i|^\beta},
\tag{5.20}
$$

where the vectors $\mathbf{p}_i = \mathbf{m} - \mathbf{r}_i$ determine the position of the observed particle with respect to the remaining particles. They are introduced primarily to simplify notation. We have

---

[2]We shall later give a precise definition of what is meant by a "stable balance position" of a particle.

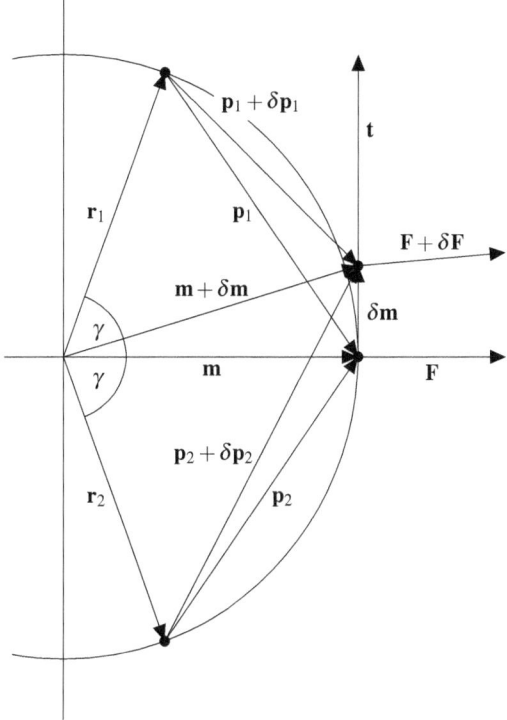

Fig. 5.15 Particle movement in the two-dimensional case

$$\mathbf{p}_1 = \begin{bmatrix} 1 - \cos\gamma & -\sin\gamma \end{bmatrix},$$
$$\mathbf{p}_2 = \begin{bmatrix} 1 - \cos\gamma & \sin\gamma \end{bmatrix},$$
$$|\mathbf{p}_1| = |\mathbf{p}_2| = 2\sin\frac{\gamma}{2}, \tag{5.21}$$

and by using some simple trigonometric identities we obtain

$$\mathbf{F} = \left(2\sin\frac{\gamma}{2}\right)^{-\beta+2}\mathbf{m}. \tag{5.22}$$

We need to show that, if the observed particle is in some neighbourhood of the stable balance position ($\mathbf{m}$), then the presented algorithm will make it converge towards this position, provided that (5.12) holds. Assume that the particle is moved by an infinitesimally small amount along the circle $\delta\mathbf{m} = \delta m\,\mathbf{t}$, where $\delta m = |\delta\mathbf{m}|$, and $\mathbf{t} = \begin{bmatrix} 0 & 1 \end{bmatrix}$ is the tangent unit vector at $\mathbf{m}$ (see Fig. 5.15). The vectors determining the position of this particle with respect to the other particles are now $\mathbf{p}_i + \delta\mathbf{p}_i = \mathbf{m} + \delta\mathbf{m} - \mathbf{r}_i$, i.e.

$$
\begin{aligned}
\mathbf{p}_1 + \delta\mathbf{p}_1 &= \begin{bmatrix} 1 - \cos\gamma & -\sin\gamma + \delta m \end{bmatrix}, \\
\mathbf{p}_2 + \delta\mathbf{p}_2 &= \begin{bmatrix} 1 - \cos\gamma & \sin\gamma + \delta m \end{bmatrix}.
\end{aligned}
\tag{5.23}
$$

and hence, the force exerted on the particle at this position is

$$
\begin{aligned}
\mathbf{F} + \delta\mathbf{F} = \quad & |\mathbf{p}_1 + \delta\mathbf{p}_1|^{-\beta} \cdot \begin{bmatrix} 1 - \cos\gamma & -\sin\gamma + \delta m \end{bmatrix} \\
& + |\mathbf{p}_2 + \delta\mathbf{p}_2|^{-\beta} \cdot \begin{bmatrix} 1 - \cos\gamma & \sin\gamma + \delta m \end{bmatrix},
\end{aligned}
\tag{5.24}
$$

where

$$
\begin{aligned}
|\mathbf{p}_1 + \delta\mathbf{p}_1|^{-\beta} &= \left((1 - \cos\gamma)^2 + (-\sin\gamma + \delta m)^2 \right)^{-\frac{\beta}{2}}, \\
|\mathbf{p}_2 + \delta\mathbf{p}_2|^{-\beta} &= \left((1 - \cos\gamma)^2 + (\sin\gamma + \delta m)^2 \right)^{-\frac{\beta}{2}}.
\end{aligned}
\tag{5.25}
$$

Note that, since δm is an infinitesimally small quantity, any differentiable expression depending on δm can be linearized by developing it into a Maclaurin series and disregarding all higher order terms

$$
f(\delta m) \approx f(\delta m)\Big|_{\delta m = 0} + \frac{\mathrm{d}f(\delta m)}{\mathrm{d}\delta m}\Big|_{\delta m = 0} \delta m.
\tag{5.26}
$$

By linearizing the right-hand sides of (5.25) in this way, we obtain

$$
\begin{aligned}
\left((1 - \cos\gamma)^2 + (\pm\sin\gamma + \delta m)^2 \right)^{-\frac{\beta}{2}} &= \\
= \left(2\sin\frac{\gamma}{2} \right)^{-\beta} &\pm \beta \left(2\sin\frac{\gamma}{2} \right)^{-\beta-1} \left(\cos\frac{\gamma}{2} \right) \delta m,
\end{aligned}
\tag{5.27}
$$

which, together with (5.24), gives

$$
\mathbf{F} + \delta\mathbf{F} = \left[\left(2\sin\tfrac{\gamma}{2} \right)^{-\beta+2} \quad 2 \left(2\sin\tfrac{\gamma}{2} \right)^{-\beta} \left(1 - \beta\cos^2\tfrac{\gamma}{2} \right) \delta m \right].
\tag{5.28}
$$

We also have (again ignoring higher order infinitesimals)

$$
|\mathbf{F} + \delta\mathbf{F}| = \left(2\sin\frac{\gamma}{2} \right)^{-\beta+2} = |\mathbf{F}|.
\tag{5.29}
$$

Therefore, if the particle finds itself at the position $\mathbf{m} + \delta\mathbf{m}$, the force it will feel is given by (5.28). Recall that in this case the algorithm works as follows: We move the particle along the direction of the force and then project it back onto the circle. The direction of the force is determined by the vector

$$\frac{\mathbf{F} + \delta\mathbf{F}}{|\mathbf{F} + \delta\mathbf{F}|} = \left[1 \quad \frac{1 - \beta \cos^2 \frac{\gamma}{2}}{2 \sin^2 \frac{\gamma}{2}} \delta m \right] = \mathbf{G} + \delta\mathbf{G}, \tag{5.30}$$

where $\mathbf{G} = \frac{\mathbf{F}}{|\mathbf{F}|} = \mathbf{m}$ is the normalized force at the stable balance position, and $\delta\mathbf{G} = \frac{\delta\mathbf{F}}{|\mathbf{F}|}$. We now need to move the particle for $\alpha(\mathbf{G} + \delta\mathbf{G})$, and then to project it onto the circle, i.e. normalize its radius vector. The new position is therefore

$$\mathbf{m} + \delta\mathbf{m}' = \frac{\mathbf{m} + \delta\mathbf{m} + \alpha(\mathbf{G} + \delta\mathbf{G})}{|\mathbf{m} + \delta\mathbf{m} + \alpha(\mathbf{G} + \delta\mathbf{G})|}. \tag{5.31}$$

We have

$$\mathbf{m} + \delta\mathbf{m} + \alpha(\mathbf{G} + \delta\mathbf{G}) = \left[1 + \alpha \quad \left(1 + \alpha \frac{1 - \beta \cos^2 \frac{\gamma}{2}}{2 \sin^2 \frac{\gamma}{2}} \right) \delta m \right] \tag{5.32}$$

and

$$|\mathbf{m} + \delta\mathbf{m} + \alpha(\mathbf{G} + \delta\mathbf{G})| = 1 + \alpha, \tag{5.33}$$

so that

$$\mathbf{m} + \delta\mathbf{m}' = \left[1 \quad \frac{1 + \alpha b}{1 + \alpha} \delta m \right], \tag{5.34}$$

where

$$b = \frac{1 - \beta \cos^2 \frac{\gamma}{2}}{2 \sin^2 \frac{\gamma}{2}}. \tag{5.35}$$

Thus the new distance of the particle from the stable balance position (after the execution of one step of the algorithm) is

$$\delta m' = |\delta\mathbf{m}'| = \left| \frac{1 + \alpha b}{1 + \alpha} \right| \delta m. \tag{5.36}$$

We would like this displacement to be as small as possible ($\delta m' \approx 0$) so that the particle converges quickly to the stable balance position, but in general it is enough that

$$\left| \frac{1 + \alpha b}{1 + \alpha} \right| < 1 \tag{5.37}$$

to ensure convergence. In other words, the optimal choice is $\alpha = \frac{-1}{b}$. Unfortunately, expression (5.35) cannot be generalized to the higher dimensional case in a straightforward way. We therefore find a simpler expression by lower bounding b,

$$b > \frac{-\beta \cos^2 \frac{\gamma}{2}}{2 \sin^2 \frac{\gamma}{2}} > \frac{-\beta}{2 \sin^2 \frac{\gamma}{2}} = \frac{-2\beta}{\mu}, \tag{5.38}$$

where

$$\mu = d_{\min}^2, \tag{5.39}$$

and $d_{\min} = \min_i |\mathbf{p}_i| = 2 \sin \frac{\gamma}{2}$ is the minimum Euclidean distance from the observed particle to any of the other particles. Now from $\alpha = \frac{-1}{b}$, and by using the above bound instead of b, we get

$$\alpha = \frac{\mu}{2\beta}. \tag{5.40}$$

It is easy to show that this value satisfies (5.37) for all $\beta > 1$, thus completing the proof of the claim.

General Case

Let us now consider the general case with N particles on the D-dimensional unit sphere. The idea of the proof is the same as in the 2D case. Namely, we again observe a particle in a stable balance position $\mathbf{m}$, and consider what happens if this particle is moved slightly in any direction.

Definition 1.1 Let some configuration of particles be specified, and observe the particle at the point $\mathbf{m}$. Let $\mathbf{t}$ be a unit vector in the tangent hyperplane at $\mathbf{m}$. If the particle is moved by an infinitesimal amount in the direction of $\mathbf{t}$, i.e. $\delta\mathbf{m} = \delta m \, \mathbf{t}$, the new tangent vector (lying in the plane defined by $\mathbf{m}$ and $\mathbf{t}$) is $\mathbf{t} + \delta\mathbf{t}$, and the force exerted on the particle is $\mathbf{F} + \delta\mathbf{F}$. We say that $\mathbf{m}$ is a stable balance position for the observed particle if

$$(\mathbf{F} + \delta\mathbf{F}) \cdot (\mathbf{t} + \delta\mathbf{t}) < 0. \tag{5.41}$$

Intuitively, this condition means that the force pushes the particle back to $\mathbf{m}$. See Fig. 5.16 for an illustration.

It is easy to show that the above condition implies that $\mathbf{F} = |\mathbf{F}|\mathbf{m}$, i.e. that the force exerted on the particle in a stable balance position is orthogonal to the surface of the sphere. Consequently, $\mathbf{F} \cdot \mathbf{t} = 0$. We shall use these facts later.

Let us now prove a generalization of Theorem 1.

Theorem 2 *Observe a configuration of N points on a D-dimensional sphere, and let one of the particles be in an infinitesimally small neighbourhood of a stable balance position. Then the above algorithm will make it converge towards this position.*

Proof The force exerted on the observed particle located at $\mathbf{m}$ is

$$\mathbf{F} = \sum_i \frac{\mathbf{m} - \mathbf{r}_i}{|\mathbf{m} - \mathbf{r}_i|^\beta} = \sum_i \frac{\mathbf{p}_i}{|\mathbf{p}_i|^\beta}, \tag{5.42}$$

where $\mathbf{p}_i = \mathbf{m} - \mathbf{r}_i$, and the $\mathbf{r}_i$'s are the positions of the other particles, as before. Since this is by assumption a stable balance position, we have $\mathbf{F} = |\mathbf{F}|\mathbf{m}$. Assume now that the particle is displaced by $\delta \mathbf{m}$, i.e. that its new position is $\mathbf{m} + \delta \mathbf{m}$. (For now we allow all directions of the displacement, i.e. we do not assume that $\delta \mathbf{m}$ is tangential to the sphere at $\mathbf{m}$.) Then its position relative to the particle $\mathbf{r}_i$ is changed by $\delta \mathbf{p}_i = \delta \mathbf{m}$. The force it feels at this position is

$$
\begin{aligned}
\mathbf{F} + \delta \mathbf{F} &= \sum_i |\mathbf{p}_i + \delta \mathbf{p}_i|^{-\beta} (\mathbf{p}_i + \delta \mathbf{p}_i) \\
&= \sum_i |\mathbf{p}_i + \delta \mathbf{m}|^{-\beta} (\mathbf{p}_i + \delta \mathbf{m}).
\end{aligned}
\tag{5.43}
$$

We have

$$
\begin{aligned}
|\mathbf{p}_i + \delta \mathbf{m}|^{-\beta} &= \left((\mathbf{p}_i + \delta \mathbf{m}) \cdot (\mathbf{p}_i + \delta \mathbf{m}) \right)^{-\frac{\beta}{2}} \\
&= (\mathbf{p}_i \cdot \mathbf{p}_i + 2\mathbf{p}_i \cdot \delta \mathbf{m})^{-\frac{\beta}{2}},
\end{aligned}
\tag{5.44}
$$

where we have disregarded higher order infinitesimals, as usual. (We denote by $\mathbf{x} \cdot \mathbf{y}$ the inner product of vectors $\mathbf{x}$ and $\mathbf{y}$. Note that, since we have adopted the row-vector notation, we could also write $\mathbf{x}\mathbf{y}^\mathsf{T}$ for the inner product.) As in the proof of Theorem 1, we can linearize expressions depending on infinitesimally small quantities, but we now need the multivariate form,

$$
f(\delta \mathbf{m}) \approx f(\delta \mathbf{m}) \bigg|_{\delta \mathbf{m} = 0} + \sum_{u=1}^{D} \frac{\partial f(\delta \mathbf{m})}{\partial \delta m_u} \bigg|_{\delta \mathbf{m} = 0} \delta m_u,
\tag{5.45}
$$

Fig. 5.16 Particle movement in the general case

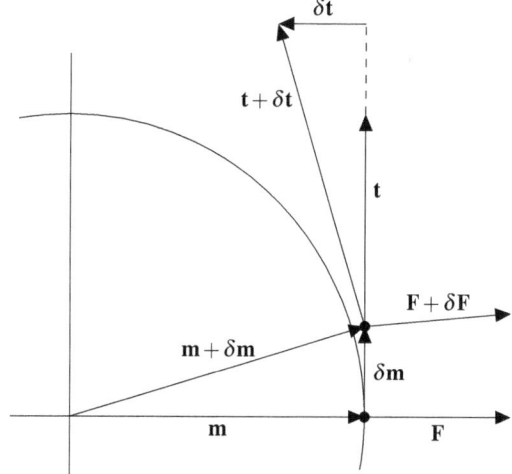

where $\delta\mathbf{m} = \begin{bmatrix} \delta m_1 & \cdots & \delta m_D \end{bmatrix}$. By linearizing expression (5.44) in this way, we obtain

$$|\mathbf{p}_i + \delta\mathbf{m}|^{-\beta} = |\mathbf{p}_i|^{-\beta} - \beta|\mathbf{p}_i|^{-\beta-2}\mathbf{p}_i \cdot \delta\mathbf{m} \qquad (5.46)$$

and hence the force can be expressed as

$$
\begin{aligned}
\mathbf{F} + \delta\mathbf{F} &= \sum_i \left(|\mathbf{p}_i|^{-\beta} - \beta|\mathbf{p}_i|^{-\beta-2}\mathbf{p}_i \cdot \delta\mathbf{m} \right) (\mathbf{p}_i + \delta\mathbf{m}) \\
&= \sum_i \left(|\mathbf{p}_i|^{-\beta}\mathbf{p}_i + |\mathbf{p}_i|^{-\beta}\delta\mathbf{m} - \beta|\mathbf{p}_i|^{-\beta-2}(\mathbf{p}_i \cdot \delta\mathbf{m})\mathbf{p}_i \right),
\end{aligned}
\qquad (5.47)
$$

from which we conclude that the differential of the force at the stable balance position is

$$
\begin{aligned}
\delta\mathbf{F} &= \sum_i \left(|\mathbf{p}_i|^{-\beta}\delta\mathbf{m} - \beta|\mathbf{p}_i|^{-\beta-2}(\mathbf{p}_i \cdot \delta\mathbf{m})\mathbf{p}_i \right) \\
&= \sum_i |\mathbf{p}_i|^{-\beta} \left(\delta\mathbf{m} - \beta(\mathbf{q}_i \cdot \delta\mathbf{m})\mathbf{q}_i \right),
\end{aligned}
\qquad (5.48)
$$

where $\mathbf{q}_i = \frac{\mathbf{p}_i}{|\mathbf{p}_i|}$. We further have

$$
\begin{aligned}
|\mathbf{F} + \delta\mathbf{F}|^{-1} &= \left((\mathbf{F} + \delta\mathbf{F}) \cdot (\mathbf{F} + \delta\mathbf{F}) \right)^{-\frac{1}{2}} \\
&= (\mathbf{F} \cdot \mathbf{F} + 2\mathbf{F} \cdot \delta\mathbf{F})^{-\frac{1}{2}} \\
&= |\mathbf{F}|^{-1} - |\mathbf{F}|^{-3}\mathbf{F} \cdot \delta\mathbf{F},
\end{aligned}
\qquad (5.49)
$$

where the last expression is obtained by linearization, as before. Now,

$$
\begin{aligned}
|\mathbf{F} + \delta\mathbf{F}|^{-1}(\mathbf{F} + \delta\mathbf{F}) &= |\mathbf{F}|^{-1}\mathbf{F} + |\mathbf{F}|^{-1}\delta\mathbf{F} - |\mathbf{F}|^{-3}(\mathbf{F} \cdot \delta\mathbf{F})\mathbf{F} \\
&= \mathbf{G} + \delta\mathbf{G},
\end{aligned}
\qquad (5.50)
$$

where $\mathbf{G} = |\mathbf{F}|^{-1}\mathbf{F} = \mathbf{m}$ is the normalized force at the stable balance position, and $\delta\mathbf{G} = |\mathbf{F}|^{-1}\delta\mathbf{F} - |\mathbf{F}|^{-1}(\mathbf{m} \cdot \delta\mathbf{F})\mathbf{m}$. Notice that

$$\delta\mathbf{G} \cdot \mathbf{m} = |\mathbf{F}|^{-1}\delta\mathbf{F} \cdot \mathbf{m} - |\mathbf{F}|^{-1}(\mathbf{m} \cdot \delta\mathbf{F})(\mathbf{m} \cdot \mathbf{m}) = 0. \qquad (5.51)$$

Now, the new position of the particle, after moving and projecting back onto the sphere, is

$$
\begin{aligned}
\mathbf{m} + \delta\mathbf{m}' &= \frac{\mathbf{m} + \delta\mathbf{m} + \alpha(\mathbf{G} + \delta\mathbf{G})}{|\mathbf{m} + \delta\mathbf{m} + \alpha(\mathbf{G} + \delta\mathbf{G})|} \\
&= \frac{(1+\alpha)\mathbf{m} + \delta\mathbf{m} + \alpha\delta\mathbf{G}}{|(1+\alpha)\mathbf{m} + \delta\mathbf{m} + \alpha\delta\mathbf{G}|}.
\end{aligned}
\qquad (5.52)
$$

Let us calculate the new displacement $\delta\mathbf{m}'$. Taking (5.51) into account, we have

$$|(1+\alpha)\mathbf{m} + \delta\mathbf{m} + \alpha\delta\mathbf{G}|^{-1} =$$

$$= \left(\left((1+\alpha)\mathbf{m} + \delta\mathbf{m} + \alpha\delta\mathbf{G}\right) \cdot \left((1+\alpha)\mathbf{m} + \delta\mathbf{m} + \alpha\delta\mathbf{G}\right)\right)^{-\frac{1}{2}}$$

$$= \left((1+\alpha)^2 \mathbf{m} \cdot \mathbf{m} + 2(1+\alpha)\mathbf{m} \cdot \delta\mathbf{m}\right)^{-\frac{1}{2}} \tag{5.53}$$

$$= \left((1+\alpha)^2 + 2(1+\alpha)\mathbf{m} \cdot \delta\mathbf{m}\right)^{-\frac{1}{2}},$$

which after linearization becomes

$$|(1+\alpha)\mathbf{m} + \delta\mathbf{m} + \alpha\delta\mathbf{G}|^{-1} = \frac{1}{1+\alpha} - \frac{1}{(1+\alpha)^2}\mathbf{m} \cdot \delta\mathbf{m}. \tag{5.54}$$

Plugging this back into (5.52) we get

$$|(1+\alpha)\mathbf{m} + \delta\mathbf{m} + \alpha\delta\mathbf{G}|^{-1}\left((1+\alpha)\mathbf{m} + \delta\mathbf{m} + \alpha\delta\mathbf{G}\right) =$$

$$= \mathbf{m} + \frac{1}{1+\alpha}\delta\mathbf{m} + \frac{\alpha}{1+\alpha}\delta\mathbf{G} - \frac{1}{1+\alpha}(\mathbf{m} \cdot \delta\mathbf{m})\mathbf{m}. \tag{5.55}$$

It follows that the new distance (after the execution of one step of the algorithm) from the stable balance position is given by

$$\delta\mathbf{m}' = \frac{1}{1+\alpha}\delta\mathbf{m} + \frac{\alpha}{1+\alpha}\delta\mathbf{G} - \frac{1}{1+\alpha}(\mathbf{m} \cdot \delta\mathbf{m})\mathbf{m}$$

$$= \frac{1}{1+\alpha}(\delta\mathbf{m} - (\mathbf{m} \cdot \delta\mathbf{m})\mathbf{m}) + \frac{\alpha}{1+\alpha}\left(|\mathbf{F}|^{-1}\delta\mathbf{F} - |\mathbf{F}|^{-1}(\mathbf{m} \cdot \delta\mathbf{F})\mathbf{m}\right), \tag{5.56}$$

or equivalently by

$$\delta\mathbf{m}' = \frac{1}{1+\alpha}(\delta\mathbf{m} - (\mathbf{m} \cdot \delta\mathbf{m})\mathbf{m})$$

$$+ \frac{\alpha}{1+\alpha}|\mathbf{F}|^{-1}\left(\sum_i |\mathbf{p}_i|^{-\beta}(\delta\mathbf{m} - \beta(\mathbf{q}_i \cdot \delta\mathbf{m})\mathbf{q}_i)\right.$$

$$\left. - \left(\mathbf{m} \cdot \sum_i |\mathbf{p}_i|^{-\beta}(\delta\mathbf{m} - \beta(\mathbf{q}_i \cdot \delta\mathbf{m})\mathbf{q}_i)\right)\mathbf{m}\right) \tag{5.57}$$

$$= \frac{1}{1+\alpha}\delta\mathbf{m_t}$$

$$+ \frac{\alpha}{1+\alpha}|\mathbf{F}|^{-1}\sum_i |\mathbf{p}_i|^{-\beta}\left(\delta\mathbf{m_t} - \beta(\mathbf{q}_i \cdot \delta\mathbf{m})(\mathbf{q}_i - (\mathbf{q}_i \cdot \mathbf{m})\mathbf{m})\right).$$

In Eq. (5.57) we have denoted by $\delta\mathbf{m_t} = \delta\mathbf{m} - (\mathbf{m} \cdot \delta\mathbf{m})\mathbf{m}$ the tangential component of the vector $\delta\mathbf{m}$ (see Fig. 5.17), which can also be written as

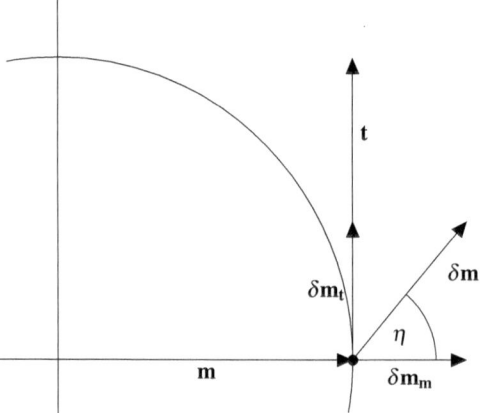

Fig. 5.17 Particle displacement decomposition

$$\delta \mathbf{m_t} = \delta \mathbf{m}(\mathrm{I} - \mathbf{m}^{\mathsf{T}}\mathbf{m}), \tag{5.58}$$

where I is the $D \times D$ identity matrix, and $\mathbf{m}^{\mathsf{T}}\mathbf{m}$ is the $D \times D$ matrix of the projection operation (onto $\mathbf{m}$). Observe from the figure that $\delta \mathbf{m}$ can be decomposed as

$$\delta \mathbf{m} = \delta m(\cos \eta \, \mathbf{m} + \sin \eta \, \mathbf{t}) = \delta \mathbf{m_m} + \delta \mathbf{m_t}, \tag{5.59}$$

so we can write

$$\begin{aligned}
\delta \mathbf{m}' = \;& \frac{1}{1+\alpha}\delta \mathbf{m_t} \\
& + \frac{\alpha}{1+\alpha}|\mathbf{F}|^{-1}\sum_i |\mathbf{p}_i|^{-\beta}\Big(\delta \mathbf{m_t} - \beta\big(\mathbf{q}_i \cdot (\delta \mathbf{m_m} + \delta \mathbf{m_t})\big)\big(\mathbf{q}_i - (\mathbf{q}_i \cdot \mathbf{m})\mathbf{m}\big)\Big).
\end{aligned} \tag{5.60}$$

From (5.60) we can deduce that $\delta \mathbf{m}' \cdot \mathbf{m} = 0$, i.e. that the displacement $\delta \mathbf{m}'$ has only the tangential component, as expected (after projecting, the particle is displaced by an infinitesimal amount in the tangent hyperplane at the point $\mathbf{m}$). Plugging (5.59) into (5.60) we obtain

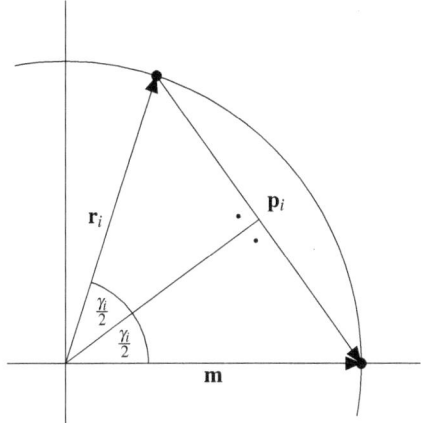

Fig. 5.18 $\mathbf{r}_i - \mathbf{m}$ plane

$$
\delta\mathbf{m}' = \frac{\delta m}{1+\alpha}\Bigg(\quad \sin\eta\,\mathbf{t}
$$
$$
+ \alpha|\mathbf{F}|^{-1}\sum_i |\mathbf{p}_i|^{-\beta}\Big(\quad \sin\eta\,\mathbf{t}
$$
$$
- \beta\cos\eta\,(\mathbf{q}_i\cdot\mathbf{m})\mathbf{q}_i
$$
$$
+ \beta\cos\eta\,(\mathbf{q}_i\cdot\mathbf{m})^2\mathbf{m} \tag{5.61}
$$
$$
- \beta\sin\eta\,(\mathbf{q}_i\cdot\mathbf{t})\mathbf{q}_i
$$
$$
+ \beta\sin\eta\,(\mathbf{q}_i\cdot\mathbf{t})(\mathbf{q}_i\cdot\mathbf{m})\mathbf{m}\Big)\Bigg).
$$

To simplify expression (5.61) recall that $\mathbf{q}_i = \frac{\mathbf{p}_i}{|\mathbf{p}_i|}$, and observe from Fig. 5.18 that

$$
|\mathbf{p}_i| = 2\sin\frac{\gamma_i}{2},
$$
$$
\mathbf{q}_i\cdot\mathbf{m} = |\mathbf{q}_i||\mathbf{m}|\cos\left(\frac{\pi}{2}-\frac{\gamma_i}{2}\right) = \sin\frac{\gamma_i}{2} = \frac{|\mathbf{p}_i|}{2}. \tag{5.62}
$$

We further have

$$\sum_i |\mathbf{p}_i|^{-\beta}(\mathbf{q}_i \cdot \mathbf{m})^2 \mathbf{m} = \sum_i |\mathbf{p}_i|^{-\beta}\frac{|\mathbf{p}_i|}{2}(\mathbf{q}_i \cdot \mathbf{m})\mathbf{m}$$

$$= \frac{1}{2}\sum_i |\mathbf{p}_i|^{-\beta}(\mathbf{p}_i \cdot \mathbf{m})\mathbf{m}$$

$$= \frac{1}{2}\left(\left(\sum_i |\mathbf{p}_i|^{-\beta}\mathbf{p}_i\right) \cdot \mathbf{m}\right)\mathbf{m} \qquad (5.63)$$

$$= \frac{1}{2}(\mathbf{F} \cdot \mathbf{m})\mathbf{m} = \frac{1}{2}|\mathbf{F}|\mathbf{m}$$

$$= \frac{1}{2}\mathbf{F},$$

and similarly

$$\sum_i |\mathbf{p}_i|^{-\beta}(\mathbf{q}_i \cdot \mathbf{t})(\mathbf{q}_i \cdot \mathbf{m}) = \frac{1}{2}\mathbf{F} \cdot \mathbf{t} = 0, \qquad (5.64)$$

and

$$\sum_i |\mathbf{p}_i|^{-\beta}(\mathbf{q}_i \cdot \mathbf{m})\mathbf{q}_i = \frac{1}{2}\mathbf{F}. \qquad (5.65)$$

We now obtain

$$\delta\mathbf{m}' = \frac{\delta m \sin \eta}{1 + \alpha}\left(\left(1 + \alpha|\mathbf{F}|^{-1}\sum_i |\mathbf{p}_i|^{-\beta}\right)\mathbf{t} - \alpha|\mathbf{F}|^{-1}\beta\sum_i |\mathbf{p}_i|^{-\beta}(\mathbf{q}_i \cdot \mathbf{t})\mathbf{q}_i\right). \qquad (5.66)$$

This is a linear transformation, which can be written in the following form

$$\delta\mathbf{m}' = \frac{\delta m \sin \eta}{1 + \alpha}\mathbf{t}(\mathbf{I} + \alpha\mathbf{B})$$

$$= \delta\mathbf{m}_\mathbf{t}\frac{\mathbf{I} + \alpha\mathbf{B}}{1 + \alpha} \qquad (5.67)$$

$$= \delta\mathbf{m}(\mathbf{I} - \mathbf{m}^\mathrm{T}\mathbf{m})\frac{\mathbf{I} + \alpha\mathbf{B}}{1 + \alpha},$$

where I is the identity matrix, and

$$\mathbf{B} = |\mathbf{F}|^{-1}\sum_i |\mathbf{p}_i|^{-\beta}(\mathbf{I} - \beta\mathbf{q}_i^\mathrm{T}\mathbf{q}_i). \qquad (5.68)$$

Here $\mathbf{q}_i^\mathrm{T}\mathbf{q}_i$ is the matrix of the projection operation (onto $\mathbf{q}_i$). It follows now from (5.67) that

$$|\delta\mathbf{m}'| \le |\delta\mathbf{m}|\, \|\mathbf{J}(\mathbf{m})\|, \qquad (5.69)$$

where

$$J(\mathbf{m}) = \left(I - \mathbf{m}^\mathsf{T}\mathbf{m}\right) \frac{I + \alpha B}{1 + \alpha}, \tag{5.70}$$

and $\|\cdot\|$ denotes the matrix norm (defined by $\|S\| = \max_{|\mathbf{u}|=1} |\mathbf{u}S|$). Notice the similarity of (5.69) with the expression obtained in the two-dimensional case, (5.36). (The factor $\left(I - \mathbf{m}^\mathsf{T}\mathbf{m}\right)$ which appears above is a consequence of $\delta\mathbf{m}$ being arbitrary, not necessarily tangential as was assumed in the proof of Theorem 1. By (5.58), the vector $\delta\mathbf{m}\left(I - \mathbf{m}^\mathsf{T}\mathbf{m}\right)$ is precisely the tangential component of $\delta\mathbf{m}$.)

Now, to prove that the particle converges towards its stable balance position, i.e. that the displacement decreases in every step, we need to show that $\|J(\mathbf{m})\| < 1$. Note that the matrix $J(\mathbf{m})$ is symmetric, and recall that the norm of a symmetric matrix is equal to its spectral radius, i.e. the largest absolute value of its eigenvalues. It can readily be shown (by using Cauchy–Schwartz inequality, for example) that the norm of a symmetric matrix S can also be expressed as

$$\|S\| = \max_{|\mathbf{u}|=1} |\mathbf{u}S \cdot \mathbf{u}|. \tag{5.71}$$

Actually, in the case of $J(\mathbf{m})$ the maximization can be done only over the unit vectors $\mathbf{t}$ orthogonal to $\mathbf{m}$, i.e. vectors lying in the tangential hyperplane at $\mathbf{m}$. This follows from the fact that the vectors $\mathbf{u}J(\mathbf{m})$ have only tangential components, as is easily concluded from (5.66) and (5.64). Therefore, we can write

$$\|J(\mathbf{m})\| = \max_{|\mathbf{t}|=1, \mathbf{t}\cdot\mathbf{m}=0} |\mathbf{t}J(\mathbf{m}) \cdot \mathbf{t}|. \tag{5.72}$$

Observe that the quadratic form defining the norm of our matrix can be expressed as

$$\mathbf{t}J(\mathbf{m}) \cdot \mathbf{t} = \mathbf{t}\frac{I + \alpha B}{1 + \alpha} \cdot \mathbf{t} = \frac{1 + \alpha b}{1 + \alpha}, \tag{5.73}$$

with

$$b = |F|^{-1} \sum_i |\mathbf{p}_i|^{-\beta}\left(1 - \beta(\mathbf{q}_i \cdot \mathbf{t})^2\right). \tag{5.74}$$

We therefore need to show that $\left|\frac{1+\alpha b}{1+\alpha}\right| < 1$.

Recall that for $\delta\mathbf{m} = \delta m\, \mathbf{t}$ we must have $(\mathbf{F} + \delta\mathbf{F}) \cdot (\mathbf{t} + \delta\mathbf{t}) < 0$ because $\mathbf{m}$ is by assumption a stable balance position. This can further be written as

$$(\mathbf{F} + \delta\mathbf{F}) \cdot (\mathbf{t} + \delta\mathbf{t}) = \mathbf{F} \cdot \delta\mathbf{t} + \delta\mathbf{F} \cdot \mathbf{t}$$

$$= \mathbf{F} \cdot (-\delta m)\mathbf{m} + \delta m \sum_i |\mathbf{p}_i|^{-\beta}\big(\mathbf{t} - \beta(\mathbf{q}_i \cdot \mathbf{t})\mathbf{q}_i\big) \cdot \mathbf{t}$$

$$= -\delta m|\mathbf{F}| + \delta m \sum_i |\mathbf{p}_i|^{-\beta}\big(1 - \beta(\mathbf{q}_i \cdot \mathbf{t})^2\big)$$

$$= \delta m|\mathbf{F}|\Big(-1 + |\mathbf{F}|^{-1} \sum_i |\mathbf{p}_i|^{-\beta}\big(1 - \beta(\mathbf{q}_i \cdot \mathbf{t})^2\big)\Big) < 0,$$

$$(5.75)$$

where we have used the following facts: $\mathbf{F} \cdot \mathbf{t} = 0$, $\delta\mathbf{t} = -\delta m\,\mathbf{m}$ (see Fig. 5.16), and $\delta\mathbf{F} \cdot \delta\mathbf{t}$ is disregarded as a higher order infinitesimal. The condition for the particle being in a stable balance position is therefore equivalent to

$$|\mathbf{F}|^{-1} \sum_i |\mathbf{p}_i|^{-\beta}\big(1 - \beta(\mathbf{q}_i \cdot \mathbf{t})^2\big) < 1. \tag{5.76}$$

The left-hand side of (5.76) is precisely b (see (5.74)) and hence we have shown that $b < 1$, which further implies that $\frac{1+\alpha b}{1+\alpha} < 1$. It is left to prove that also $\frac{1+\alpha b}{1+\alpha} > -1$. Observe that

$$b \geq |\mathbf{F}|^{-1} \sum_i |\mathbf{p}_i|^{-\beta}(1 - \beta) \tag{5.77}$$

because $\mathbf{q}_i \cdot \mathbf{t} \leq 1$. Also, we have

$$|\mathbf{F}| = \mathbf{F} \cdot \mathbf{m} = \sum_i |\mathbf{p}_i|^{-\beta}\mathbf{p}_i \cdot \mathbf{m} = \frac{1}{2} \sum_i |\mathbf{p}_i|^{-\beta}|\mathbf{p}_i|^2, \tag{5.78}$$

where the last equality follows from (5.62). From (5.77) and (5.78) (and the fact that $1 - \beta < 0$) we now get

$$b \geq (1 - \beta)\frac{\sum_i |\mathbf{p}_i|^{-\beta}}{\frac{1}{2}\sum_i |\mathbf{p}_i|^{-\beta}|\mathbf{p}_i|^2}$$

$$\geq (1 - \beta)\frac{\sum_i |\mathbf{p}_i|^{-\beta}}{\frac{1}{2}\sum_i |\mathbf{p}_i|^{-\beta} \min_k |\mathbf{p}_k|^2} \tag{5.79}$$

$$= (1 - \beta)\frac{2}{\mu},$$

where

$$\mu = \min_i |\mathbf{p}_i|^2 \tag{5.80}$$

is the squared minimum distance from the observed particle to any of the other particles, as already introduced in (5.39). We now finally obtain (recall that the parameter of the algorithm α is defined as $\frac{\mu}{2\beta}$)

$$\alpha b = \frac{\mu}{2\beta} b \geq \frac{\mu}{2\beta}(1 - \beta)\frac{2}{\mu} > -1, \tag{5.81}$$

which implies $\frac{1+\alpha b}{1+\alpha} > -1$. We have thus shown that $\left|\frac{1+\alpha b}{1+\alpha}\right| < 1$, and consequently (see (5.73)) that

$$\|J(\mathbf{m})\| < 1. \tag{5.82}$$

This, as we have already discussed (see (5.69)), means that the particle converges to a stable balance position according to our algorithm whenever it finds itself in an infinitesimal neighbourhood of such a position.

We now wish to extend the above result to the case of a *finite* neighbourhood of a single particle. A couple of definitions are needed first.

By an ϵ-neighbourhood of $\mathbf{m}$ we will understand an open ball around $\mathbf{m}$ of radius ϵ, i.e. $\{\mathbf{x} \in \mathbb{R}^D : |\mathbf{x} - \mathbf{m}| < \epsilon\}$. In the following proof, we will assume that the particle's initial position is in some neighbourhood of a stable balance position, and consequently, that it is not necessarily on the sphere. This assumption yields a slightly more general claim than is needed, but actually simplifies the proof.

Assume that the observed particle is located at $\mathbf{x}$ and one step of the algorithm is executed. We can define a mapping $\mathbf{f}$ such that $\mathbf{f}(\mathbf{x})$ represents the new position of the particle after the execution of one step of the algorithm, assuming that the positions of the remaining $N - 1$ particles are kept fixed. (Clearly, this mapping depends on the positions of the other particles, but to keep the notation simple we do not make this dependence explicit.) Note that $\mathbf{f}(\mathbf{m}) = \mathbf{m}$ because if the particle is in the stable balance position, then $\mathbf{F} = |\mathbf{F}|\mathbf{m}$ and hence the algorithm will not move the particle. In the following proof, we will also need the derivative (Jacobian matrix) of this mapping at the point $\mathbf{m}$, which has in fact already been calculated in (5.67). Namely, using the notation from the proof of Theorem 2, we have $\delta\mathbf{m}' = \mathbf{f}(\mathbf{m} + \delta\mathbf{m}) - \mathbf{f}(\mathbf{m})$, and then it follows from (5.67) that the Jacobian matrix of the mapping $\mathbf{f}$ at the point $\mathbf{m}$ is precisely

$$J(\mathbf{m}) = (I - \mathbf{m}^{\mathsf{T}}\mathbf{m})\frac{I + \alpha B}{1 + \alpha}. \tag{5.83}$$

By (5.82) we also know that $\|J(\mathbf{m})\| < 1$. We can now prove the desired claim.

Theorem 3 *Let some configuration of N points on a D-dimensional unit sphere be given, and let one of the particles be in a stable balance position $\mathbf{m}$. Then there exists a neighbourhood of $\mathbf{m}$ such that, if the particle is moved to any position belonging to this neighbourhood, the algorithm will make it converge towards $\mathbf{m}$.*

Proof We need to prove that if $\mathbf{x}$ is in some sufficiently small neighbourhood of the stable balance position $\mathbf{m}$, then the new position $\mathbf{f}(\mathbf{x})$ is closer to $\mathbf{m}$ than the initial position

$$|\mathbf{f}(\mathbf{x}) - \mathbf{m}| < |\mathbf{x} - \mathbf{m}|, \tag{5.84}$$

which means that the algorithm "pushes" the particle towards the stable balance position $\mathbf{m}$. By the mean value theorem for vector-valued functions we can write

$$\mathbf{f}(\mathbf{x}) - \mathbf{f}(\mathbf{m}) = \left(\int_0^1 J(\mathbf{m} + t(\mathbf{x} - \mathbf{m})) \, dt \right) \cdot (\mathbf{x} - \mathbf{m}) \qquad (5.85)$$

and consequently

$$|\mathbf{f}(\mathbf{x}) - \mathbf{f}(\mathbf{m})| \leq J_{\max}(\epsilon) \, |(\mathbf{x} - \mathbf{m})| , \qquad (5.86)$$

where

$$J_{\max}(\epsilon) = \sup_{\mathbf{y} : |\mathbf{y} - \mathbf{m}| < \epsilon} \|J\mathbf{y}\|. \qquad (5.87)$$

Now, since $\|J(\mathbf{m})\| < 1$ by (5.82) and the Jacobian is a continuous function, there must exist some $\epsilon^* > 0$ such that $J_{\max}(\epsilon^*) < 1$. We then conclude from (5.86) that (5.84) holds whenever the initial position $\mathbf{x}$ of the particle is in an ϵ^*-neighbourhood of the stable balance position $\mathbf{m}$. This completes the proof.

A final remark is appropriate. Theorem 3 ensures that the iterative procedure converges when all the particles except one are fixed and when the one that is being moved is in some neighbourhood of its stable balance position. This is only a necessary condition but not a sufficient one for convergence of the procedure when all the particles are being moved simultaneously, since in that case their stable balance positions also move. Despite this, experimental data indicates that numerical convergence is achieved for arbitrary initial positions.

References

1. Adams PG (1997) A numerical approach to Tamme's problem in Euclidean n-space. MSc thesis, Oregon State University
2. Bachoc C, Vallentin F (2009) Optimality and uniqueness of the $(4, 10, 1/6)$ spherical code. J Comb Theory Ser A116(1):195–204
3. Ballinger B, Blekherman G, Cohn H, Giansiracusa N, Kelly E, Schürmann A (2009) Experimental study of energy-minimizing point configurations on spheres. Exp Math 18(3):257–283
4. Bannai E, Sloane NJA (1981) Uniqueness of certain spherical codes. Can J Math 33(2):437–449
5. Blagojević V, Bojić D, Bojović M, Cvetanović M, Đorđević J, Đurđević Đ, Furlan B, Gajin S, Jovanović Z, Milićev D, Milutinović V, Nikolić B, Protić J, Punt M, Radivojević Z, Stanisavljević Ž, Stojanović S, Tartalja I, Tomašević M, Vuletić P (2017) A systematic approach to generation of new ideas for PhD research in computing. Adv Comput 104:1–31
6. Böröczky KJ, Glazyrin A (2017) Stability of optimal spherical codes, arXiv:1711.06012
7. Burr AG (1989) Spherical codes for M-ary code shift keying. In: Proceedings of the second IEE national conference on telecommunications, New York, United Kingdom, pp 67–72
8. Buzan T (1974) Use your head. BBC Books, London
9. Clare BW, Kepert DL (1991) The optimal packing of circles on a sphere. J Math Chem 6(1):325–349
10. Cohn H, Kumar A (2007) Universally optimal distribution of points on spheres. J Am Math Soc 20(1):99–148
11. Cohn H, Kumar A (2007) Uniqueness of the $(22, 891, 1/4)$ spherical code. New York J Math 13:147–157
12. Cohn H, Jiao Y, Kumar A, Torquato S (2011) Rigidity of spherical codes. Geom Topol 15(4):2235–2273

13. De Bono E (1992) Serious creativity: using the power of lateral thinking to create new ideas. Harper Business, New York
14. Gao J, Rudolph L, Hartmann C (1988) Iteratively maximum likelihood decodable spherical codes and a method for their construction. IEEE Trans Inf Theory 34(3):480–485
15. Hamkins J, Zeger K (1997) Asymptotically dense spherical codes-part I: wrapped spherical codes. IEEE Trans Inf Theory 43(6):1774–1785
16. Hamkins J, Zeger K (1997) Asymptotically dense spherical codes-part II: laminated spherical codes. IEEE Trans Inf Theory 43(6):1786–1798
17. Hamkins J, Zeger K (2002) Gaussian source coding with spherical codes. IEEE Trans Inf Theory 48(11):2980–2989
18. Jovanović Ž, Milutinović V (2012) FPGA accelerator for floating-point matrix multiplication. IET Comput Digit Tech 6:249–256
19. Karlof JK (1993) Decoding spherical codes for the Gaussian channel. IEEE Trans Inf Theory 39(1):60–65
20. Kotlar M, Babović Z, Milutinović V (2016) Implementation of perceptron algorithm using dataflow paradigm. In: Proceedings of the 13th symposium on neural networks and applications (NEUREL), Belgrade, Serbia, pp 1–5
21. Kotlar M, Milutinović V (2018) Comparing controlflow and dataflow for tensor calculus: speed, power, complexity, and MTBF. In: Proceedings of the fourth international workshop on communication architectures for HPC, big data, deep learning and clouds at extreme scale (ExaComm), Frankfurt, Germany
22. Kottwitz DA (1991) The densest packing of equal circles on a sphere. Acta Crystallogr Sect A 47:158–165
23. Lazić DE (1980) Class of block codes for the Gaussian channel. IET Electron Lett 16(5):185–186
24. Lazić DE, Drajić DP, Šenk V (1982) A table of some small-size three-dimensional best spherical codes. In: Proceedings of the IEEE international symposium on information theory (ISIT). Les Arcs, France
25. Lazić DE, Šenk V, Šeškar I (1987) Arrangements of points on a sphere which maximize the least distance. Bull Appl Math (Technical University of Budapest) 47(479/87):7–21
26. Lazić DE, Šenk V, Zamurović R (1988) An efficient numerical procedure for generating best spherical arrangements of points. In: Proceedings of the international AMSE conference "Modelling & Simulation", vol 1C, Istanbul, Turkey, pp 267–278
27. Milutinović V, Kotlar M, Stojanović M, Dundić I, Trifunović N, Babović Z (2017) DataFlow supercomputing essentials: algorithms, applications and implementations. Springer, Berlin
28. Musin OR, Tarasov AS (2012) The strong thirteen spheres problem. Discret Comput Geom 48(1):128–141
29. Musin OR, Tarasov AS (2015) The Tammes problem for $N = 14$. Exp Math 24(4):460–468
30. Nguyen HQ, Goyal VK, Varshney LR (2009) On concentric spherical codes and permutation codes with multiple initial codewords. In: Proceedings of the IEEE international symposium on information theory (ISIT), Seoul, Korea, pp 2038–2042
31. Nurmela KJ (1995) Constructing spherical codes by global optimization methods. Helsinki University of Technology, Digital systems laboratory, Research report A32
32. Orloff MA (2006) Inventive thinking through TRIZ, a practical guide. Springer, Berlin
33. Osborn AF (1963) Applied imagination: principles and procedures of creative problem solving, Third Revised edn. Charles Scribner's Sons, New York
34. Savransky SD (2001) Engineering of creativity: introduction to TRIZ methodology of inventive problem solving. CRC Press, New York
35. Silverstein D, DeCarlo N, Slocum M (2008) Insourcing innovation: how to achieve competitive excellence using TRIZ. CRC Press, New York
36. Sloane NJA (1981) Tables of sphere packings and spherical codes. IEEE Trans Inf Theory 27(3):327–338
37. Tarnai T, Gáspár Zs (1983) Improved packing of equal circles on a sphere and rigidity of its graph. Math Proc Camb Philos Soc 93(2):191–218

38. Utkovski Z, Lindner J (2006) On the construction of non-coherent space time codes from high-dimensional spherical codes. In: Proceedings of the ninth IEEE international symposium on spread spectrum techniques and applications, Manaus-Amazon, Brazil, pp 327–331
39. Wang J (2009) Finding and investigating exact spherical codes. Exp Math 18(2):249–256
40. http://en.wikipedia.org/wiki/Binary64
41. http://neilsloane.com/packings/index.html

Part III
Applications in Image Understanding, Biomedicine, Physics Simulation, and Business

Chapter 6
Face Recognition Using Maxeler DataFlow

**Tijana Sustersic, Aleksandra Vulovic, Nemanja Trifunovic,
Ivan Milankovic and Nenad Filipovic**

Abstract Face recognition has its theoretical and practical value in daily life. In this chapter, we will present face recognition application and discuss its implementation using the Maxeler DataFlow paradigm. We first give theoretical background and overview of the existing solutions in the area of algorithms for face recognition. Maxeler card is based on FPGA technology and therefore a brief explanation of the main FPGA characteristics are given and the comparison is made with GPU technology. After that, we analyze one of the PCA algorithms called Eigenface for its application, first on PC and then on Maxeler card. The results show that this algorithm is suitable for implementing on Maxeler card using the dataflow paradigm. By analyzing aforementioned algorithm, it could be seen that execution timing could be reduced, which is especially important when working with large databases. We could conclude that the use of the Maxeler DataFlow paradigm provides advantages in comparison to PC application, resulting in reduction in memory access latency and increase in power efficiency, due to the execution of instructions in natural sequence as data propagates through the algorithm. Since there are many technical challenges

T. Sustersic · A. Vulovic · I. Milankovic · N. Filipovic (✉)
Faculty of Engineering, University of Kragujevac, Sestre Janjic 6, Kragujevac, Serbia
e-mail: fica@kg.ac.rs

T. Sustersic
e-mail: tijanas@kg.ac.rs

A. Vulovic
e-mail: aleksandra.vulovic@kg.ac.rs

I. Milankovic
e-mail: ivan.milankovic@kg.ac.rs

T. Sustersic · A. Vulovic · I. Milankovic · N. Filipovic
Research and Development Center for Bioengineering (BioIRC),
Prvoslava Stojanovica 6, Kragujevac, Serbia

N. Trifunovic
School of Electrical Engineering, University of Belgrade, Bulevar kralja
Aleksandra 73, Belgrade, Serbia
e-mail: nemanja@maxeler.com

© Springer Nature Switzerland AG 2019
V. Milutinovic and M. Kotlar (eds.), *Exploring the DataFlow
Supercomputing Paradigm*, Computer Communications and Networks,
https://doi.org/10.1007/978-3-030-13803-5_6

(e.g., viewpoint, lightening, facial expression, different haircut, presence of glasses, hats, etc.) affecting successful recognition, this area is to be further examined and algorithms could be adapted for dataflow implementation.

6.1   Introduction

Different body characteristics such as voice, face, gait cycle, etc., have been used for centuries in order to recognize people. Face recognition can be considered quite intuitive for people, but it is rather complex to be applied for a machine vision system [1]. In previous decades, a noticeable advancement was achieved regarding development of person recognition, using these biometric characteristics.

According to Stan and Anil [2], face recognition is a visual pattern recognition problem. A complete face recognition system has to be able to automatically detect faces present in images and videos and identify whether a certain face is a known face or an unknown face. It can operate in either one or both of the two modes:

1. face verification (or authentication)
2. face identification (or recognition)

Face verification mode is a one-to-one match when a query face image is compared to a template face image in order to confirm identity that is being claimed. Face identification mode includes a one-to-many match when a query face image is compared against the images in the database in order to identify the query face [3]. The first step in the face recognition process is detection and segmentation of the face areas from the whole image [3]. If the video is used for face recognition, the detected faces may need to be tracked using a face tracking component [3]. With the face detection successfully done, one of the available or developed new algorithms is applied. The applied algorithm should make decision whether a face belongs to the known face in the available database or to the unknown face. In both cases, the current database needs to be updated. If the face is unknown, a new folder with person's information is added to the database, or if it is a known face, the database needs to be updated with new images. The results displayed to the user are followed by appropriate information (Fig. 6.1).

One of the approaches for face recognition is to use feature extraction. The choice of the feature extractor is very important for creation of high recognition rate system. There are two main approaches [4]:

1. statistical based approach
2. approach based on extracting structural features

These approaches have been intensively studied [5, 6]. The first approach considers features extracted from the whole image, whilst the second approach is focused on extraction of local structural features such as shapes of the eyes, mouth or nose [7, 8]. The problem with the first approach is the use of global data, which means that some information is insignificant for recognition process (shoulders, hair …),

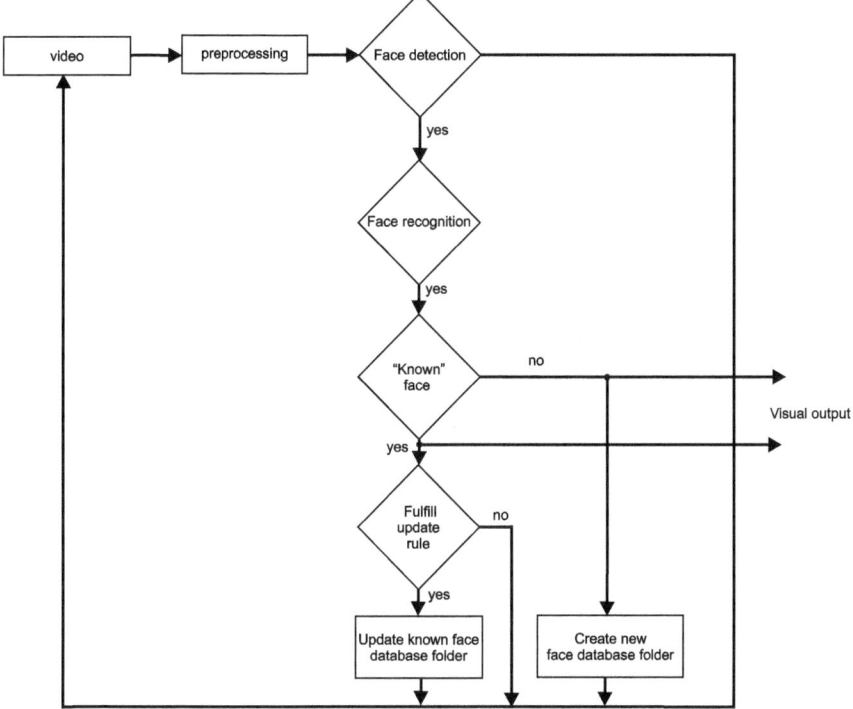

Fig. 6.1 Phases in face recognition

while the problem with the second approach is variation of facial appearance and environmental conditions [9].

Authors have used different methods to reduce the amount of irrelevant information. Authors Turk and Pentland [10] tried to multiply the input image by a 2D Gaussian window that was centered on the face in order to reduce the effects of the non-face portion.

Another method for reducing the amount of irrelevant data involves neural networks. This approach has shown equivalent or slightly better results compared to conventional classifiers for a number of different classification problems [11]. The more feature elements are used, the more complex structure of the neural network classifier is needed, which affects the convergence in the learning phase.

Eigenfaces is an algorithm that falls under the previously mentioned approaches that use feature extraction, where the features are human faces. Eigenface algorithm is a basic facial recognition introduced by M. Turk and A. Pentland in 1991 [10].

The recognition process consists of initialization and face classification. During initialization, the eigenface basis is established; while during face classification, a new image is projected onto the "face space" and the resulting image is categorized by the weight patterns as a known face, an unknown face or a non-face image.

Image1 Pixel1	Image2 Pixel1	Image3 Pixel1	....	ImageM Pixel1	Image1 Pixel1
Image1 Pixel2	Image2 Pixel2	Image3 Pixel2	....	ImageM Pixel2	Image1 Pixel2
Image1 Pixel3	Image2 Pixel3	Image3 Pixel3	....	ImageM Pixel3	Image1 Pixel3
....	....	....	....	....	....
Image1 PixelN	Image2 PixelN	Image3 PixelN	....	ImageM PixelN	Image1 PixelN

Fig. 6.2 Conversion of images into column vectors

The first step in this process is to obtain a training set of **M** grayscale face images $\mathbf{I}_1, \mathbf{I}_2, \ldots, \mathbf{I}_M$. These images should have the following characteristics:

- to be normalized (every pixel has a value between 0 and 255) and
- to have the same size

Pixels of one image are put in an array and placed in a corresponding column of the matrix. After this, the matrix should look like the one shown in Fig. 6.2.

Each column of the previous matrix is called face vector Γ_i.

Eigenface algorithm calculates the eigenvalue decomposition of a data covariance matrix or singular value decomposition of a data matrix, usually after mean centering the data for each attribute.

The algorithm can be divided into the following two steps:

1. identification
2. recognition

Steps in identification can be divided into the following substeps:

- compute the average face
- subtract the average face from the training faces
- compute covariance matrix
- determine eigenvectors and eigenvalues by using one of the methods
- choose a certain number of eigenvectors with highest eigenvalues
- project training faces into Facespace

In order to classify faces, we need to analyze the distribution of faces in the face space. The first step is to compute average (mean) face vector Ψ. This vector represents the center of gravity of faces in a database. Average face vector is calculated by averaging the values of each pixel across all our face images.

$$\Psi = \frac{1}{M} \sum_{i=1}^{M} \Gamma_i; \qquad (6.1)$$

The result of this process is a single column which contains the mean pixels (Fig. 6.3). Cell N contains averaged value for Nth pixel for all faces in the database.

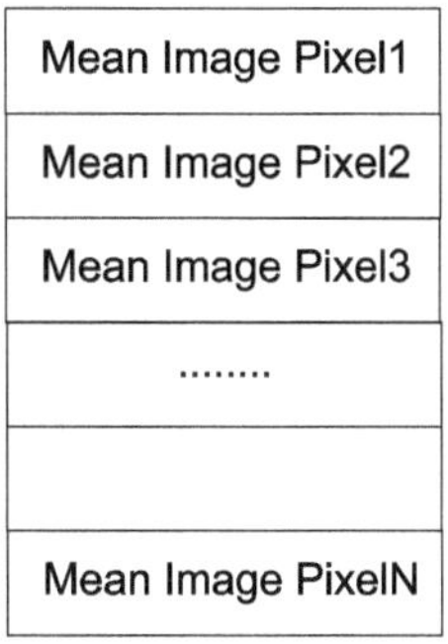

| Mean Image Pixel1 |
| Mean Image Pixel2 |
| Mean Image Pixel3 |
| .... |
| |
| Mean Image PixelN |

Fig. 6.3 Mean face column

Each eigenface represents the differences from the calculated average face.

After this, the mean face is subtracted from each face vector Γ_i, to obtain a set of vectors Φ_i. The reason for this subtraction is to remove common features and to be left with only the distinguishing features from each face.

$$\Phi_i = \Gamma_i - \Psi; \qquad (6.2)$$

Then we find the Covariance matrix C.

$$C = AA^T; \qquad (6.3)$$

where A = $[\Phi_1, \Phi_2,..., \Phi_M]$.

The next step is to calculate Eigenvectors u_i of matrix C. For this purpose, we have used the Jacobi method.

The generalized eigenvalue problem for two $N \times N$ matrices A and B is to find a set of eigenvalues λ and eigenvectors x $\neq 0$ such that

$$Ax = \lambda Bx; \qquad (6.4)$$

If B is the identity matrix, this equation reduces to the eigenvalue problem [12]

$$Ax = \lambda x; \qquad (6.5)$$

In numerical linear algebra, the Jacobi eigenvalue algorithm is an iterative method for the calculation of the eigenvalues and eigenvectors of a real symmetric matrix (a process known as diagonalization).

This method is a fairly robust way to extract all of the eigenvalues and eigenvectors of a symmetric matrix. The method is based on a series of rotations, called Jacobi or Givens rotations. The rotations are chosen in order to eliminate off-diagonal elements while retaining the eigenvalues. Accumulating the products of the explained transformations, we obtain the eigenvectors of the matrix [13, 14].

Eigenvector associated with the highest eigenvalue reflects the highest variance, and the one associated with the lowest eigenvalue reflects the smallest variance. Eigenvalues decrease exponentially so that about 90% of the total variance is contained in the first 5–10% eigenvectors [15].

Based on the mentioned information, we have chosen eigenvectors value for the current database to be 20. The same approach can be used for other databases. First, eigenvalues and eigenvector should be calculated, and then those values used to decide on the number of vectors.

When converted to matrix, obtained eigenvectors u_i have a face-like appearance. Because of this, these eigenvectors are called Eigenfaces.

Each face in the training set (when the mean is removed) — Φ_i can be represented as a linear combination of these eigenvectors u_i.

$$\Phi_i = \sum_{j=1}^{K} \omega_j u_j; \tag{6.6}$$

where K is the number of eigenvectors and u_j are eigenfaces.

Weights (ω_j) can be calculated as

$$\omega_j = u_j^T \Phi_i; \tag{6.7}$$

where $i = 1, 2,..., M$ (M—number of training set images)

Each normalized training face Φ_i is represented in this basis by a vector:

$$\Omega_i = \begin{bmatrix} w_1^i \\ w_2^i \\ \vdots \\ w_K^i \end{bmatrix}, i = 1, 2,..., M \tag{6.8}$$

Steps in Recognition can be divided into substeps:

- Subtract the average face from the face
- Compute the projection into Facespace
- Compute the distance in the face space between the face and all known faces
- Compute the threshold

If we assume that the unknown image is Γ and it has the same size as other images in the database. First, we need to normalize it as

$$\Phi = \Gamma - \Psi; \tag{6.9}$$

Then, we project this normalized image onto the Eigenspace (the collection of Eigenvectors/faces) to find out the weights.

Table 6.1 Advantages and disadvantages of eigenface algorithm

Advantages	Disadvantages
Fast	Lighting variations affect results
robust	$>15°$ head rotation reduces accuracy

$$\omega_j = u_j^T \Phi_i$$

where Φ_i can be presented as

$$\Omega = \begin{bmatrix} w_1 \\ w_2 \\ \vdots \\ w_K \end{bmatrix} \tag{6.10}$$

After the weight vector for the unknown image is found, the next step is to classify it. For the classification task, we use distance measures. Classification is performed in the following way:

- Find $e_r = \min\| \Omega - \Omega_i \|$
- If $e_r < \Theta$, where Θ is chosen experimentally, we can say that unknown image is recognized (this image has the smallest distance to the unknown face in the face space.)
- If $e_r > \Theta$, then unknown image does not belong to the database.

For distance measures, the most commonly used measure is the Euclidean Distance.

$$||x - y||_e = \sqrt{| x_i - y_i |^2}; \tag{6.11}$$

It is necessary to set a threshold that allows us to determine whether a person is in the current database. There is no formula for calculating the threshold. According to Gupta et al. [16], the value is chosen arbitrarily or as some factor of maximum value of minimum Euclidean distances of each image from other images. The threshold value is taken as 0.8 times of the maximum value of minimum Euclidean distances of each image from other images i.e. $\Theta = 0.8*\max(e)$.

Advantages and disadvantages of this algorithm are given in Table 6.1.

6.2 Application of Face Recognition

Face recognition has been an attractive research topic in recent years. The main reason for this is a broad range of possible applications, such as access control systems, image, and film processing, model-based video coding, criminal identification and authentication in secure systems [4].

The application of face recognition identification can be categorized into the following two main fields:

- law enforcement application and
- commercial application.

Law enforcement applications include identifying individuals at airports and in transportation (train stations, airports, border crossings, etc. have become prime locations for catching wanted individuals (i.e., terrorists). The sheer volume of travelers, however, has made it difficult to effectively identify these individuals). There is also an application in public security systems (e.g., criminal identification, digital driver license, etc.) [1]. Commercial applications range from static matching of photographs on credit cards, drivers licenses, ATM cards, and photo ID to real-time matching with still images or video image sequences for access control. Each application presents different constraints in terms of processing [17].

One interesting combination of both law enforcement and commercial application is in the case of car theft protection, where the system can automatically learn its owner's face in different views and keep updating the database. When someone else enters the car without the key, the system is switched into the alarm mode. It can help by sending a corresponding warning message to the owner who can decide whether to report to the police. The face shots of the unknown driver are stored in the database. Since GPS and mobile communication systems are more and more popular in cars, the mugshot as well as the car's current position can be sent using the in-car mobile phone to the car owner if the alarm is verified by the car owner. If the owner thinks it is a false alarm, he/she can use the mobile phone to switch the alarm mode off. In the normal mode, the system learns new faces, for example, of the spouse, relatives or friends of the owner [1].

We use a large number of passwords everyday; for example, to withdraw money from an ATM machine, to unlock a computer or phone, or to access different web sites. Although biometric personal identification has been used for decades (fingerprint scan or retinal scan), the problem with this type of identification is that it depends on the willingness of participants. The advantage of face recognition is that the participant is not required to cooperate. Some of the advantages/disadvantages of different biometrics are described in [18]. Table 6.2 lists some of the applications of face recognition. Face Recognition systems can be grouped into two categories depending on whether they use images or video. Different fields are involved in the process of face recognition—image processing, analysis and pattern recognition and the complexity of the problem depends on whether the input into Face Recognition system is an image or video, image quality, background noise, variability of the images [19].

There is little information on which software is used in European countries in the area of law enforcement—public law and order, as well as in courts. One of the mentioned companies that is said to work also in Europe is NEC Corporation of America (NEC). They provide a solution that law enforcement officials can rely upon to manage physical access, identify criminals, as well as exonerate the innocent. In addition to its facial recognition system, NEC is able to provide its products and

Table 6.2 Common application of face recognition [19]

Areas	Specific application
Entertainment	Video game, virtual reality, training programs
	Human–robot interaction, human–computer interaction
Smart cards	Drivers' licenses, entitlement programs
	Immigration, national ID, passports, voter registration
	Welfare fraud
Information security	TV Parental control, personal device logon, desktop logon
	Application security, database security, file encryption
	Intranet security, Internet access, medical records
	Secure trading terminals
Law enforcement and surveillance	Shoplifting, suspect tracking and investigation
	Advanced video surveillance, CCTV control
	Portal control, postevent analysis

Table 6.3 Some of the available products on the market

Company	Product	Website
Luxand	Luxand FaceSDK^{TM}	https://www.luxand.com/
FaceFirst	FaceFirst®platform	https://www.facefirst.com/
Cognitec	FaceVACS-DBScan	http://www.cognitec.com/
Vocord	VOCORD FaceControl and VOCORD FaceControl 3D	http://www.vocord.com/
Face4 systems	Face Recognition System (FRS)	http://www.face4systems.com/
IdentyTech solutions^{TM}	GateKeeper^{TM}	http://www.identytech.com/
FaceKey^{TM}	FaceKey	https://www.facekey.com/
Passfaces^{TM}	Passfaces Web Access	http://www.realuser.com/
BioID®	BioID®PhotoVerify	https://www.bioid.com/

applications for voice, data and video communications solutions. NEC is also said to have recently agreed with its master distributor for German SKM Skyline and is actively looking to recruit additional resellers to exploit the market interest in Germany and other European markets. However, on a global scale, there are quite a number of face recognition products available on the market (Table 6.3).

6.3 Issues and Technical Challenges

When humans are trying to recognize a face, they use a wide spectrum of stimuli obtained from our senses (visual, auditory, olfactory, tactile, etc.). It is hard to develop a face recognition system that will mimic the face recognition ability of humans.

The human brain is limited by a total number of persons that it can accurately remember. That is the biggest advantage for creating a face recognition system, as a computer is able to handle large number of face images [19]. Studies in psychology and neuroscience have effect on systems for machine recognition of faces and vice versa [20, 21]. It is believed that face recognition is a process that is different from other object recognition tasks [22].

- **Ranking of significance of facial features**: hair, eyes, mouth, and face outline have been considered to be important for perceiving and remembering faces. It has been shown by several studies that the nose has small effect on recognition. One reason for this could be that these studies have used frontal images. If we consider profile mugshots, the nose is more important than mouth. Attractiveness and beauty were analyzed, and it was concluded that attractive faces have the best recognition rate; the least attractive faces come next, followed by the mid-range faces.
- **Distinctiveness**: studies have shown that distinctive faces are better and faster recognized than typical faces [23].
- **The role of spatial frequency analysis**: earlier studies concluded that information in low spatial frequency bands plays a dominant role in face recognition. It was concluded that low, band pass, and high-frequency components play different roles. Based on the low-frequency components only, we can successfully perform gender classification, while high-frequency components are needed for identification.
- **Viewpoint-invariant recognition**: some experiments have suggested that memory for faces is highly viewpoint-dependent. Generalization even from one profile viewpoint to another is poor, though generalization from one three-quarter view to the other is very good.
- **Effect of lighting change**: It was observed that photographic negatives of faces are difficult to recognize [24, 25]. It was demonstrated that bottom lighting makes identification of familiar faces harder and later Hill and Bruce [25], showed the importance of top lighting for face recognition.
- **Movement and face recognition**: It was observed that movement helps with face recognition of familiar faces shown under different types of degradation [24, 26]. Unfortunately, experiments with unfamiliar faces showed no benefit when these faces are displayed in moving sequences compared to still image.
- **Facial expressions**: Neurophysiological studies indicate that analysis of facial expressions is carried out at the same time as face recognition. Patients with difficulties in identifying familiar faces have no problem in recognizing facial expressions of emotions. Patients with "organic brain syndrome" are capable to recognize faces, but have problem with expression analysis.

Evaluation reports and other independent studies indicate that accuracy of many face recognition methods and speaker identification systems depend on many factors. With respect to face recognition, accuracy deteriorates with changes in lighting, illumination and other factors [27]. Whereas shape and reflectance are intrinsic properties of a face object, appearance of a face is also subject to several other factors, including facial pose (or, equivalently, camera viewpoint), facial expression, pres-

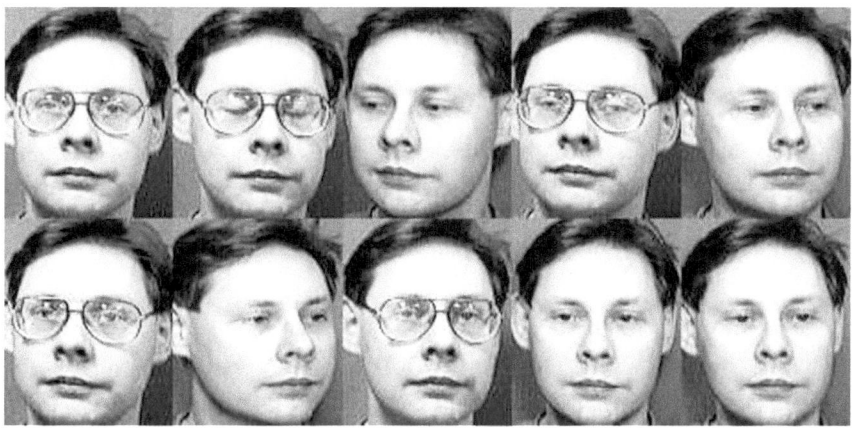

Fig. 6.4 Example of different facial poses and facial expressions [28]

ence/absence of glasses, etc. Examples of such variations in one person can also be found in AT&T publicly available database, which will be used later on for testing purposes of our algorithm [28] (Fig. 6.4).

In addition to these, various imaging parameters, such as aperture, exposure time, lens aberrations, and sensor spectral response also increase intrasubject variations [2]. Nevertheless, the number of examples per person available for learning phase is usually much smaller than the dimensionality of the image space; a system trained on not so many examples may not represent well unseen instances of the face [2].

6.4 Existing Solutions

Face recognition methods can be divided into two groups:

1. appearance-based and
2. model-based algorithms (Fig. 6.5)

Appearance-based methods consider face to be raw intensity image [28]. An image is considered as a high-dimensional vector. Usually, statistical techniques are used in order to derive a feature space from the image distribution. The sample image is compared to the training set. Appearance methods can be classified as linear or non-linear, while model-based methods can be 2D or 3D [28]. Linear appearance-based methods perform a linear dimension reduction. The face vectors are projected to the basis vectors, while the projection coefficients are used as the feature representation of each face image. Some examples of these methods are Linear Discriminant Analysis (LDA), Principal Component Analysis (PCA) or Independent Component Analysis (ICA). Nonlinear appearance methods are more complex. The model-based approach tries to model a human face. The new sample is fitted to the model, and the parameters of the model are used to recognize the image. On the other hand,

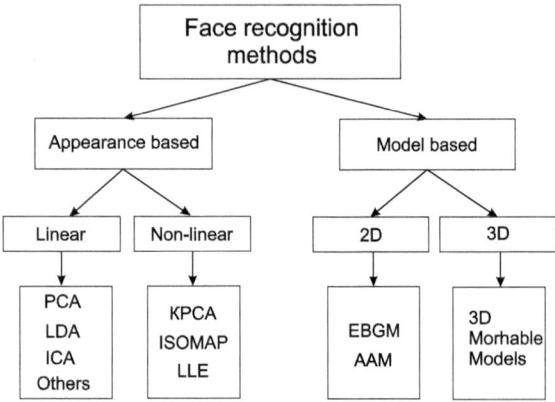

model-based approaches can be two-dimensional or three-dimensional. These models are often morphable, which means they are able to classify faces even when pose changes are present. Three-dimensional models are more complicate, as they try to capture the three-dimensional nature of human faces. Examples of this approach are KernelPCA (KPCA) [29], Elastic Bunch Graph Matching (EBGM) [30] or 3D Morphable Models [31–33].

In addition to previously mentioned methods, artificial neural networks are a popular tool in face recognition. They have been used in pattern recognition and classification. Neural networks-based approaches are learned from the example-images and rely on the techniques from machine learning to find relevant characteristics of face images. Kohonen [34] was the first to demonstrate that a neural network could be used to recognize aligned and normalized faces. Many different methods based on neural network have been proposed since then. The main goal of the paper by Vulovic et al. [35] was to compare different training algorithms for neural networks and to use them for face recognition problem as it is a nonlinear problem. With all used training functions, percentage of recognition was above 90%, where training functions *traincgb* and *trainrp* performed best in given conditions.

Field-Programmable Gate Array (FPGA) is technology that can be used for accuracy and speed improvement. The main goal of the paper by Sustersic et al. [36] was to load photos from files to FPGA and display, as well as describe implementation of the Eigenface algorithm using DE2 Altera board. They showed that DE2 Altera board could be used for reading databases and photos from crime scenes and discussed how Eigenface algorithm could be implemented on this board in order to speedup the process of face recognition [36]. Another paper by Sustersic et al. [37] researched how an algorithm for face recognition, based on Fast Fourier Transform (FFT), could be implemented on a FPGA chip. Due to the lack of memory on the chip, the implementation of the whole system could not be achieved and the results were discussed based on the results obtained by the simulation. The implemented part of the system included displaying images on a VGA monitor and the result of the algorithm was shown on a seven-segment display. The authors pointed out 79%

accuracy. The advantage of the presented system lies in the possibility of working with images in real time [37].

Recently, field-programmable gate array (FPGA) has attracted lot of attention. The reason for that is its high flexibility and superior performance. FPGA presents an array of gates that can be redesigned by the designer. Programmable logic devices (PLDs) and the logic cell array (LCA) concept were used for FPGA creation [38]. It can implement a wide range of arithmetic and logic functions by adding a two-dimensional array of configurable logic blocks (CLBs) and programming the interconnection that connects the configurable resources [39]. Compared to other popular IC technologies, FPGA has the following advantages [39–42]:

- **Performance**: FPGA has the ability to exceed the speed limit of sequential execution technologies. Moreover, it is able to process data at a much higher speed than DSP processors and microprocessors, which performance is determined by the system clock rate. FPGA can achieve higher performance in applications that require large arithmetic resources.
- **Cost**: FPGA is a cost-effective solution for system development due to its reprogrammable nature. Power consumption can be reduced with design. It is customizable and can be easily reconfigured. The non-recurring engineering cost is relatively low.
- **Reliability**: The FPGA resources can be separated by task requirements. The tasks interact through well-defined interfaces between one another, which greatly reduces unexpected interaction. If one task changes, it will not affect the execution of the other tasks. This high isolation and high parallelism mechanism not only minimize the reliability concerns, but also reduce the deterministic hardware dedicated to every task.
- **Flexibility**: Since the functionality of FPGA is not fixed by the manufacturer, but defined by the designer, it allows easy system maintenance and quick upgrade with simple prototype changes. By reconfiguration even on site, FPGA requires very low prototyping costs and avoid the compatibility problem.
- **Time to market**: Reconfiguration, programming flexibility, and fast prototyping made it is easy to develop, verify and debug the devices. Products using FPGA are getting to market quickly thanks to the different development tools that have shorten developing period.

Due to all this, FPGA has become important candidate for implementation of OFDM technology for wireless and mobile communication systems. The challenge with DataFlow computing is identification and modification of algorithms that are based on the DataFlow architecture platform properties [43].

6.5 Essence of the DataFlow Implementation

The Data flow Programming Model represents somewhat an opposite approach to the conventional control flow programming model. In general, each step in a control flow model depends on the completion of the previous step, which means that tasks are executed in series. To be more specific, a sequence of instructions which represent the source code are loaded to the memory and guide the processor to allow reading and writing data from and to the memory in order to complete specific commands. Despite the improvements and optimization techniques (e.g., forwarding and prediction logic) to speedup the process, the control flow programming model remains sequential. Therefore, its latency depends on the number of memory accesses and the time required for a CPU clock cycle. On the contrary, data flow computing is based on the fact that order of execution is controlled only by flow of the data, which makes this model a simple, yet powerful for parallel computing. To further explain the working process of data flow, instructions are performed immediately after the required resources necessary for the execution are available. Additionally, data is considered as streams (require independent processing) rather than batches (set of data which are processed as a unit). This means that dataflow programming models comprise multiple components that can process input data at the same time, unlike control flow modeling. Terminology also specifies the smallest unit of dataflow models as a component, whereas the smallest unit of control flow models is called the task [43].

Data on a dataflow engine (DFE) is streamed from memory onto a chip. Data moves directly from one functional unit to another, until the process is completed [44]. After the process is completed, it is written off to the chip memory. Calculation can be done using a dataflow core or control flow core [45]. A dataflow core enables computation to be laid out spatially on the chip while control flow core computes on the same functional unit at different points in time. The DFE itself represents the computation, which means that there is no need for instructions, branch prediction, memory decode logic, or out of order scheduling. This means that the chip is able to assign all its resources for computations. In order to create a computer system at cluster level, this engine (DFE) must be unified with conventional CPUs, storage and network. MPC-C and MPC-X are two types of Maxeler Maximum Performance Computing (MPC) dataflow compute nodes that are being used. The first compute node (MPC-C) allows for a standalone deployment of dataflow technology with the use of a fixed combination of coupled DFEs and CPUs. The second compute node MPC-X can balance different compute technologies, as the balance of DFEs and CPUs is flexible at runtime. The DFEs are managed by the system software, which allocates them to different applications.

The computers based on von Neumann architecture or control flow architecture fetch the instructions and data from memory, perform operations specified by instructions and write the results back to the memory. The main drawback of this architecture is that each execution of an instruction requires a cyclic access for memory, which results in a large number of data transfers between CPU and memory.

The dataflow computing paradigm is fundamentally different from the standard control flow computing. In dataflow computing, the high level language is used to generate a dataflow graph which is directly mapped to the DFE. Each node in this graph performs specific operation and outputs the result of that operation to another node in the graph, thus the cyclic access for memory is avoided.

The dataflow computing paradigm as such has certain advantages and certain disadvantages compared to control flow computing. The main advantages are that the instructions are executed in natural sequence as data propagate through the algorithm. It also reduces the effect of memory access latency because all of the data travel through the graph, and they are attached to their nodes. The disadvantage is the lack of central control because each node is activated only when all of its inputs are available. Furthermore, the nodes are forced to use data when they are available even if they are not needed at that time.

The dataflow computing is nowadays used in a large number of applications. It does not only accelerate the applications, but it makes them more energy efficient compared to sequential computing because of low processing power consumption of DFEs. Energy efficiency is achieved with low DFE's frequency and it is well known fact that power consumption is directly proportional to frequency. DFE's frequency can go up to a few hundreds of MHz, while frequency of nowadays processors goes up to a few GHz.

DataFlow computing is used to accelerate a wide range of applications (i.e., finite difference solvers, financial tree-based PDE solvers, Monte Carlo simulations...). Multiple DFEs can be used for a single problem by dividing the domain into subdomains and assigning them to each DFE. This dataflow computing method is used in automation applications [46], digital signal processing [47, 48], mathematics for solving system of equations [49] and floating-point matrix multiplication [50], financial derivatives pricing [51], artificial neural networks [52] and many more.

Maxeler has developed DFEs which use high-speed, large area FPGA chip as a computational unit. The general architecture of these DEFs is shown in Fig. 6.6. Each DFE consists of one or more kernels, manager, fast memory (FMem), large memory (LMem) and is connected to the CPU via PCI Express bus.

Kernels represent hardware implementation of an algorithm and their main task is to perform computation as data flows through DFE. Manager has a task to define the way the data flows between kernels, memories and host processor. The DFE has two types of memory: fast memory (FMem) and large memory (LMem). The FMem represents on-chip memory and can store several megabytes of data with terabytes/second of access bandwidth. The LMem much larger than the FMem and represents off-chip memory. It can store many gigabytes of data. The cost of the greater capacity LMem is payed with much lower access bandwidth.

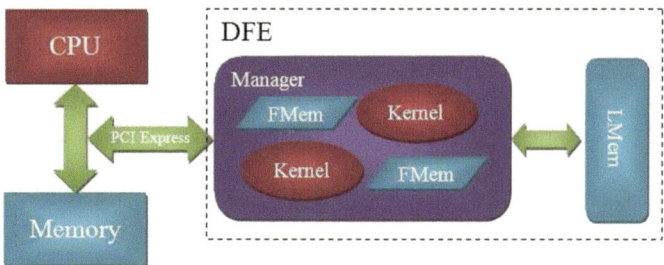

Fig. 6.6 Maxeler's DFE architecture

Maxeler has also developed its own hardware description language called MaxJ. The MaxJ is Java-based language which enables a user without significant expertise in hardware design to write high-performance applications for FPGA with higher level of abstraction from hardware than a HDL. MaxCompiler uses MaxJ code and in its runtime generates VHDL which is built into a bitstream. Voss et al. [53] showed on the gzip design example that using MaxJ takes only one person and a period of one month to develop an application and achieve better performance than the similar work created in Verilog and OpenCL. Maxeler has also provided simulation and debugging tools for designs to be tested before creating a real DFE, which means it provides much faster development than it is the case with low-level hardware description languages such as VHDL and Verilog.

Contemporary computer architectures can be divided into:

- general-purpose computing and
- problem-specific computing.

Although the first group is the most common, the second one is attracting more attention with appearance of challenging computational problems. The first special-purpose architectures were very effective since they had a focus on mapping custom solutions into hardware. The problem was that they were not adaptable to changes and were dependent on the hardware they were designed for, which means that they were expensive and not very convenient for transferring to another hardware. Field-Programmable Gate Arrays changed that fact, since they can be thought of as programmable ASICs. FPGAs are reconfigurable logic chips that can be reprogrammed in seconds to implement customized designs. Their main drawback is their inability to reach the same high level of performance in terms of speed and energy efficiency. Their main advantage is flexibility and user-friendliness due to the fact that they are programmable. They are also highly parallel and accelerated implementations, but their major drawback is a slow clock [43]. FPGAs form a fundamental component of the Maxeler technologies.

Each Data flow Engine (DFE) is a reconfigurable chip with a lot of memory. Their engines do not perform calculations in time, but rather in space as they are designed with the FPGAs. The movement of the data, which are streamed on the chip of the DFE from some memory, is specific and driven from one functional unit to the other. All this happens without requiring the interference of an external off-chip memory, all the way to the completion of the execution. The DFE here is only is used for computation purposes and one or more Kernels and a Manager are used in order to reach maximum performance in terms of speed and efficiency. A Kernel contains the computation that will be performed, whereas the Manager controls the data movement within the DFE. Then, the MaxCompiler utilizes the information provided by the aforementioned components to produce the corresponding Data flow implementations. These implementations can be directly called from the CPU by using the SLiC (Simple Live CPU) interface that is automatically built based on the current Data flow program and allows an easy access to DFEs from the attached CPU. Furthermore, the DFE has access to two types of Memories-FMem (on-chip memory), which can store up to several megabytes of data, and LMem (off-chip memory), which can store several gigabytes of data. In addition, FMem offers a broad bandwidth that can reach up to 21 TB/s within the chip. This is a significant advantage in comparison to traditional CPUs with multilevel caches since at traditional CPUs the fastest cache memory level is the only memory close to the computational unit. In that process data are duplicated through the cache hierarchy. DFEs are controlled by the system manager, who acts as the one that can assign one or more DFEs for a specific application and sets them free whenever once they have completed their task [43].

Maxeler's DFEs are widely used in many fields. Gan et al. [54] summarize their experience of using Maxeler DFEs to eliminate the main bottlenecks and obtain higher efficiencies in solving geoscience problems. They managed to achieve better results in both the performance and power efficiency over traditional multicore and many-core architectures. Grull et al. [55] showed acceleration of 200 compared to an Intel i5 450 CPU for localization microscopy, and an acceleration of 5 over an Nvidia Tesla C1060 for electron tomography while maintaining full accuracy using Maxeler's DFEs. Gan et al. [56] used Maxeler's DFEs to find the solution of global shallow water equations (SWEs), which is one of the most essential equation sets describing atmospheric dynamics. They achieved speedup of 20 over a fully optimized design on a CPU rack (with two eight-core processors) and speedup of 8 over the fully optimized Kepler GPU design. They also managed to have 10 times more power efficiency than a Tianhe-1A supercomputer node. Weston et al. [51] achieved acceleration over 270 times faster than a single Intel Core for a multi-asset Monte Carlo model. Milankovic et al. [57] presented a few DFE configurations for acceleration of image segmentation algorithm for breast mammogram images and they achieved acceleration up to seven times for the best DFE configuration. It must be also emphasized that the use of dataflow in tensor problems is of great importance. Since the problem of face recognition is a tensor problem in many aspects (image is a 3rd order tensor by definition, face images database is a 6th order tensor), dataflow paradigm has great potential in this field [58]. Kotlar et al. also discussed a nontrivial problem of computing eigenvalues and eigenvectors. They proposed dataflow implementation for calculating eigenvalues and eigenvectors, based on the QR decomposition. The proposed dataflow solution implements the QR decomposition using two different methods Gram–Schmidt method and Householder method [58].

6.6 Comparison of GPU and FPGA Characteristics

The main advantages of the FPGA compared to other types of conventional architectures (GPUs, multicore CPU, ASICs) are [43]:

- Flexibility (much greater than the one offered by ASICs)
- Potential to optimize low-level elements
- Power efficiency compared to multicore CPU and GPUs

Huge processing power of GPUs on one side, and the great power efficiency of the FPGA technology on the other side, show the competitiveness of these two technologies. When comparing floating-point processing of GPUs and FPGAs, the total floating-point operations per second of the best GPUs are higher than that of FPGA's with the maximum DSP capabilities, which means GPUs have better floating-point processing. This is also true for processing/€ as mid-class devices can be compared within the same order of magnitude, but GPU wins in the category money per GFLOP. However, timing latency of FPGAs is one order of magnitude less than GPUs, since FPGAs have deterministic timing. Measuring GFLOPS per watt, FPGAs are 3–4 times better. Although still far away, latest GPU products are dramatically improving the power consumption. Similar situation is happening when interfaces are concerned, since GPUs interface via PCIe, while FPGA flexibility allows connection to any other device via almost any physical standard or custom interface. regarding backward compatibility, GPUs are better than FPGAs, since software developed for older GPUs can work in new devices, and FPGA HDL can be moved to newer platforms, but it needs reworking. FPGA also lacks flexibility to modify hardware implementation of the synthesized code, which is not a problem for GPUs developers. Size is definitely the advantage of FPGAs, which implement the solution in smaller dimensions. Finally, it should be said that today many algorithms are designed directly for GPUs and there are not many FPGA developers, which increases the cost of their hiring.

GPUs performance is mainly dependent on the commercial graphic cards characteristics. Some of the widely used graphic cards, along with the comparison of their main characteristics and prices, are given in Tables 6.4 and 6.5. GPUs price ranges from less than 400 € are given in Table 6.4, while those above 400 € are given in Table 6.5.

Although Altera and Xilinx are the two main competitive companies in production of FPGA chips, we will show only characteristics of a selection of Xilinx 7-Series

Table 6.4 Compared characteristics of different graphic cards—bellow 400 [60]

		Nvidia GeForce GT 730	AMD RadeonR7 360	Nvidia GeForce GTX 970
Price (approx.)		80 €	120 €	250 €
Processing power	Single	693 GFLOPS	1,612 GFLOPS	3,494 GFLOPS
	Double	32 GFLOPS	100 GFLOPS	109 GFLOPS
Technology		28 nm	28 nm	28 nm
GPU		GK208 (Kepler)	Tobago (GCN 1.1)	GM204 (Max-well)
Core clock		902 MHz	1050 MHz	1050 MHz
Power consumption stress test		93 W	100 W	242 W
Price efficiency		0.10 €/GFLOPS	0.07 €/GFLOPS	0.07 €/GFLOPS
Power efficiency		7 GFLOPS/W	16 GFLOPS/W	14 GFLOPS/W

Table 6.5 Compared characteristics of different graphic cards—above 400 € [51]

		Sapphire Radeon R9 390	Sapphire Radeon R9 Fury X	Nvidia GeForce GTX 980 Ti
Price (approx.)		400 €	600 €	700 €
Processing power	Single	5,120 GFLOPS	7,168 GFLOPS	5,632 GFLOPS
	Double	640 GFLOPS	448 GFLOPS	176 GFLOPS
Technology		28 nm	28 nm	28 nm
GPU		Grenada (GCN 1.1)	Fiji (GCN 1.2)	GM200
Core clock		1000 MHz	1050 MHz	1000 MHz
Power consumption stress test		323 W	358 W	250 W
Price efficiency		0.08 €/GFLOPS	0.08 €/GFLOPS	0.12 €/GFLOPS
Power efficiency		16 GFLOPS/W	20 GFLOPS/W	23 GFLOPS/W

Table 6.6 Compared characteristics of Xilinx SoCs and FPGAs [60]

	Zynq SoC Z-7020	Zynq SoC Z-7010	Artix-7 200T	Kintex-7 480T	Virtex-7 690T
Price (approx.)	100 €	3000 €	190 €	2500 €	11200 €
Dual ARM®Cortex^{TM}-A9 MPCore	Yes	Yes	–	–	–
Programmable logic cells	85000	444000	215360	477000	693120
Programmable DSP slices	220	2020	740	1920	3600
Peak DSP performance	276 GMACs	2622 GMACs	929 GMACs	2845 GMACs	5335 GMACs
Processing power—single	180 GFLOPS	1560 GFLOPS	648 GFLOPS	1800 GFLOPS	3120 GFLOPS
Technology	28 nm	28 nm	28 nm	28 nm	28 nm
PCIe interface	–	x8 Gen2	x4 Gen2	x8 Gen2	x8 Gen3
Power consumption	2.5 W	20 W	9 W	25 W	40 W
Price efficiency-single	0.56 €/GFLOPS	1.92 €/GFLOPS	0.29 €/GFLOPS	1.39 €/GFLOPS	3.59 €/GFLOPS
Power efficiency-single	72 GFLOPS/W	78 GFLOPS/W	72 GFLOPS/W	72 GFLOPS/W	78 GFLOPS/W

devices (Table 6.6). Excellent power efficiency (higher than 70 GFLOPS/W) and decent total processing power enable the implementation of the FPGAs devices into small and efficient hardware [59].

6.7 Details of the Implementation

The database used in the implementation was AT&T [28] database which consists of grayscale photos of 40 people, where each person was photographed 10 times (400 photos in total). These 10 photos of each person were made at different times, varying certain parameters such as the lighting, facial expressions (with open/closed eyes, smiling/not smiling) and certain facial details (with glasses/without glasses). All the images were taken against a dark homogeneous background with the subjects in an upright, frontal position (with tolerance for some side movement). Each image had the size of 112 x 92 pixels with 256 levels of grey. The algorithm for face recognition that was implemented was one of the Principal Component Analysis algorithms (PCA)-Eigenface algorithm. This implementation included working matrices with larger dimension in the following manner:

1. Start with $U = 1$
2. Determine the off-diagonal element A_{ij} that is the largest in absolute value and compute the rotation angle α from A_{ij} and the corresponding diagonal elements:

$$\alpha = \frac{1}{2} \arctan \left(\frac{2A_{ij}}{A_{jj} - A_{ii}} \right) \qquad (6.12)$$

3. Construct a matrix V that is identical to the unit matrix except for

$$V_{ii} = V_{jj} \cos \alpha \qquad (6.13)$$

$$V_{ij} = \sin \alpha \qquad (6.14)$$

$$V_{ji} = -\sin \alpha \qquad (6.15)$$

4. Then compute the matrix product $A' = V^T A V$ and $U' = U V$; A'_{ij} becomes zero by this operation, the other elements in rows and columns i and j are changed.
5. If the largest absolute value of the diagonal elements A_{ij} is greater than threshold, repeat the procedure from step 2 with A' instead of A and U' instead of U. After convergence, A' contains the eigenvalues and U' the eigenvectors.

With the use of this algorithm, the number of calculations is greatly reduced—from the order of the number of pixels in the images (N^2) to the order of the number of images from the training set (M). It should be the case that the training set of face images will be relatively small ($M \ll N^2$), and this way calculations become quite manageable. However, since the algorithm itself assumes that the problem is linear, authors [60] show that the times for obtaining results are around 36 s for training (using AT&T database) using Windows OS and around 0.8 s for recognition per image. When using larger databases, times for training will become larger, and that could be the motivation for implementing the same algorithm on Maxeler chip.

6.8 Some Performance Indicators

The results are obtained using PC running 64-bit Windows 10 with 8GB of RAM memory with Code Blocks 16.01. There are two possibilities:

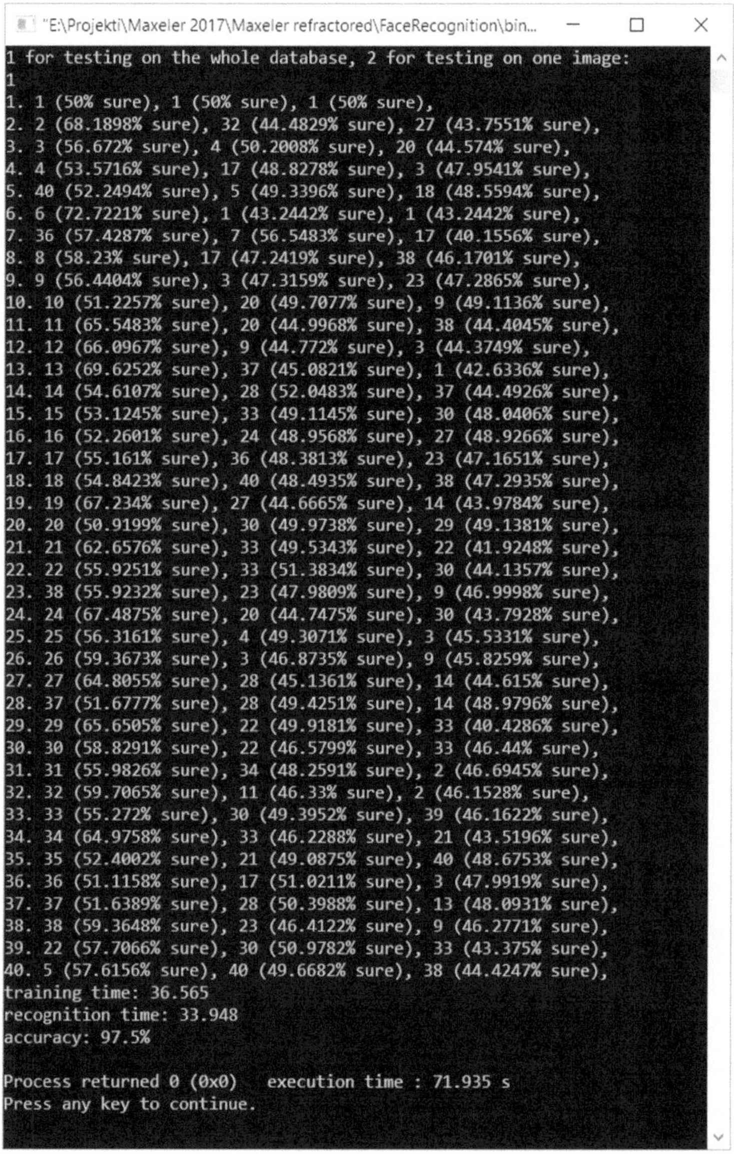

```
"E:\Projekti\Maxeler 2017\Maxeler refractored\FaceRecognition\bin...   —   □   ×
1 for testing on the whole database, 2 for testing on one image:
1
1. 1 (50% sure), 1 (50% sure), 1 (50% sure),
2. 2 (68.1898% sure), 32 (44.4829% sure), 27 (43.7551% sure),
3. 3 (56.672% sure), 4 (50.2008% sure), 20 (44.574% sure),
4. 4 (53.5716% sure), 17 (48.8278% sure), 3 (47.9541% sure),
5. 40 (52.2494% sure), 5 (49.3396% sure), 18 (48.5594% sure),
6. 6 (72.7221% sure), 1 (43.2442% sure), 1 (43.2442% sure),
7. 36 (57.4287% sure), 7 (56.5483% sure), 17 (40.1556% sure),
8. 8 (58.23% sure), 17 (47.2419% sure), 38 (46.1701% sure),
9. 9 (56.4404% sure), 3 (47.3159% sure), 23 (47.2865% sure),
10. 10 (51.2257% sure), 20 (49.7077% sure), 9 (49.1136% sure),
11. 11 (65.5483% sure), 20 (44.9968% sure), 38 (44.4045% sure),
12. 12 (66.0967% sure), 9 (44.772% sure), 3 (44.3749% sure),
13. 13 (69.6252% sure), 37 (45.0821% sure), 1 (42.6336% sure),
14. 14 (54.6107% sure), 28 (52.0483% sure), 37 (44.4926% sure),
15. 15 (53.1245% sure), 33 (49.1145% sure), 30 (48.0406% sure),
16. 16 (52.2601% sure), 24 (48.9568% sure), 27 (48.9266% sure),
17. 17 (55.161% sure), 36 (48.3813% sure), 23 (47.1651% sure),
18. 18 (54.8423% sure), 40 (48.4935% sure), 38 (47.2935% sure),
19. 19 (67.234% sure), 27 (44.6665% sure), 14 (43.9784% sure),
20. 20 (50.9199% sure), 30 (49.9738% sure), 29 (49.1381% sure),
21. 21 (62.6576% sure), 33 (49.5343% sure), 22 (41.9248% sure),
22. 22 (55.9251% sure), 33 (51.3834% sure), 30 (44.1357% sure),
23. 38 (55.9232% sure), 23 (47.9809% sure), 9 (46.9998% sure),
24. 24 (67.4875% sure), 20 (44.7475% sure), 30 (43.7928% sure),
25. 25 (56.3161% sure), 4 (49.3071% sure), 3 (45.5331% sure),
26. 26 (59.3673% sure), 3 (46.8735% sure), 9 (45.8259% sure),
27. 27 (64.8055% sure), 28 (45.1361% sure), 14 (44.615% sure),
28. 37 (51.6777% sure), 28 (49.4251% sure), 14 (48.9796% sure),
29. 29 (65.6505% sure), 22 (49.9181% sure), 33 (40.4286% sure),
30. 30 (58.8291% sure), 22 (46.5799% sure), 33 (46.44% sure),
31. 31 (55.9826% sure), 34 (48.2591% sure), 2 (46.6945% sure),
32. 32 (59.7065% sure), 11 (46.33% sure), 2 (46.1528% sure),
33. 33 (55.272% sure), 30 (49.3952% sure), 39 (46.1622% sure),
34. 34 (64.9758% sure), 33 (46.2288% sure), 21 (43.5196% sure),
35. 35 (52.4002% sure), 21 (49.0875% sure), 40 (48.6753% sure),
36. 36 (51.1158% sure), 17 (51.0211% sure), 3 (47.9919% sure),
37. 37 (51.6389% sure), 28 (50.3988% sure), 13 (48.0931% sure),
38. 38 (59.3648% sure), 23 (46.4122% sure), 9 (46.2771% sure),
39. 22 (57.7066% sure), 30 (50.9782% sure), 33 (43.375% sure),
40. 5 (57.6156% sure), 40 (49.6682% sure), 38 (44.4247% sure),
training time: 36.565
recognition time: 33.948
accuracy: 97.5%

Process returned 0 (0x0)   execution time : 71.935 s
Press any key to continue.
```

Fig. 6.7 C++ performance results when recognition is tested on all images of different persons

1. Give path to all the images in order to perform training and validation

   • 97.5% accuracy
   • Result in a format: 2. 2, 32, 27 (Fig. 6.7) which means
     2. — searched image
     2, 32, 27 — recognized persons

Looking for...

| Match no 1 | Match no 2 | Match no 3 |
| 67.4875% sure | 44.7475% sure | 43.7928% sure |

Fig. 6.8 Face recognition result where the correct face is the first choice

```
■ "E:\Projekti\Maxeler 2017\Maxeler refractored\FaceRecognition...   —   □   ×
Please enter 1 or 2:
1 for testing on the whole database, 2 for testing on one image:
2
2. 2 (68.1898% sure), 32 (44.4829% sure), 27 (43.7551% sure),
training time: 35.429
recognition time: 0.839

Process returned 0 (0x0)    execution time : 37.636 s
Press any key to continue.
```

Fig. 6.9 C++ performance results when recognition is tested on one image of the person

Figure 6.8 shows another recognition example, where the correct face is the first choice, while the second and third one are similar looking faces.

2. Give path to specified image for recognition

   • Result in the same format (Fig. 6.9)

Table 6.7 shows benchmark results for the first possibility when path is given to all the images in order to perform training and validation.

Table 6.7 Benchmark results for the first possibility

Training average	36.488s
Recognition average for all images	34.27466s
Recognition average per image	0.8568s

Figure 6.10 shows another recognition example, where the correct face is the second choice, and not the first one.

Looking for...

Match no 1	**Match no 2**	**Match no 3**
57.4287% sure	**56.5483% sure**	**40.1556% sure**

Fig. 6.10 Face recognition result where the correct face is the second choice

Training average time is the average time necessary for training on 390 images. Recognition average for all images is the time necessary for matching 40 images (verification stadium), which gives us recognition average per image, which is the time necessary for finding one person image match in the database.

6.9 Conclusion

This chapter presented a brief review of face recognition algorithms that are in use today, and special attention was devoted to the face recognition application using Maxeler DataFlow. The practical value of face recognition not only in our daily life, but also for the use in airports, employee entries, criminal detection systems and identification based on facial features, show the motivation for speedup and accuracy improvement of these systems that implement face recognition. As there are numerous features in face or facial expression, most of these methods make trade-offs such as hardware requirements, time to update image database, time for extracting features, response time, etc. Principal Component Analysis (PCA) is a well-studied approach used to extract features by making sample projection from higher dimension to lower dimension. Therefore, special attention in this chapter is given to the Eigenface algorithm because of its simplicity, and yet, promising accuracy and computational time. The main problem with this algorithm is the time for extracting eigenvalues, which is high even with enough memory. For testing purposes, the mentioned Eigenface algorithm was tested on AT&T database of human faces using traditional approach of control flow computing on a PC. It was also discussed that the same algorithm could be implemented on Maxeler card using dataflow computing. The results show that the use of the Maxeler DataFlow paradigm provides great benefits in comparison to PC application. Since the instructions are executed in natural sequence as data propagate through the algorithm, and there is a reduction in memory access latency because of the data that travels through the graph. Power efficiency is also an advantage of Maxeler card. There are still many technical challenges (e.g., viewpoint, lightening, facial expression, different haircut, presence of glasses, hats, etc.) that influence the accuracy of face recognition, which will be the topic of interest in future research. Another area of future research is the adaptation of several face recognition algorithms for dataflow implementation.

Acknowledgements The part of this research is supported by HOLOBALANCE: HOLOgrams in an ageing population with BALANCE disorders project funded by European Union's Horizon 2020 research and innovation programme under grant agreement No 769574 and Ministry of Education, Science and Technological Development of Serbia, with projects OI174028 and III41007.

References

1. Mou D (2010) Machine-based intelligent face recognition. Springer, Berlin
2. Jain AK, Li SZ (2005) Handbook of face recognition. Springer, London
3. Draper BA, Baek K, Bartlett MS, Beveridge JR (2003) Recognizing faces with PCA and ICA. Comput Vis Image Underst 91(1–2):115–137
4. Haddadnia J, Faez K, Ahmadi M (2002) A neural based human face recognition system using an efficient feature extraction method with pseudo zernike moment. J Circuits Syst Comput 11(3):283–304
5. Yang M-H, Kriegman DJ (2002) Detecting faces in images: a survey. IEEE Trans Pattern Anal Mach Intell 24(1):34–58
6. Hjelmas E, Low BK (2001) Face detection: a survey. Comput Vis Image Underst 83(3):236–274
7. Bichsel M, Pentland AP (1994) Human face recognition and the face image set's topology. CVGIP: Image Underst 59(2):254–261
8. Lin KH, Lam KM, Siu WC (2001) Locating the eye in human face images using fractal dimensions. IEEE Proc-Vis Image Signal Processing 148(6):413–421
9. Chen LF, Liao HM, Lin J, Han C (2001) Why recognition in a statistic-based face recognition system should be based on the pure face portion: a probabilistic decision-based proof. Pattern Recognit 34(7):1393–1403
10. Turk M, Pentland A (1991) Eigenfaces for recognition. J Cogn Neurosci 3(1):71–86

11. Zhou W (1999) Verification of the nonparametric characteristics of backpropagation neural networks for image classification. IEEE Trans Geosci Remote Sens 37(2):771–779
12. Golub GH, Van Loan CF (1996) Matrix computations, 3rd edn. The John Hopkins University Press, Baltimore
13. Press WH, Flannery BP, Teukolsky SA, Vetterling WT (1992) Numerical recipes in C: the art of scientific computing, 2nd edn. Cambridge University Press, Cambridge
14. Kiusalaas J (2010) Numerical methods in engineering with Python 3, 2nd edn. Cambridge University Press, Cambridge
15. Kim K, Face Recognition Using Principle Component Analysis, Department of Computer Science, University of Maryland, College Park, USA
16. Gupta S, Sahoo OP, Goel A, Gupta R (2010) A new optimized approach to face recognition. Glob J Comput Sci Technol 10(1):15–17
17. Tolba AS, El-Baz AH, El-Harby AA (2006) Face recognition: a literature review. Int J Signal Process 2(2):88–103
18. Phillips PJ (1998) Support vector machines applied to face fecognition. Adv Neural Inf Process Syst 803–809
19. Zhao W, Chellappa R, Phillips PJ, Rosenfeld A (2003) Face recognition: a literature survey. ACM Comput Surv (CSUR) 35(4):399–458
20. Hancock PJ, Bruce V, Burton MA (1998) A comparison of two computer-based face identification systems with human perceptions of faces. Vis Res 38(15–16):2277–2288
21. Kalocsai P, Zhao W, Elagin E (1998) Face similarity space as perceived by humans and artificial systems. Automatic face and gesture recognition, Nara, Japan
22. Ellis HD (1986) Introduction to aspects of face processing: ten questions in need of answers. Aspects of face processing. Springer, Dordrecht, pp 3–13
23. Bruce V, Burton MA, Dench N (1994) What's distinctive about a distinctive face? Q J Exp Psychol 47(1):119–141
24. Bruce V, Hancock PJ, Burton AM (1998) Human face perception and identification. Face recognition. Springer, Berlin, pp 51–72
25. Hill H, Bruce V (1996) The effects of lighting on the perception of facial surfaces. J Exp Psychol: Hum Percept Perform 22(4):986
26. O'Toole AJ, Roark DA, Abdi H (2002) Recognizing moving faces: a psychological and neural synthesis. Trends Cogn Sci 6(6):261–266
27. Hassaballah M, Aly S (2015) Face recognition: challenges, achievements and future directions. IET Comput Vis 9(4):614–626
28. AT&T Laboratories, The database of faces, Cambridge. http://www.cl.cam.ac.uk/research/dtg/attarchive/facedatabase.html. Accessed 12 Jan 2016
29. Fernandes S, Bala J (2013) Performance analysis of PCA-based and LDA based algorithms for face recognition. Int J Signal Process Syst 1(1):1–6
30. Moghaddam B (2002) Principal manifolds and probabilistic subspaces for visual recognition. IEEE Trans Pattern Anal Mach Intell 24(6):780–788
31. Wiskott L, Fellous J-M, Kruger N, von der Malsburg C (1997) Face recognition by elastic bunch graph matching. IEEE Trans Pattern Anal Mach Intell 19(7):775–779
32. Blanz V, Vetter T (2003) Face recognition based on fitting a 3D morphable model. IEEE Trans Pattern Anal Mach Intell 25(9):1063–1074
33. Huang J, Heisele B, Blanz V (2003) Component-based face recognition with 3D morphable models, In: Proceedinds of the international conference on audio- and video-based biometric person authentication, Guildford, UK
34. Kohonen T (2012) Self-organization and associative memory. Springer, New York
35. Vulovic A, Sustersic T, Rankovic V, Peulic A, Filipovic N (2018) Comparison of different neural network training algorithms with application to face recognition. EAI Endorsed Trans Ind Netw Intell Syst 18(12):e3
36. Sustersic T, Vulovic A, Filipovic N, Peulic A (2018) FPGA implementation of face recognition algorithm. Pervasive Comput Parad Ment Health. Selected Papers from MindCare 2016, Fabulous 2016, and IIoT 2015, pp 93–99

37. Šušteršič T, Peulić A (2018) Implementation of face recognition algorithm on field programmable gate array (FPGA). J Circuits, Syst Comput, p 1950129
38. Zhou H (2013) Design and FPGA implementation of OFDM system with channel estimation and synchronization. Montral, Qubec, Canada
39. Xilinx Inc., System generator for DSP. http://www.xilinx.com/support/documentation/dt_sysgendsp_sysgen12-1.html. Accessed 10 Jan 2018
40. Cummings M, Haruyama S (1999) FPGA in the software radio. IEEE Commun Mag 37(2):108–112
41. Altera Inc., DSP builder. http://www.altera.com/products/software/products/dsp/dsp-builder.html. Accessed 10 Jan 2018
42. Altera Inc., FPGA vs. DSP design reliability and maintenance white paper. http://www.altera.com/literature/wp/wp-01023.pdf. Accessed 10 Jan 2018
43. Nikolakaki SM (2015) Real-time Stream data processing with FPGA-based supercomputer. Chania
44. Pell O (2012) Maximum performance computing with data-flow engines. https://www.maxeler.com/media/MaxelerSummaryMPCDatafowEngines.pdf. Accessed 5 Sept 2017
45. Technologies M (2017) Dataflow computing. https://www.maxeler.com/technology/dataflow-computing/. Accessed 6 Sept 2017
46. Panfilov P, Salibekyan S (2014) Dataflow computing and its impact on automation applications. Procedia Eng 69:1286–1295
47. Bhattacharya B, Bhattacharyya SS (2001) Parameterized dataflow modeling for DSP systems. IEEE Trans Signal Process 49(10):2408–2421
48. Voigt S, Baesler M, Teufel T (2010) Dynamically reconfigurable dataflow architecture for high-performance digital signal processing. J Syst Arch 56(11):561–576
49. Morris GR, Abed KH (2013) Mapping a Jacobi Iterative solver onto a high-performance heterogeneous computer. IEEE Trans Parallel Distrib Syst 24(1):85–91
50. Jovanovic Z, Milutinovic V (2012) FPGA accelerator for floating-point matrix multiplication. IET Comput Digit Tech 6(4):249–256
51. Weston S, Spooner J, Racaniere S, Mencer O (2012) Rapid computation of value and risk for derivatives portfolios. Concurr Comput: Pract Exp 24(8):880–894
52. Li WXY, Chaudhary S, Cheung RCC, Matsumoto T, Fujita M (2013) Fast simulation of digital spiking silicon neuron model employing reconfigurable dataflow computing. In: International conference on field-programmable technology (FPT). Kyoto, Japan
53. Voss N, Becker T, Mencer O, Gaydadjiev G (2017) Rapid development of Gzip with MaxJ. In: International symposium on applied reconfigurable computing. Delft, The Netherlands
54. Gan L, Fu H, Mencer O, Luk W, Yang G (2017) Chapter four data flow computing in geoscience applications. Creativity in computing and dataflow supercomputing, vol 104. Academic Press, Cambridge, pp 125–158
55. Grull F, Kebschull U (2014) Biomedical image processing and reconstruction with dataflow computing on FPGAs. In: 24th international conference on field programmable logic and applications (FPL). Munich, Germany
56. Gan L, Fu H, Luk W, Yang C, Xue W, Huang X, Zhang Y, Yang G (2015) Solving the Global Atmospheric Equations through Heterogeneous Reconfigurable Platforms. ACM Trans Reconfigurable Technol Syst (TRETS)-Spec Sect FPL 2013 8(2):11
57. Milankovic IL, Mijailovic NV, Filipovic ND, Peulic AS (2017) Acceleration of image segmentation algorithm for (Breast) Mammogram Images using High-Performance Reconfigurable Dataflow Computers. Comput Math Methods Med
58. Kotlar M, Milutinovic V (2018) Comparing controlflow and dataflow for tensor calculus: speed, power, complexity, and MTBF. In: High performance computing - ISC high performance 2018 international workshops, ExaComm, Frankfurt, Germany, June 24–28
59. BERTEN Digital Signal Processing, GPU vs FPGA performance comparison 19 May 2016. http://www.bertendsp.com/pdf/whitepaper/BWP001_GPU_vs_FPGA_Performance_Comparison_v1.0.pdf. Accessed 12 Dec 2017
60. Agarwal M, Jain N, Kumar M, Agrawal H (2010) Face recognition using eigen faces and artificial neural network. Int J Comput Theory Eng 2(4):624–629

Chapter 7
Biomedical Images Processing Using Maxeler DataFlow Engines

Aleksandar S. Peulic, Ivan Milankovic, Nikola V. Mijailovic
and Nenad Filipovic

Abstract Image segmentation is one of the most common procedures in medical imaging applications. It is also very important task in breast cancer detection. Breast cancer detection procedure based on mammography can be divided into several stages. The first stage is the extraction of region of interest from a breast image, after which the identification of suspicious mass regions, their classification, and comparison with the existing image database follows. It is often the case that already existing image databases have large sets of data for which processing requires a lot of time, thus the acceleration of each of the processing stages in breast cancer detection is a very important issue. Image filtering is also one of the most common and important tasks in image processing applications. It is, in most cases, preprocessing procedure for 3D visualization of an image stack. In order to achieve high-quality 3D visualization of a 2D image stack, it is of particular importance that all the images of the input stack are clear and sharp, thus their filtering should be executed carefully. There are also many algorithms for 3D visualization, so it is important to choose the right one which will execute fast enough and produce satisfied quality. In this chapter, the implementation of the already existing algorithm for region-of-interest-based image segmentation for mammogram images on High-Performance Reconfigurable DataFlow Computers (HPRDC) is proposed. As a DataFlow Engine (DFE) of such HPRDC, Maxeler's acceleration card is used. The experiments for examining the acceleration of that algorithm on the Reconfigurable DataFlow Computers (RDC)

A. S. Peulic · I. Milankovic · N. V. Mijailovic · N. Filipovic (✉)
Faculty of Engineering, University of Kragujevac, Sestre Janjic 6, Kragujevac, Serbia
e-mail: fica@kg.ac.rs

A. S. Peulic
e-mail: aleksandar.peulic@kg.ac.rs

I. Milankovic
e-mail: ivan.milankovic@kg.ac.rs

N. V. Mijailovic
e-mail: nmijailovic@kg.ac.rs

I. Milankovic · N. V. Mijailovic · N. Filipovic
Research and Development Center for Bioengineering (BioIRC),
Prvoslava Stojanovica 6, Kragujevac, Serbia

© Springer Nature Switzerland AG 2019
V. Milutinovic and M. Kotlar (eds.), *Exploring the DataFlow
Supercomputing Paradigm*, Computer Communications and Networks,
https://doi.org/10.1007/978-3-030-13803-5_7

are performed with two types of mammogram images with different resolutions. There were, also, several DFE configurations and each of them gave different acceleration of algorithm execution. Those accelerations are presented and experimental results have been shown good acceleration. Also, image processing using a mean filtering algorithm combined with thresholding and binarization algorithms and 3D visualization of murine lungs using marching cubes method are explained. These algorithms are mapped on the Maxeler's DFE to significantly increase calculation speed. Optimal algorithm calculation speed was up to 20-fold baseline calculation speed.

7.1 Introduction

The image segmentation algorithms are one of the most used procedures in medical imaging analysis. It is also one of the most important tasks in image processing [1]. In computer vision, the image segmentation is the process of partitioning the image into multiple segments. This technique or group of techniques refers to dividing images into regions with similar attributes. The attributes are often gray levels, colors, edges, texture characteristics, or spectral properties. The main task of image segmentation is to represent the image in a simpler way, into something that is more meaningful and because of that it is easier to be analyzed in the image processing tasks.

There are various applications in medical diagnostics for which the process of image segmentation is one of the main parts. There are several algorithms for image segmentation which are used for detecting microcalcification in the mammography images [2]and these algorithms have a major role in the opportune detection and treatment of a lesion. Kallergi Maria [3] created automated computed tools for microcalcification detection which use artificial neural networks and are based on wavelet filters. Edge detection operators are used for tumor region extraction and enhancement of classification of mammographic images [4]. In order to segment dense areas of the breast and the existence of mass and to visualize other breast regions, the graph cuts (GC) segmentation technique is used [5]. Aghdam Hamed Habibi et al. [6] proposed a probabilistic approach for breast boundary extraction in mammograms.

In most cases, detection of microcalcification is performed with preprocessed images. Oliver Arnau et al. [7] created huge dictionary database by filtering images with different kinds of filters. New images were compared with database where every pixel of breast image is a center of the patch. The results of that processing are probability images where more brightness pixel corresponds to more reliance to be microcalcification. The final result of the segmentation process is a uniform region. The segmentation quality is rated by the ratio of uniformity and homogeneity of the estimated region [8]. The regions need to be shell free and edges of regions are smooth and space accurate. The segmentation of images has a significant application in digital mammography. The main tasks of scientists and researchers are to detect suspicious mass regions of the breast for which they are trying to develop sophisticated image analysis tools. This process begins with extraction of Regions of Interest (ROIs) of breast images. Then, the detection of suspicious regions and their classification is performed, after which the comparison with the existing image database starts. For medical diagnostic decision systems, it is very important to provide a large training data set. For preparation of this data, a large database of medical images can be very helpful. The processing time can be a limiting parameter here. To accelerate the processing time, the multicore, many-core, and FPGA-based dataflow architectures can be used. Alongside with image segmentation algorithms, the image filtering algorithms are commonly used in image processing to remove noise from images, emphasize significant information, and prepare images for further processing such as image segmentation, edge detection, object tracking [9], etc. Median filter, mean filter, Gaussian filter, and other convolution filters are among the most popular image filtering algorithms. A general introduction to the field can be found in standard image processing textbooks such as [1], while the specific issues of processing 3D-images are described, e.g., in [10]. There are many studies which aim to improve image filtering algorithms. Banks et al. [11] used a form of evolutionary computation called genetic programming to optimize the sequence in which image noise filters should be applied to remove two types of image noise and one type of communication noise. Another study focused on improving adaptive image filtering by combining the advantages of mean and median filters [12]. Their proposed adaptive filter effectively filtered out both the Gaussian noise and salt and pepper noise while preserving image edge and detailed information. Horentrup and Schlosser used an extension of the guided image filter to significantly increase the quality of filter output using confidence information associated with the filter input [9]. The speed of image filtering algorithms can be problematic for real-time applications and large datasets.

Once the filtering of 2D image stack is finished, it could be visualized. There are many methods for 3D visualization such as marching cubes algorithm which creates triangle models of constant density surfaces from 2D image stack [13]. This algorithm has been widely used. Ma et al. [14] improved this algorithm to better meet the reconstruction of molecular face in order to use it for its visualization. Viceconti et al. [15] compared standard marching cube and the discretized marching cube algorithms in terms of local accuracy when used to reconstruct the geometry of a human femur and as a result the standard method was found to be more accurate. Wang et al. [16] featured tetrahedronization scheme that incorporates marching cubes surface smoothing together with a smooth-distortion factor in order to develop a technique for the automatic generation of voxel-based patient-specific finite element models.

In this chapter, dataflow architecture provided by Maxeler is used to accelerate region-of-interest-based image segmentation algorithm for (breast) mammogram images which is developed by Milosevic et al. [17]. Also, we have proposed mapping of the mean filter in combination with thresholding onto Maxeler's DataFlow Engine (DFE) to decrease the preprocessing time of large 2D image sequences of murine lungs. The filtered images are then used for 3D visualization of mice lungs.

7.2   Background

A great increase in calculation speed in many algorithms can be achieved by combining High-Performance Computing (HPC) systems and Reconfigurable DataFlow Computing (RDC) systems [18]. This combination is usually called High-Performance Reconfigurable DataFlow Computing (HPRDC). Consisting of some variation of reconfigurable computing and general-purpose processing, HPRDC improves performance and decreases latency by transferring a performance demanding part of some application onto a reconfigurable dataflow computer. In such a way, the whole application will be divided into two parts: the simpler function is executed on the HPC and the performance demanding function is executed on an RDC system such as the Maxeler's. By familiarizing with the RDC systems and clarifying speedup possibilities, one can achieve additional performance advantages over other single-core solutions. The RDC systems represent the combination of reconfigurable logic and, in most cases, are based on FPGA architecture.

7.2.1   Field-Programmable Gate Arrays

Field-programmable gate arrays are integrated circuits designed in such a manner so that they can be configured by the designer to implement some digital computations. They were invented in the 1980s by Ross Freeman and Bernard Vonderschmitt [19]. Their main goal is to accelerate certain calculation tasks in respect to the single processing unit. They also have low processing power consumption. They achieve the best results in speedup with algorithms with limited data dependencies and significant scope for parallel execution.

In order to implement complex computational tasks, the modern FPGAs have a large number of logic blocks, I/O pads, and routing channels. Process of mapping an algorithm into FPGA requires paying a special attention on the available amount of recourses. The number of the logic blocks and I/O pads can be easily determined from the design, but the number of routing channels may vary significantly.

To define the behavior of the FPGA, the designer can use Hardware Description Language (HDL) or some other specialized computer language. The design, described using some of those languages, is transformed into technology-mapped netlist via an electronic design automation tool. That netlist is then mapped to the actual FPGA architecture. At each point of the process of defining the behavior of the FPGA, the functionality of the design is validated via timing analysis, simulation, and other verification methodologies. When the process of validating is done and it is concluded that functionality of the design is correct, the binary file is generated. This binary file is then used to configure the FPGA. In theory, any algorithm can be mapped into FPGA, but, in practice, the main constraints are the already mentioned available recourses, clock rates, and available I/O pads.

7.2.2 DataFlow Computing

Today, most computers are based on von Neumann architecture. This architecture transforms the program source into a list of instructions which are loaded into the memory. These instructions alongside with appropriate data are fetched from the memory into the processor core where operations specified by instructions are performed and the results of those operations are written back to the memory. Modern processors contain many levels of caching, forwarding, and prediction logic to improve the efficiency of this architecture, but still, the main drawback of this architecture is that each execution of an instruction requires a cyclic access for memory, which results in a large number of data transfers between processor core and memory.

The dataflow computing architecture is fundamentally different from the von Neumann architecture. In a dataflow computing architecture, the program source is transformed into a dataflow graph rather than into a set of instructions. DataFlow graph, as shown in Fig. 7.1, consists of a lot of dataflow cores and is directly mapped to the DFE.

Each dataflow core performs specific operation on input data and outputs the result of that operation to another dataflow core in the dataflow graph without writing to the off-chip memory until the chain of processing is complete. DataFlow core executes only when all inputs are available. Thus, the data can be streamed from memory

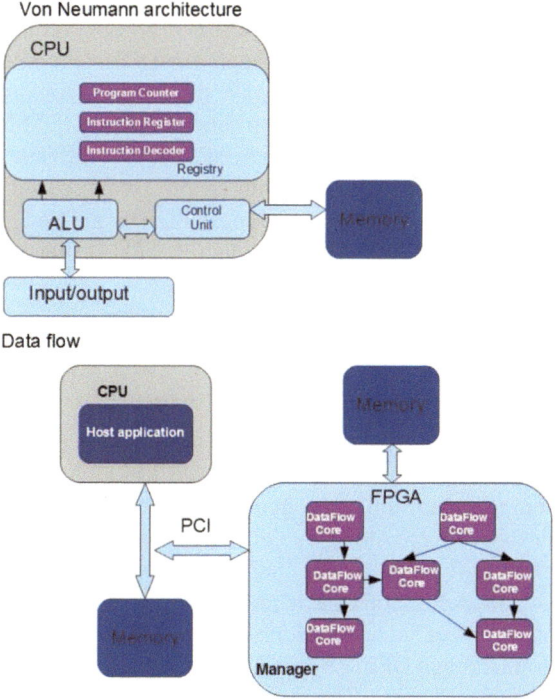

Fig. 7.1 Comparison between von Neumann architecture and dataflow computing architecture

into the DFE where all of the processing is performed and only the results of that processing are outputted to the memory. In this way, the cyclic access for memory is avoided.

The main advantage of dataflow architecture in comparison to von Neumann architecture is that each operation is given its own part of the DFE, and all operations execute with true parallelism. This means that DFE is essentially computing multiple results at a time. On each tick, a new result is calculated, even though the number of operations required for computing may be very large. This is achieved because the instructions are executed in natural sequence, in space rather than in time, as data propagate through the algorithm. DFE also reduces the effect of memory access latency because all data travel through the graph and they are attached to their own nodes. Thus, DFE architecture is especially suitable for big data applications where very low latency is needed. Also, DFE runs on much lower frequency (up to few 100 MHz) than computers based on von Neumann architecture (few GHz), thus it makes applications more energy efficient.

DataFlow computing is currently used in a large number of applications and is the subject of many current studies. Veljovic [20] described discrepancy reduction between the topology of dataflow graph and the topology of FPGA structure on the example of moving from two input adders to three input adders. Riebler et al. [21] studied the design and development of DFEs methodology. Another study [22] demonstrated that the key search algorithm calculation speed could be improved with DFEs up to 205-fold. The development of flexible and customizable long-length LDPC decoders for DFEs was proposed by [23]. They compared the performance of the DFE approach with state-of-the-art parallel computing architectures and showed that for the real-time throughputs the DFE solution is much more power efficient. Li et al. showed that DFEs have the potential to conduct large-scale and fast simulation of the Digital Spiking Silicon Neuron (DSSN) model-based network [24]. Pell et al. presented the framework for finite difference wave propagation modeling using DFEs [25] demonstrating that the proposed solution was up to 30 times more energy efficient than conventional CPUs. Oriato et al. were able to increase execution speed up to 74-fold compared to x86 CPU in their dataflow implementation of the meteorological limited area model [26].

7.2.3 Maxeler's DataFlow Engines

Maxeler has developed DFEs which use high-speed, large area FPGA chip as a computational unit. The general architecture of these DEFs is shown in Fig. 7.2. Each DFE consists of one or more kernels, manager, fast memory (FMem), large memory (LMem), and is connected to the CPU via PCI Express bus.

Architecture of DFE is presented in Fig. 7.2. Kernels represent hardware implementation of some algorithm and their main task is to perform computation as data flows through DFE. Each DFE design can have multiple kernels which communicate among each other. The manager has a task to define the way of data flow between

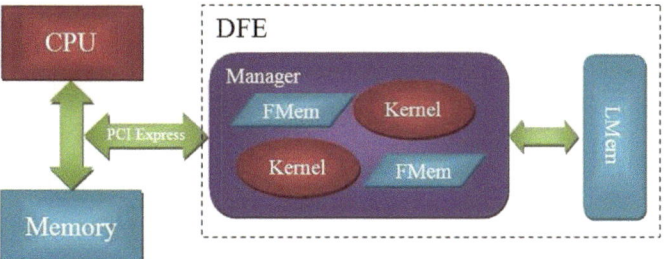

Fig. 7.2 Maxeler's DFE architecture

kernels, memories, and host processor. The DFE has two types of memory: fast memory and large memory. The FMem is on-chip memory and can store several megabytes of data with terabytes/second of access bandwidth. The LMem is off-chip memory and it is substantially larger than the FMem. It can store many gigabytes of data. The cost of the greater capacity LMem pays with much lover access bandwidth.

Maxeler has also developed its own hardware description language called MaxJ. The MaxJ is Java-based language which enables a user without significant expertise in hardware design to write high-performance applications for FPGA with higher level of abstraction from hardware than an HDL. MaxCompiler uses MaxJ code and in its runtime generates VHDL which is built into bitstream using FPGA vendor's toolchain. Different generations of Maxeler's DFEs have different FPGA vendors which are either Xilinx of Altera. Voss et al. [27] showed on the gzip design example that using MaxJ takes only one person and a period of 1 month to develop application and achieve better performance than the related work created in Verilog and OpenCL. Maxeler has also provided simulation and debugging tools which allow designs to be tested before building for a real DFE which provides much faster development than with using low-level hardware description languages such as Verilog and VHDL.

Maxeler's DFEs are widely used in many fields. Niu et al. [28] proposed DFE implementation of stencil-based algorithms and achieved linear speedup using Maxeler's MPC-C500 computing system with four Virtex-6 SX475T FPGAs. Gan et al. [29] summarize their experiences of using Maxeler DFEs to eliminate the main bottlenecks and obtain higher efficiencies in solving geoscience problems. They managed to achieve better results in both the performance and power efficiency over traditional multicore and many-core architectures. Grull et al. [30] showed acceleration of 200 compared to an Intel i5 450 CPU for localization microscopy, and an acceleration of 5 over an Nvidia Tesla C1060 for electron tomography while maintaining full accuracy using Maxeler's DFEs. Gan et al. [31] used Maxeler's DFEs to find the solution of global Shallow Water Equations (SWEs), which is one of the most essential equation sets describing atmospheric dynamics. They achieved speedup of 20 over a fully optimized design on a CPU rack with two 8-core processors and speedup of 8 over the fully optimized Kepler GPU design. They also managed to have 10 times more power efficiency than a Tianhe-1A supercomputer node. Weston et al. [32] achieved acceleration over 270 times faster than a single Intel Core for a multi-asset Monte Carlo model.

7.3 Region-of-Interest-Based Image Segmentation Algorithm for (Breast) Mammogram Images on Maxeler's DFE

The method for mammogram ROI detection [17] is composed of pectoral muscle removal and background removal which represent any artifact present outside the breast area, such as patient markings [33]. The example of those two is presented in Fig. 7.3. There are two views of breast mammogram images: left sided and right sided. For the sake of simplicity, both algorithms, the background removal and the pectoral muscle removal, will be explained for the right-sided view of breast mammogram image. The algorithms for the left-sided view are very similar to the algorithms for the right-sided view, thus there is no need to explain both algorithms (Table 7.1).

7.3.1 Background Partition Removal Algorithm

The basic idea of background partition removal algorithm is to find the largest area of connected nonblack pixels and then set all other pixels to black. The algorithm is as follows:

Fig. 7.3 Pectoral muscle and background

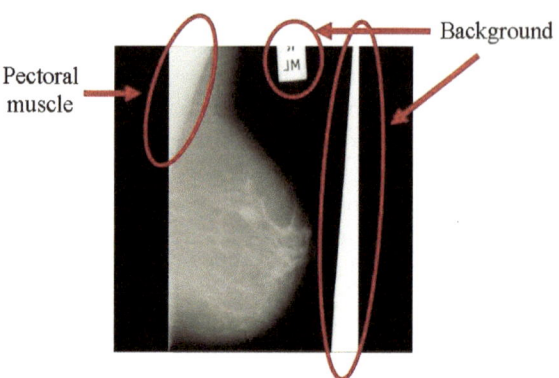

Pectoral muscle

Background

Table 7.1 Background partition removal algorithm

Step 1	Start with first row;
Step 2	Scan from left to right side;
Step 3	While pixel is black go to the next pixel and after that go to step 4;
Step 4	While pixel is not black go to the next pixel and after that go to step 5;
Step 5	If it is a first row then set all other pixels in that row to black and go to step 7, otherwise go to step 6;
Step 6	If the above pixel is black then set the current pixel to black and go to the next pixel and repeat step 6, otherwise go to the next pixel and repeat step 6. If there are no more pixels in the current row then go to step 7;
Step 7	Repeat steps 2 to 6 for the next row

Table 7.2 Pectoral muscle removal algorithm

Step 1	Start with first row;
Step 2	Scan from left to right side;
Step 3	While pixel value is less than the threshold value go to the next pixel;
Step 4	If the pixel belongs to the first tenth part of the mammogram, then the pixel value is greater than or equal to the threshold value; set the current pixel value to black and go to the next pixel and repeat step 4, otherwise go to step 6;
	If the pixel does not belong to the first tenth part of the mammogram anymore go to step 5;
Step 5	If the pixel value is greater than or equal to the threshold value and the above pixel is black, then set the current pixel value to black and go to the next pixel and repeat step 5, otherwise go to step 6;
Step 6	Repeat steps 2 to 5 for the next row

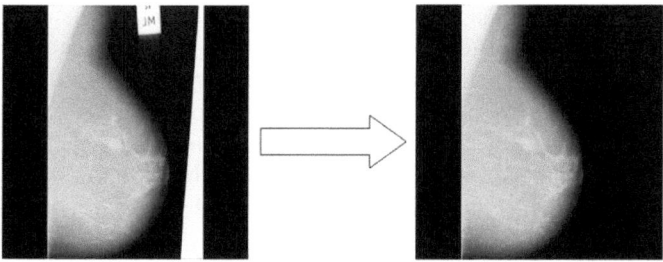

Fig. 7.4 Background partition removal algorithm result

The result of the background partition removal algorithm is shown in Fig. 7.4. As it can be noticed, the unnecessary background has been removed successfully (Table 7.2).

After the background partition removal has been done, the next task is pectoral muscle removal.

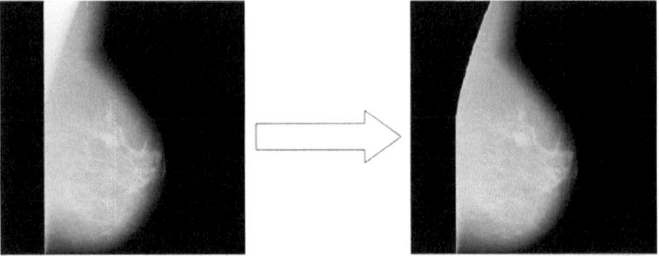

Fig. 7.5 Pectoral muscle removal algorithm result

7.3.2 Pectoral Muscle Removal

Pectoral muscle tissue is usually denser than the rest of the breast. Therefore, pectoral muscle and the central part of the breast can be extracted by applying local threshold operation with appropriate threshold value. The algorithm for pectoral muscle removal is as follows:

The result of the pectoral muscle removal algorithm is shown in Fig. 7.5. As it can be noticed, the unwanted pectoral muscle has been removed successfully.

7.3.3 Mapping the Region-of-Interest-Based Image Segmentation Algorithm for (Breast) Mammogram Images on DFE

As DFE platform for mapping the region-of-interest-based image segmentation algorithm the Maxeler's DFE is used. In the case of this algorithm, the graph that represents it consists of two kernels, one for background partition removal and the other one for pectoral muscle removal. The manager is responsible for getting the data about mammogram images from the host processor, streaming them to the input of the kernel for background partition removal, getting the output of this kernel and streaming it to the input of the kernel for pectoral muscle removal and streaming the output of this kernel back to the host processor.

After implementing this graph on DFE, there were still a lot of unused FPGA resources left on it. Because of that, as it is shown in Fig. 7.6, this graph is multiplied eight times, so that the DFE can process eight mammogram images at the same time. There cannot be more than eight graphs on the DFE because it is limited to eight input and eight output streams. The host processor streams the data about one or more, but, up to eight mammogram images ($M_{(1-8)}$) to the DFE. The manager on the DFE collects these data and steams them to the kernels $K_{(1-8)}$. The kernels process these

Fig. 7.6 Block diagram for the DFE design

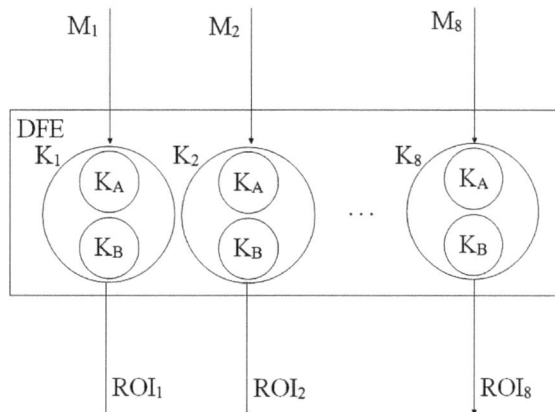

data and stream region of interest images ($ROI_{(1-8)}$) to the manager, as a result of that processing. The manager collects these output data from the kernels and streams them to the host processor. The host processor writes these data to the memory. In this way, the process of region-of-interest-based image segmentation on the DFE is done.

On the block diagram that represents the DFE design, there are eight kernels $K_{(1-8)}$ which can process eight mammogram images at the same time. Each kernel $K_{(1-8)}$ is constructed by two kernels K_A and K_B. The kernel K_A implements the background partition removal algorithm, whereas the kernel K_B implements the pectoral muscle removal algorithm.

Both of these kernels are defined with separate graphs which describe their functionality. The simplified versions of these two graphs are presented in Figs. 7.6 and 7.7. In these figures, some variables like "first_white" and "above_pixel" are presented as input streams, but in the final application they are calculated.

For the purpose of understanding the graph, they can be presented as input streams because their calculation does not have an effect on anything else and their name clearly describes what they are used for. The variable "first_white" is used to determine the first appearance connected nonblack pixels for the current row of the breast mammogram image. The variable "above_pixel" holds the value of the pixel that is in the same column as the current pixel of the breast mammogram image, but it is in the upper row. In this way, much more simplified and clearer graphs are derived.

Also, both graphs need to meet some conditions from the DFE usage point of view. They need to be designed in that way so that they use the least possible number of nodes, but also to meet the requirements of algorithms for background partition removal and pectoral muscle removal.

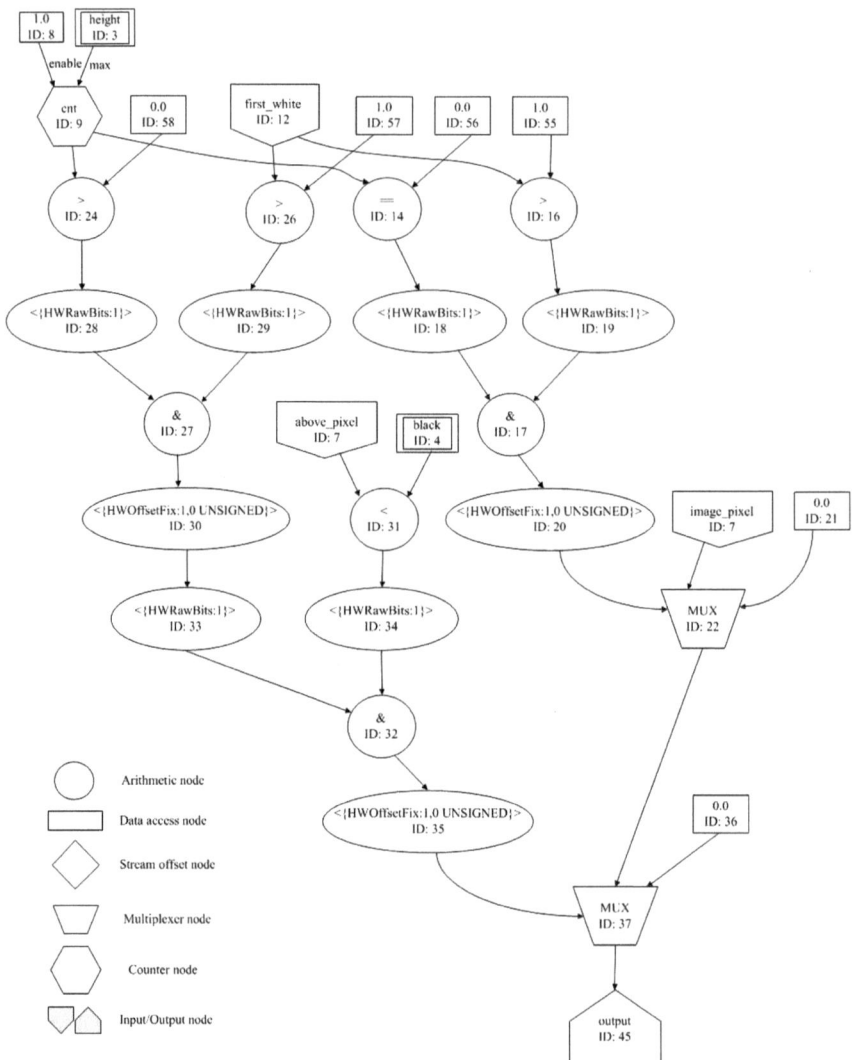

Fig. 7.7 Graph for background partition removal kernel

7.3.3.1 Graph for Background Partition Removal Kernel

The main task that the background partition removal kernel needs to accomplish is to remove the unnecessary background from the breast mammogram image. In Fig. 7.7, the graph designed for the background partition removal algorithm is shown.

The graph designed for this kernel consists of two multiplexers, few arithmetic nodes, one counter node which is used to count from 0 to the number of rows of mammogram image minus 1 with step 1, few input streams and scalar inputs and one output stream.

The main parts of this graph are two multiplexers with IDs 22 and 37. Depending on certain conditions, they stream out the value of current image pixel or set that value to the black and stream it to the output.

The first multiplexer (ID: 22) checks two conditions: whether the current pixel belongs to the first row of the mammogram image and whether it does not belong to the first connected part of the nonblack pixels of the first row. If these two conditions are satisfied, it sets the current pixel to the black, in which way the background partition pixel is removed, and streams it to the second multiplexer (ID: 37) and further to the output. Otherwise, if the conditions are not satisfied, the first multiplexer streams the unchanged current image pixel to the second multiplexer for processing, because it does not belong to the background partition part of the mammogram image.

The second multiplexer (ID: 37) checks whether the current pixel does not belong to the first row and to the first connected part of the nonblack pixels of the current row of the mammogram image and whether the above pixel value is smaller than the predefined value "black". The predefined value "black" is used as a threshold for determining the color of the pixel: black or nonblack. If all the conditions are satisfied, the second multiplexer sets the current pixel to black, in which way the background partition is removed, and streams it to the output. Otherwise, if the conditions are not satisfied, the second multiplexer streams the unchanged current image pixel to the output, because it does not belong to the background partition part of the mammogram image.

The output of the kernel for the background partition removal is connected to the input of the kernel for the pectoral muscle removal.

7.3.3.2   Graph for Pectoral Muscle Removal Kernel

The main task that pectoral muscle removal kernel needs to accomplish is to remove the part of mammogram image that represents pectoral muscle tissue. In Fig. 7.8, the graph designed for the pectoral muscle removal algorithm is shown.

The graph for background partition removal kernel consists of almost the same nodes as the graph for background partition removal. It consists of two multiplexers, several arithmetic nodes, one counter node which counts from 0 to the number of rows of mammogram image minus 1 with step 1, few input streams and scalar inputs and one output stream.

The main parts of this graph are two multiplexers with IDs 19 and 35. Depending on certain conditions, they stream out the value of the current image pixel or set that value to black and stream it to the output.

The first multiplexer (ID: 19) checks whether the current pixel belongs to the row that is in the first tenth part of the mammogram image and whether the current image pixel value is greater than or equal to the predefined variable "threshold". The predefined value "threshold" is used as threshold for determining whether the

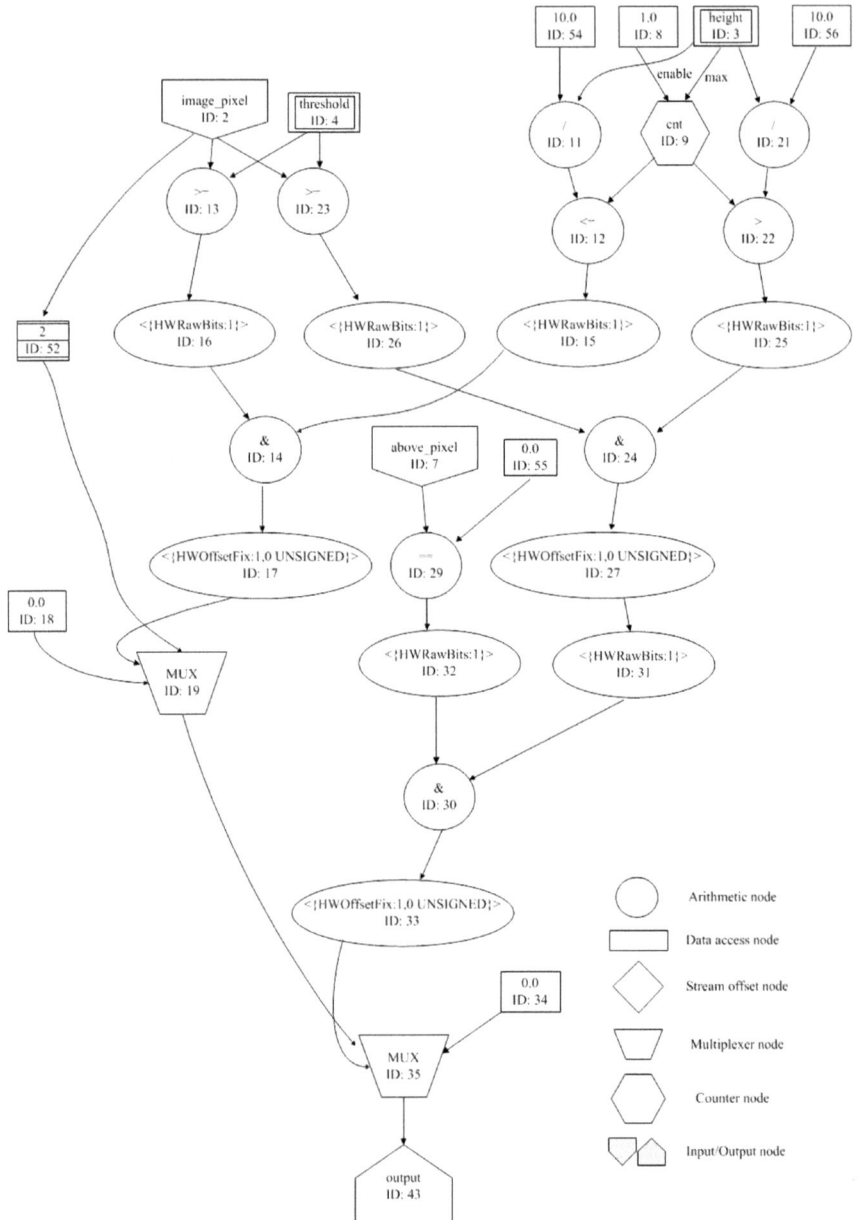

Fig. 7.8 Graph for pectoral muscle removal kernel

current pixel belongs to the pectoral muscle tissue part of the mammogram image. If all of these conditions are satisfied, the first multiplexer sets the current pixel to black, in which way the pectoral muscle part is removed, and streams it to the other multiplexers (ID: 35) and further to the output with no dependencies with the conditions for the other multiplexer. Otherwise, if the conditions are not satisfied, the first multiplexer streams the unchanged current image pixel to the second multiplexer for processing, because it does not belong to the pectoral muscle tissue part of the mammogram image.

The second multiplexer (ID: 35) checks whether the current pixel does not belong to the row that is in the first tenth part of the mammogram image, if the current image pixel value is greater than or equal to the predefined variable "threshold" and if the above pixel is black. If all those conditions are satisfied, the second multiplexer sets the current pixel to black, in which way the pectoral muscle is removed, and streams it to the output. Otherwise, if the conditions are not satisfied, the second multiplexer streams the unchanged current image pixel to the output, because it does not belong to the pectoral muscle tissue part of the mammogram image.

The output of the kernel for the pectoral muscle removal is streamed to the host processor which writes it to the memory. With storing these data into the memory, the process of ROI extraction is done.

7.3.3.3   Resource Usage

All inputs and outputs to both above kernels have 32-bits width. Multiplexers in both kernels are 2-to-1 multiplexers with 32-bits width inputs and outputs and 1-bit width selection signal. The counters have also 32-bits width. Arithmetic nodes used for comparisons ($<$, $>$, $==$, $>=$, $<=$) have 32-bits width inputs and 1-bit width output, AND arithmetic node has 1-bit width input and output and deviation arithmetic node is used for 32-bits width deviation of unsigned integers.

The FPGA resource usage per each operator according to the MaxCompiler's resource annotation [26] is presented in Table 7.3.

Table 7.3 FPGA resource usage per operator

Operator	LUTs	FFs	BRAMs	DSPs
Counters	114	99	0	0
Comparison nodes	17	1	0	0
Deviation nodes	1225	1187	0	0
AND	1	1	0	0

7.3.4 Implementation Results and Discussions

As it is already mentioned, the algorithm for region-of-interest-based image segmentation is mapped on Maxeler's DFE. Maxeler's DFE was chosen because, in literature, it showed better performances and energy efficiency against desktop processors and computing servers. On the other hand, simple array of FPGAs is not chosen because languages such as Verilog and VHDL which are widely used to design FPGA require significant expertise and considerable design efforts which is opposed to Maxeler solution which requires knowledge of Java-based MaxJ only.

The DFE is attached to the host processor via PCI Express bus and it is configured with two kernels and a manager. The Maxeler dataflow computer can be understood as a combination of two programming paradigms: control-flow and dataflow. Before one begins programming DFE, he/she must split the whole algorithm into its control-flow and dataflow parts. For instance, in this case, the control-flow part consists of reading the mammogram images from the memory and writing the processed images back to the memory, whereas the dataflow part relates to the whole image segmentation algorithm.

The execution speeds of the algorithm for region-of-interest-based image segmentation on general-purpose processor and Maxeler's DFE are compared. The Maxeler's DFE which was used is MAX2336B which contains Xilinx Virtex 5 XC5VLX330T FPGA chip. The comparisons were made on various configurations of DFE and with two types of mammogram images: with resolution of 1024×1024 pixels and 4800×6400 pixels.

The general-purpose processor that was used is Intel Core i3-3240 which works on frequency of 3.40 GHz. The operating system of the machine with this processor was running is CentOS release 5.10. The code was written in C programming language and compiled with GCC compiler.

The DFE was, as it is already mentioned, configured in various ways. It was configured to work with only one picture at the time and with two and more, but up to eight pictures at the same time. This was accomplished by mapping only one kernel for region-of-interest-based image segmentation, then mapping two and more, but up to eight kernels on the DFE. The DFE was also configured to work with different frequencies: 75 MHz, which is the default frequency and 200 MHz. The code for Maxeler's DFE was written in MaxIDE development environment and was compiled using MaxCompiler [34]. The resource usage of the DFE for the case of eight kernels and frequency of 200 MHz is shown in Table 7.4. As it can be noticed from the table, there are still a lot of unused FPGA resources on the DFE.

In Fig. 7.9, the diagram is shown that presents the amount of acceleration for two types of images, with resolutions of 1024×1024 pixels and 4800×6400 pixels, and the different number of processing images at the same time.

Figure 7.9 displays that the acceleration is much greater if DFE processes larger mammogram images. Also, with the increase in the number of processing images at the same time, the acceleration increases to some point for both mammogram image

Table 7.4 DFE resource usage

	Total available resources	Total resources used	Used by kernels	Used by manager	Stray resources
LUTs	207360	44784 (21.60%)	26006 (12.54%)	18081 (8.72%)	11 (0.01%)
FFs	207360	52447 (25.29%)	26660 (12.86%)	24664 (11.89%)	94 (0.05%)
BRAMs	324	84 (25.93%)	11 (3.4%)	71 (21.91%)	0 (0%)
DSPs	192	21 (10.94%)	21 (10.94%)	0 (0%)	0 (0%)

Fig. 7.9 Acceleration analysis with various configurations of DFE with frequency of 75 MHz

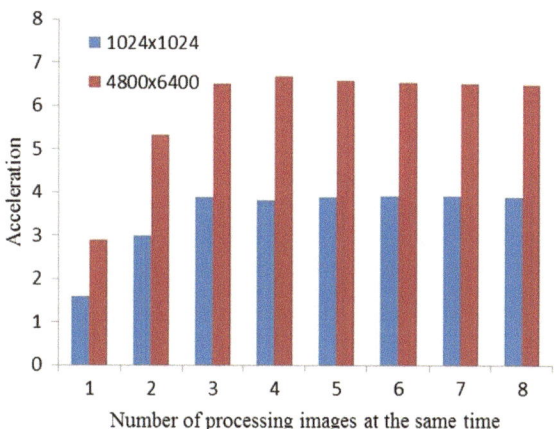

types. From that point onward, the acceleration is approximately constant. This is the point where acceleration gets bound with PCI Express bandwidth.

The same diagram as in Fig. 7.8 is presented in Fig. 7.9, but the frequency is greater and it is set to 200 MHz. As it can be noticed, the acceleration is still much greater if DFE processes larger mammogram images.

Compared to the acceleration results in Fig. 7.9 with frequency of 75 MHz, the acceleration results in Fig. 7.10 with frequency of 200 MHz in the area of one and two processing mammogram images at the same time are much better, but outside of that area the results are pretty much the same. The reason for this is also that this is the point where acceleration gets bound with PCI Express bandwidth.

The point where acceleration gets bound with PCI Express bandwidth is pretty easy to calculate. It is the point where the time required to stream data to/from DFE starts to be greater than the time required to execute the real processing of the data. The time to process all the data can be calculated using the following formula:

$$T_{processing} = \frac{cycles}{frequency}. \tag{7.1}$$

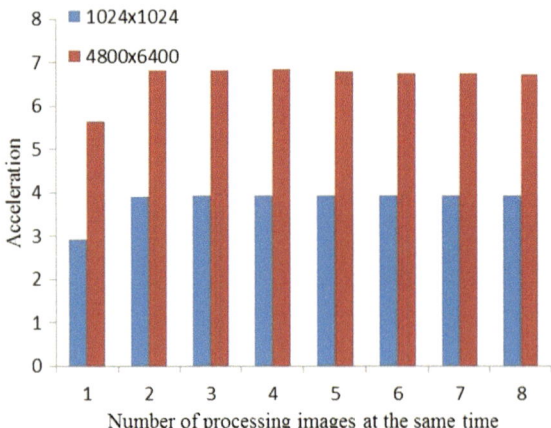

Fig. 7.10 Acceleration analysis with various configurations of DFE with frequency of 200 MHz

The cycles is the number of cycles required to do all processing and the frequency is the frequency on which DFE is running. The time to stream the data to/from DFE can be calculated using next formula which is given below:

$$T_{PCIe} = max \left(\frac{Bytes\,In}{Bandwidth\,In}, \frac{Bytes\,Out}{Bandwidth\,Out} \right). \tag{7.2}$$

From the point where acceleration gets bound with PCI Express bandwidth, the only thing that can be done is to try to compress input/output data. If this can be achieved, then it would make sense to add more pipes into the design.

7.4 Filtering and 3D Visualization of Murine Lungs on Maxeler's DFE

Maxeler's DFEs are also used for filtering and 3D visualization of murine lungs [35]. The lungs of 12-days-old C57B1/6 mice were used for 3D visualization. Image acquisition has been performed at Paul Scherrer Institute (PSI), Switzerland. Lung scans, using high definition synchrotron radiation micro-tomography, were converted to 2D image stacks with a total of 2160 slices, each with a resolution of 2560 by 2560 pixels. Each pixel size was 0.65 by 0.65 μm and the distance between two neighbor slices was also 0.65 μm. Based on this, the total field of view, which was scanned, was 1664 by 1664 by 1404 μm, i.e., approximately 3.888 mm^3. A quantitative evaluation of intussuscepted, compensatory lung growth based on these volume images has recently appeared [36].

To precisely reconstruct these 2D image stacks in 3D, there must be a high signal-to-noise ratio. However, as it can be seen in Fig. 7.11, the original tomographs have significant noise.

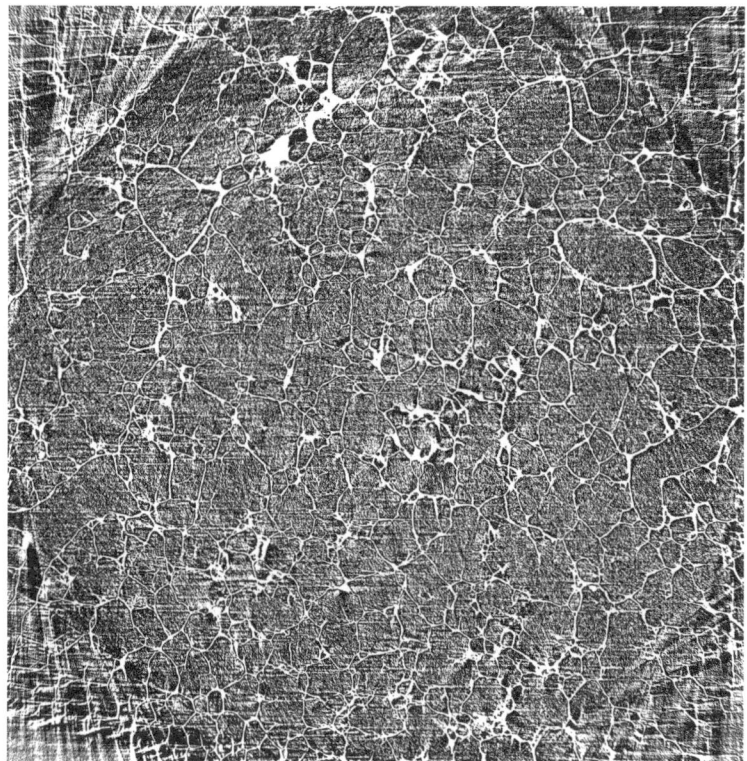

Fig. 7.11 Original 2D slice of murine lungs from the scanner output

In order to remove the noise from the original image and produce high-quality 3D visualization, the algorithm presented in Fig. 7.12 is proposed. This algorithm consists of three parts: thresholding, binarization, and filtering, followed by a quality check. The whole process of image preprocessing was repeated until the quality of 2D images is satisfactory. The criteria to be fulfilled for the satisfactory quality are that conducting airway walls should be smooth as much as possible and the pleural surface should be closed.

The first part of image preprocessing is thresholding and binarization. These are simple algorithms that determine the band of pixel values. They were represented in white and set all the other pixels to black, assigning each pixel to either fore- or background. The thresholding levels were determined empirically, the conducting airway walls were smooth and the pleural surface was closed. As shown in Fig. 7.13, the resulting output of thresholding and binarization was improved but not of satisfactory quality, thus it had to be processed furthermore.

The second part of image preprocessing is filtering. For this purpose, the mean filter is used. Mean filtering replaces each pixel value in an image with the average

Fig. 7.12 Algorithm for
image preprocessing and
quality-driven 3D
visualization

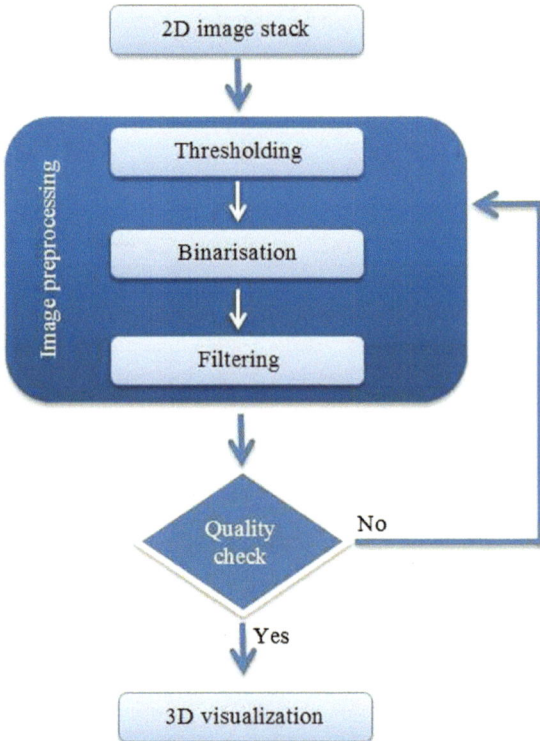

value of its neighbor pixels, including itself. Figure 7.14 shows the resulting output
of filtering, demonstrating sufficient quality for 3D visualization.

The process of preparing images for 3D visualization is often a time-consuming
process as several repetitions with different thresholding parameters are required
before images of sufficient quality are produced. Thus, it is necessary to accelerate
this process and one of the ways to accelerate it is by using DFEs.

The algorithm for processing 2D slices of murine lungs is mapped on Maxeler's
MAX2336B DFE. The Maxeler's dataflow computer can be understood as combination of two programming paradigms: control-flow which is based on von Neumann
architecture and dataflow. Before one begins programming DFE, the whole algorithm
must be split into its control-flow and dataflow parts. In this case, the control-flow
consists of reading the 2D slices of murine lungs from the memory, streaming them
to the DFE, and visualizing DFE results on the screen, whereas the dataflow relates
to the whole image processing and marching cubes algorithms. The simplified DFE
graph for the image preprocessing algorithm with one pipe is presented in Fig. 7.15.

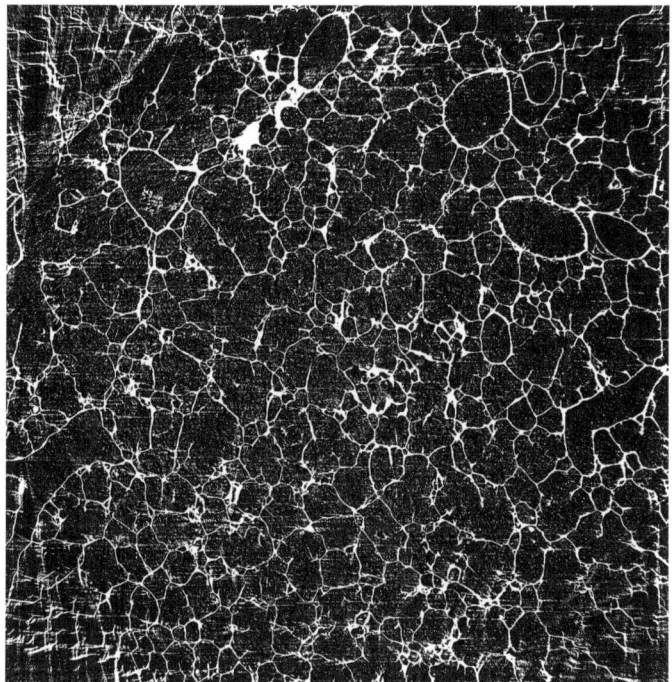

Fig. 7.13 2D slice of murine lungs after thresholding and binarization

7.4.1 Thresholding and Binarization

As shown in the above figure, the graph has nine inputs, where "curr" input represents current pixel, all "prev" inputs represent its neighbor pixels which are indexed before the current pixel and all "next" inputs represent its neighbor pixels which are indexed after current pixel. Those pixels are used in a 3 by 3 mean filter window. The "thr+bin" represents the module, which is responsible for thresholding and binarization, illustrated in Fig. 7.16.

This function compares input value with predefined down and up thresholding values. If the value is within those boundaries, the result is set to 1, otherwise the result is set to 0 which corresponds to white and black colors on the resulting image, respectively. In this way, the result always contains binarized thresholded value.

After the thresholding and binarizing, input values are summarized. In this study, we used two types of adders for this purpose and compared the calculation speeds. The first type of adders is regular BiAdder, with the DFE graph for summarization presented in Fig. 7.17.

This technique uses eight BiAdders and has a pipeline depth of four. In Fig. 7.18, the same technique is optimized for TriAdders using only four TriAdders and a pipeline depth of two.

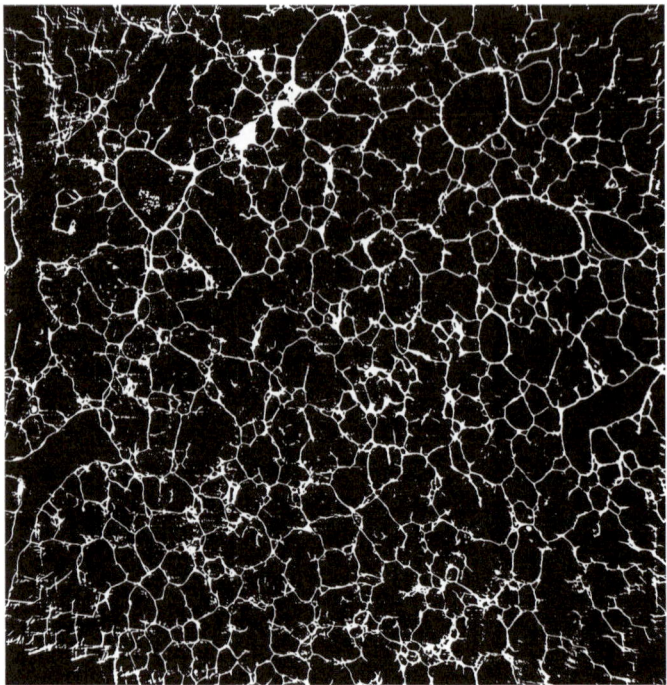

Fig. 7.14 One 2D slice of murine lungs after filtering

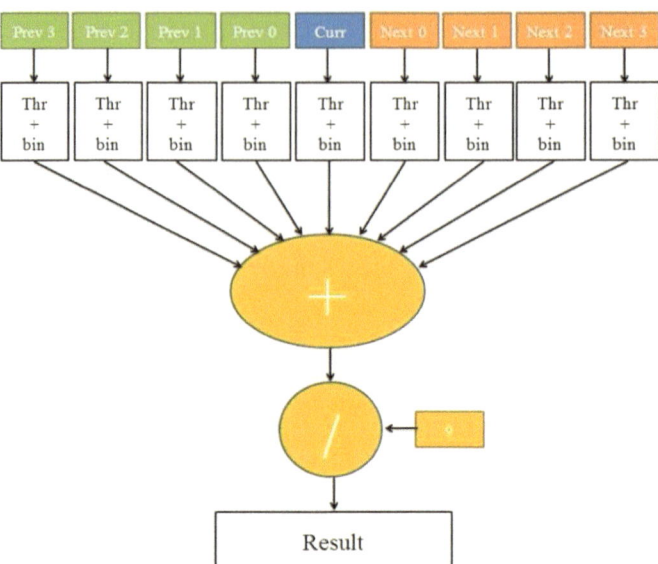

Fig. 7.15 One 2D slice of murine lungs after filtering

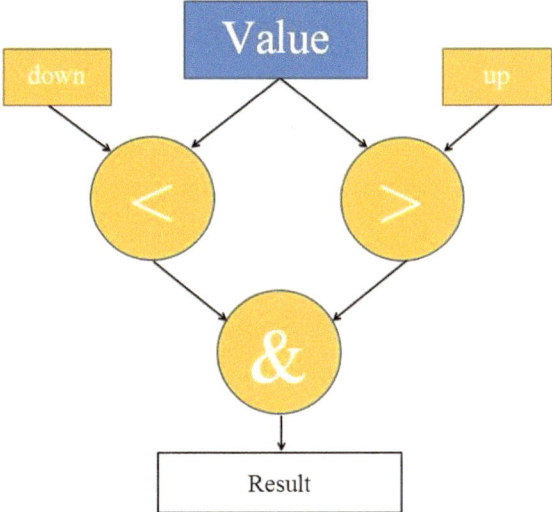

Fig. 7.16 Graph for thresholding and binarization

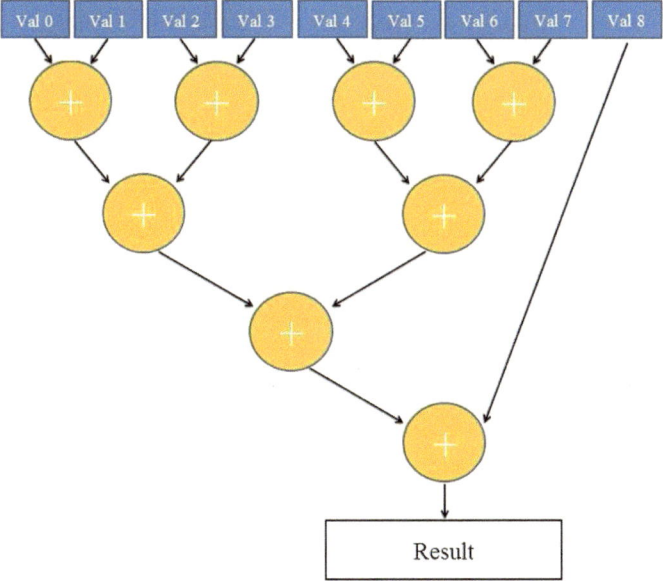

Fig. 7.17 Summarization graph using BiAdders

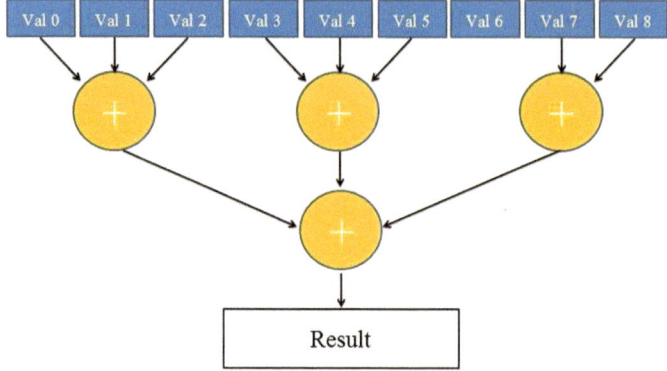

Fig. 7.18 Summarization graph using BiAdders

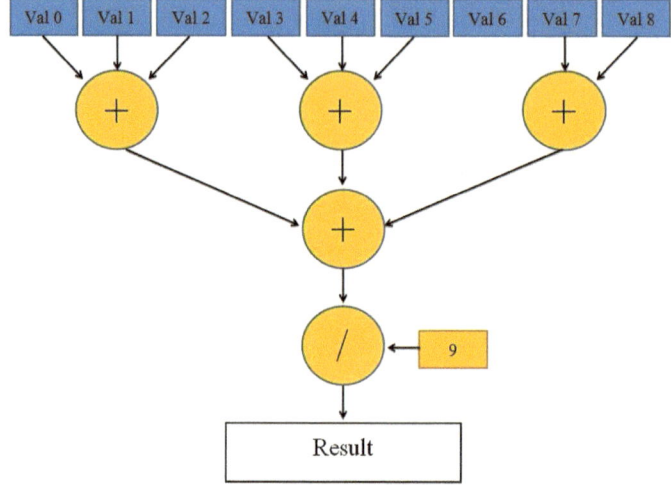

Fig. 7.19 DFE graph for median filter

7.4.2 Median Filter

The result of thresholding and binarizing is then processed with median filter. The DFE graph for median filter is shown in Fig. 7.19. The result of median filter is stored on DFE's LMem. Graphs for thresholding and binarization and median filter are multiplied 8 times so that DFE works with 8 pipes in parallel. Each pipe processes its own input image.

7.4.3   Marching Cubes Algorithm

The marching cubes algorithm divides the 3D space which should be visualized into an arbitrary number of cubes. Usually, we put cubes between two neighbor slices— two neighbor 2D images from input image stack. The problem to be solved with this algorithm is whether each vertex of the cube is inside the object that should be visualized. For each cube with all of its vertexes outside or inside the object, the surface does not pass through it. On the other hand, for each cube where some vertexes are inside and some vertexes are outside the object, the surface passes through those cubes. The surface actually intersects the edges of the cube between vertexes of opposite classification. The task of marching cubes algorithm is to draw a surface within each cube and then connect those cubes and as a result 3D visualization of an object is achieved.

Drawing the surface means triangulating the cube where filled triangles represent the surface passing through the cube. Drawing those triangles is not an easy task, but if we analyze a cube we will notice that each cube has eight vertexes out of which each vertex can be either inside or outside the object. This means that theoretically, we would have 256 different possibilities. In practice, if we include rotations and mirroring, we would actually have only 14 unique triangulations.

The DFE graph for marching cubes algorithm is shown in Fig. 7.20. This graph uses two neighbor images as input between which the cube is placed. Each of MUX_v0 – MUX_v7 represents the multiplexer whose task is to generate a key for a Lookup Table (LUT) based on the position of the corresponding vertex in the object. The key represents 8-bit binary number where binary 1 is set if the vertex is inside of the object and 0 in the opposite case. This key is used as an entry into the LUT which as a result gives one of the 14 possible triangulations. This triangulation represents one voxel of the generated 3D object. This graph is also multiplied 8 times so that DFE works with 8 pipes in parallel.

Intersection of a pleural part of the murine lung is presented in Fig. 7.21.

The acinar entrance of the acinus extracted in this case is presented in Fig. 7.22 and marked with the red arrow, while the red dashed ellipse shows the place where the acinus splits from the bronchi. Based on Bare et al. [37], the acinar entrance is determined as the transition from purely conducting to gas-exchanging airways. For its detection, the morphological criteria were used, such as thickness of the wall of the airways and appearance of alveoli.

The slice where alveolus exits from the image stack is presented in Fig. 7.23.

Marching cube algorithm on the DFE platform for murine lung is presented in Fig. 7.24. Ants represent dataflow cores and ants stream splatting algorithm loops implementation, and in the end we will read result of the dataflow computation and visualize that result to the end user.

High-quality 2D image stacks reconstructed in 3D using marching cubes algorithm are presented in Fig. 7.25.

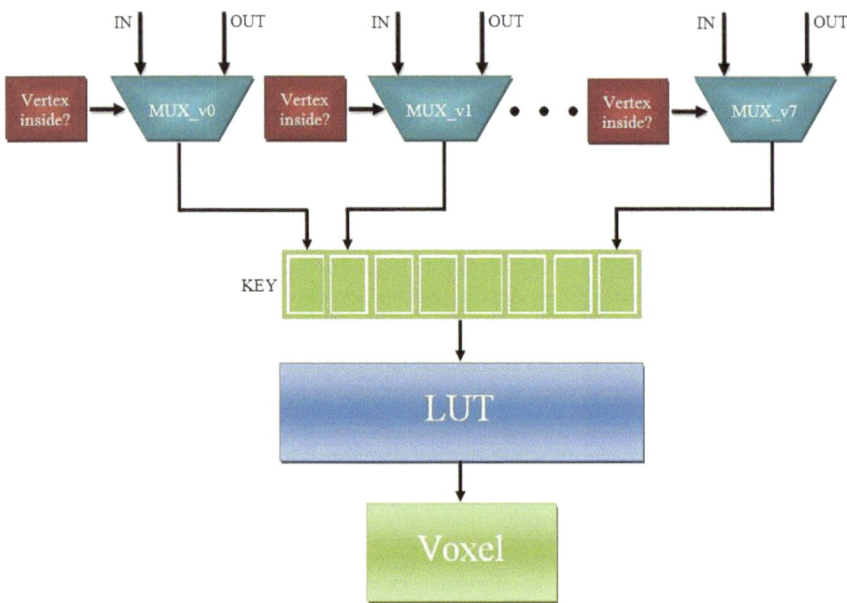

Fig. 7.20 DFE graph for median filter

Fig. 7.21 Intersection of a pleural part of the murine lung

7.4.4 *Implementation Results and Discussions*

The execution speeds of the algorithm for image preprocessing and 3D visualization are compared between the Intel Xeon CPU E5-2670 which works on frequency of 2.6 GHz and Maxeler's MAX2336B DFE which is set to work on frequency

Fig. 7.22 Acinar entrance detection

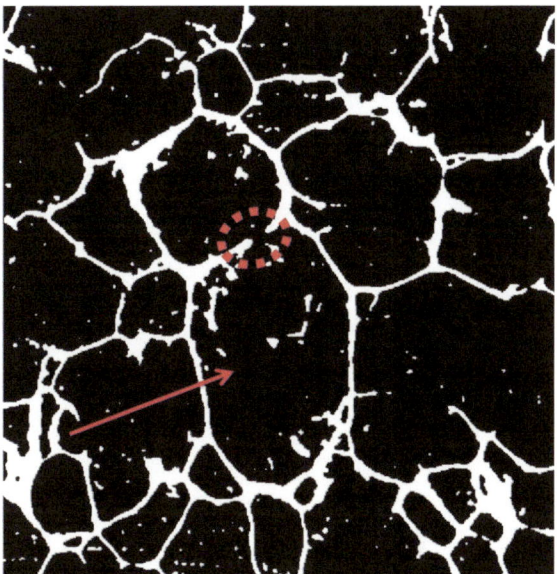

Fig. 7.23 Alveolus exits from the image stack

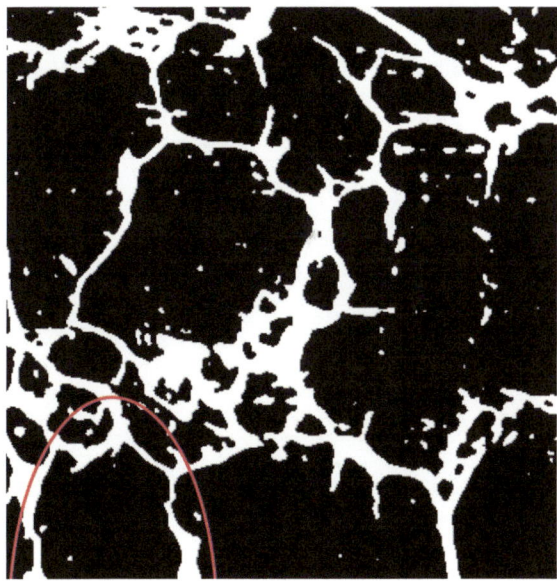

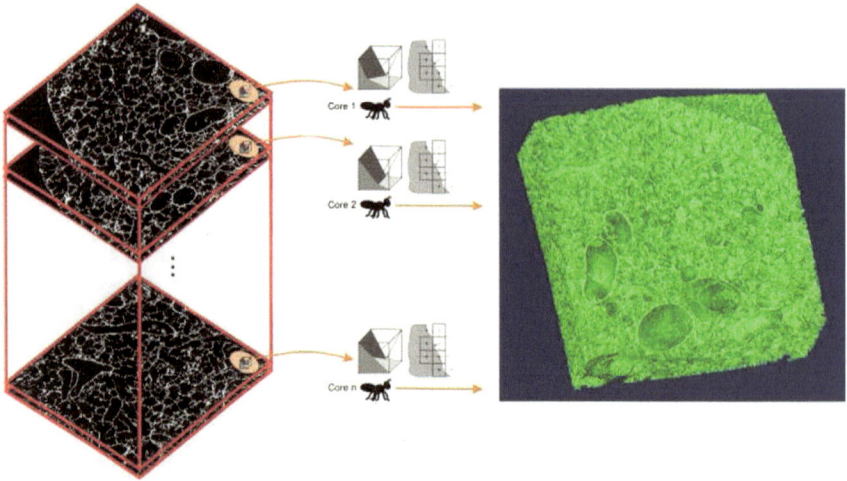

Fig. 7.24 Intersection of a pleural part of the murine lung

Fig. 7.25 Intersection of a pleural part of the murine lung

Fig. 7.26 Execution
speedup with different DFE
configurations

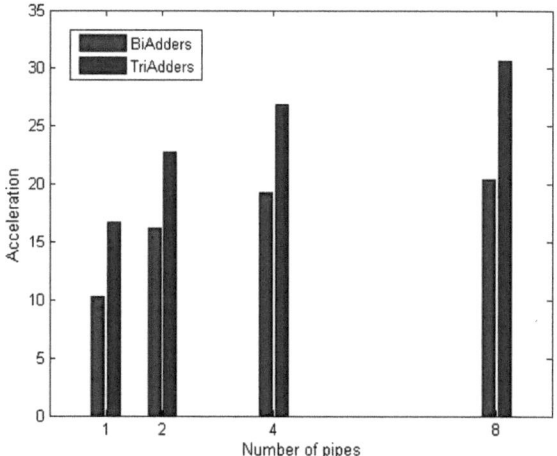

of 250 MHz. Comparisons were made on several configurations of DFE—regular
BiAdders and TriAdders and pipe line numbers of 1, 2, 4, and 8. The execution
speedup results are presented in Fig. 7.26.

7.5   Conclusion

In this chapter, the implementation of the region-of-interest-based image segmentation algorithm for breast mammograms on the DFE is proposed. The experimental
results showed that there was a significant speedup in algorithm execution on DFE
compared to the general-purpose processor. The experiments were performed on
over two types of breast mammogram images with different resolutions. The results
showed that with better image resolution, i.e., with more data per image to process,
the acceleration is greater. Also, there were several configurations of the DFE which
were implemented for testing purposes and discussed in detail. The experimental
results showed that the acceleration of algorithm execution goes near to seven times
for some DFE configurations.

Also, in this chapter, we have presented the process of filtering and 3D visualization of 2D image stack of murine lungs. For filtering, a mean filter in cooperation
with thresholding and binarization is used, while for 3D visualization, the marching
cubes algorithm is used. The implementation of those algorithms onto the Maxeler's
DFE is proposed and explained in detail. The experimental results showed that there
is a significant improvement in algorithm execution speed using DFE, compared to
an Intel Xeon processor. The experimental results showed that algorithm execution
exhibits 20-fold speed increase for the best DFE configuration.

Further work on this research may be in implementing other stages of procedure
for breast cancer detection from mammogram images on the DFE and exploring

those acceleration results. It would be interesting to try to accelerate algorithms for identification of suspicious mass regions of breast mammograms which take as input the results of the algorithm described in this paper. Also, some future work could be to port some of the skeletonization algorithms on Maxeler's DFE and check its performances.

Acknowledgements This paper was supported by the Ministry of Education, Science and Technological Development of the Republic of Serbia (projects) OI174028 and III41007.

References

1. Gonzalez RC, Woods RE (2008) Digital image processing, 3rd edn. Prentice Hall, Englewood Cliffs
2. Bankman IN, Nizialek T, Simon I, Gatewood OB, Weinberg IN, Brody WR (1997) Segmentation algorithms for detecting microcalcifications in mammograms. IEEE Trans Inf Technol Biomed 1(2):141–149
3. Maria K (2004) Computer-aided diagnosis of mammographic microcalcification clusters. Int J Med Phys Res Pract 31(2):314–326
4. Kekre HB Dr, Gharge SM (2010) Image segmentation using extended edge operator for mammographic images. Int J Comput Sci Eng 2(04):1086–1091
5. Saidin N, Mat Sakim HA, Ngah UK, Shuaib IL (2013) Computer aided detection of breast density and mass, and visualization of other breast anatomical regions on mammograms using graph cuts. Comput Math Methods Med
6. Habibi Aghdam H, Puig D, Solanas A (2013) A probabilistic approach for breast boundary extraction in mammograms. Comput Math Methods Med
7. Oliver A, Torrent A, Lladó X, Tortajada M, Tortajada L, Sentís M, Freixenet J, Zwiggelaar R (2012) Automatic microcalcification and cluster detection for digital and digitised mammograms. Knowl Based Syst 28:68–75
8. Pratt WK (2007) Digital image processing: PIKS inside. Wiley, New York
9. Horentrup J, Schlosser M (2014) Confidence-aware guided image filter. In: IEEE international conference on image processing (ICIP), Paris, pp 3243–3247
10. Ohser J, Schladitz K (2009) 3D Images of materials structures - processing and analysis. Wiley, New York
11. Banks ER, Agarwal P, McBride M, Owens C (2009) Evolving image noise filters through genetic programming DoD high performance computing modernization program users group conference (HPCMP-UGC), San Diego, CA, pp 307–312
12. Li T, Zhang X, Li C (2012) An improved adaptive image filter for edge and detail information preservation. In: International conference on systems and informatics (ICSAI), Yantai, pp 1870–1873
13. Lorensen WE, Cline HE (1987) Marching cubes: a high resolution 3D surface construction algorithm. Comput Graph 21(4)
14. Ma L, Zhao DX, Yang ZZ (2014) A software tool for visualization of molecular face (VMF) by improving marching cubes algorithm. Comput Theor Chem 1028:34–45
15. Viceconti M, Zannoni C, Testi D, Cappello A (1999) CT data sets surface extraction for biomechanical modeling of long bones. Comput Methods Programs Biomed 59(3):159–166
16. Wang ZL, Teo JCM, Chui CK, Ong SH, Yan CH, Wang SC, Wong HK, Teoh SH (2005) Computational biomechanical modelling of the lumbar spine using marching-cubes surface smoothened finite element voxel meshing. Comput Methods Programs Biomed 80(1):25–35
17. Milosevic M, Jankovic D, Peulic A (2015) Comparative analysis of breast cancer detection in mammograms and thermograms. Biomed Eng Biomed Tech 60(1):49–56

18. Smith MC, Peterson GD (2005) Parallel application performance on shared high performance reconfigurable computing resources. Perform Model Eval High-Perform Parallel Distrib Syst 60(1–4):107–125
19. Santo B (2009) 25 microchips that shook the world. IEEE Spectr 46(5):34–43
20. Veljovic D (2017) Discrepancy reduction between the topology of dataflow graph and topology of FPGA structure. In: IPSI BgD transactions on advanced research, vol 13, issue 1
21. Riebler H, Kenter T, Sorge C, Plessl C (2013) FPGA-accelerated key search for cold-boot attacks against AES. In: International conference on field-programmable technology (FPT), Kyoto, pp 386–389
22. Voros NS, Rosti A, Hübner M (2009) Dynamic system reconfiguration in heterogeneous platforms - the MORPHEUS approach, vol 40. Springer Verlag
23. Pratas F, Andrade J, Falcao G, Silva V, Sousa L (2013) Open the gates: using high-level synthesis towards programmable LDPC decoders on FPGAs. In: IEEE global conference on signal and information processing (GlobalSIP), Austin, TX, pp 1274–1277
24. Li WXY, Chaudhary S, Cheung RCC, Matsumoto T, Fujita M (2013) Fast simulation of digital spiking silicon neuron model employing reconfigurable dataflow computing. In: International conference on field-programmable technology (FPT), Kyoto, pp 478–479
25. Pell O, Bower J, Dimond R, Mencer O, Flynn MJ (2013) Finite-difference wave propagation modeling on special-purpose dataflow machines. IEEE Trans Parallel Distrib Syst 24(5):906–915
26. Oriato D, Tilbury S, Marrocu M, Pusceddu G (2012) Acceleration of a meteorological limited area model with dataflow engines. In: Symposium on application accelerators in high performance computing (SAAHPC), Chicago, IL, pp 129–132
27. Voss N, Becker T, Mencer O, Gaydadjiev G (2017) Rapid development of Gzip with MaxJ. In: Proceedings of the 13th international symposium on applied reconfigurable computing, ARC, Delft, The Netherlands, 3–7 April 2017
28. Niu X, Coutinho JGF, Luk W (2013) A scalable design approach for stencil computation on reconfigurable clusters. In: 23rd International conference on field programmable logic and applications (FPL), Porto, pp 1–4
29. Gan L, Fu H, Mencer O, Luk W, Yang G (2017) Chapter four – data flow computing in geoscience applications. Adv Comput 104:125–158
30. Grull F, Kebschull U (2014) Biomedical image processing and reconstruction with dataflow computing on FPGAs. In: 24th International conference on field programmable logic and applications (FPL)
31. Gan L, Fu H, Luk W, Yang C, Xue W, Huang X, Zhang Y, Yang G (2015) Solving the global atmospheric equations through heterogeneous reconfigurable platforms. ACM Trans Reconfigurable Technol Syst 8(2)
32. Weston S, Spooner J, Racaniere S, Mencer O (2012) Rapid computation of value and risk for derivatives portfolios. Concurr Comput: Pract Exp 24(8):880–894
33. Milankovic IL, Mijailovic NV, Filipovic ND, Peulic AS (2017) Acceleration of image segmentation algorithm for (Breast) mammogram images using high-performance reconfigurable dataflow computers. Comput Math Methods Med 2017:11. (Article ID 7909282)
34. MaxCompiler: overview, version 2011.3.1
35. Milankovic I, Peulic A, Ysasi AB, Wagner WL, Pabst AM, Ackermann M, Houdek J, Föhst S, Mentzer SJ, Konerding MA, Filipovic N, Tsuda A (2015) Acceleration of image filtering algorithms for 3D visualization of murine lungs using dataflow engines. In: IEEE 15th international conference on bioinformatics and bioengineering (BIBE), Belgrade,
36. Föhst S, Wagner W, Ackermann M, Redenbach C, Schladitz K, Wirjadi O, Ysasi AB, Mentzer SJ, Konerding MA (2015) Three-dimensional image analytical detection of intussusceptive pillars in murine lung. J Microsc 260(3):326–337
37. Barre SF, Haberthur D, Stampanoni M, Schittny JC (2014) Efficient estimation of the total number of acini in adult rat lung. Physiol Rep 2(7):1–12

Chapter 8
An Overview of Selected DataFlow Applications in Physics Simulations

Nenad Korolija and Roman Trobec

Abstract This chapter presents a wide spectrum of the dataflow implementations of applications in physics simulations. All the examples are uniformly presented, which makes possible an easy design and performance comparison among the presented dataflow algorithms.

8.1 Introduction

For many years, even in high-performance computing (HPC) applications, computers based on control-flow were the most commonly used type of machine. These computers are capable of executing any instruction at any moment. The drawback of such an approach is that the central processing unit (CPU) must be complex enough to support all the necessary functionalities [1]. Many more transistors are required than for the implementation of only a single instruction, even if it is the most complex one. Many parallel processors can increase the execution speed of HPC applications. However, most algorithms are not adequately scalable with the number of processors, because processing units have to communicate when exchanging application data and results. This reduces the number of processing units that is justifiable to use in the parallel execution of complex applications. Therefore, increasing the complexity in terms of the number of needed transistors implicitly distances processing units based on control-flow.

Another problem that is caused by having one processing unit, arithmetic logical unit (ALU), or floating-point unit (FPU) executing all of the instructions is that each instruction has to fetch the required operands at the beginning and store the results at

N. Korolija (✉)
School of Electrical Engineering, University of Belgrade, Bulevar Kralja Aleksandra 73, 11120 Belgrade, Serbia
e-mail: nenadko@gmail.com

R. Trobec
Parallel and Distributed Systems Laboratory, Department of Communication Systems, Jozef Stefan Institute (JSI), Jamova 39, 1000 Ljubljana, Slovenia
e-mail: roman.trobec@ijs.si

© Springer Nature Switzerland AG 2019
V. Milutinovic and M. Kotlar (eds.), *Exploring the DataFlow Supercomputing Paradigm*, Computer Communications and Networks, https://doi.org/10.1007/978-3-030-13803-5_8

the end. Reading instruction operands and writing results from/to the main memory would significantly slow down the instruction executions in comparison to the case where all the instruction data would already be available, e.g., in a cache memory. As a result, a reasonable portion of the chip die, dedicated only for cache memories, has to be devoted to each processor, reducing the maximum number of processors that can fit on the chip die. It could be even worse in the case of shared memory systems [2, 3].

DataFlow computing [4, 5] solves these problems by spreading the execution of the instructions in space, rather than in time; therefore, it is often referred to as spatial computing [6]. When we talk about dataflow computing, a processing unit will refer to a portion of the hardware that is needed for the execution of a single instruction. Since each processing unit should always be executing the same instruction, it could be implemented with a complexity that is several orders of magnitude lower than the control-flow ALU, if the number of transistors is counted.

Once configured, the dataflow hardware could only be used for executing dedicated preconfigured instructions, which can be considered as a limitation of the dataflow approach. In contrast, the dataflow hardware could be efficiently used for the repetitive execution of certain types of application-dedicated instructions.

A smaller number of transistors per instruction leads to a larger number of instructions that can be executed in parallel. Therefore, the dataflow approach is appropriate for computationally demanding applications, such as the simulation of natural phenomena, e.g., weather forecasting, fluid dynamics, earthquake wave propagation, etc.

8.2 DataFlow Hardware

In the 1950s, when the von Neumann paradigm for computing was born, the ratio of the time for arithmetic and logic operations and the time for communications to a system memory or to another processor were significantly larger than today. The Nobel Laureate Richard Feynman described in his lecture notes on Computing that in theory executing arithmetic and logic operations could be done with zero energy, unlike communication delays, and that the speed and energy of computing could be traded [7–9]. Based on this, we can expect that the ratio of communications to computing will increase in the future [10], which is exactly the opposite of what was the case during the times when the von Neumann paradigm was born.

DataFlow hardware is not suitable for any application. For most of the applications that run today, it would be even slower than the conventional control-flow type of processor. However, for particular algorithms, it could bring relatively large benefits in terms of the total application execution time, as well as in terms of power consumption.

The main principle of dataflow design is that each instruction could be implemented directly in the hardware. If all the instructions of an application were to be implemented in hardware, this might result in relatively complex hardware, which could consequently become quite slow. A practical implementation of such a methodology is available today in the form of the application-specific integrated circuit (ASIC), which can be directly used for relatively small-scale programs, e.g., in multimedia devices, home applications, simple controllers, etc.

However, certain complex algorithms execute the same set of instructions repeatedly on big data sets. Having that in mind, we could implement a loop where the data would be flowing through the hardware in order to have each element or group of elements parsed by the same set of instructions. However, applications rarely consist solely of instructions that would be executed over and over again. Often, a certain initialization is necessary. Therefore, it is often justifiable to combine the dataflow hardware with the conventional control-flow type of processors.

To make it possible for the dataflow hardware to be capable of executing different applications, it should be reconfigurable [11]. Reconfigurable dataflow hardware is often based on field-programmable gate arrays (FPGAs), which are relatively slow. Therefore, a configuration of the dataflow hardware is justified only for those applications that would require a significant amount of data processing.

Transferring an algorithm to the silicon is performed by an automatic compilation that could be split into two phases. First, the execution graph is generated, and then mapped onto the dataflow hardware. The execution graph can be generated by the Maxeler compiler (MaxCompiler). An important characteristic of the Maxeler framework is that it enables the simulation of the dataflow hardware that was generated after the compilation. A variety of simulation methods and computing approaches are available in [12]. Second, the execution graph is mapped onto the dataflow hardware using the synthesis tools from the FPGA device vendors. Very high speed integrated circuit (VHSIC) Hardware Description Language (VHDL) is used as an interface between these two phases.

We could expect the dataflow processing unit to be much faster than the control-flow processing unit. However, due to the need for reconfigurability, the dataflow hardware frequencies are often an order of magnitude lower than those of the control-flow type of processors. In any case, a larger number of instructions per second is achieved using dataflow hardware by running many instructions in parallel. Having simpler processing units for each instruction lowers the necessity for complex hardware (per iteration), which results in a larger number of instructions that could be run in parallel. This enables a lower power consumption, especially having in mind that dataflow hardware frequencies are much lower. Another benefit of the dataflow paradigm is in the physical size of the supercomputers, which is much smaller in the dataflow case. It is not surprising that a room with conventional control-flow type of servers could be replaced by a single dataflow hardware server. Consequently, by using less additional hardware, the reliability of the whole system is improved, because of the lower probability of component failures [13].

As the hardware technology, we have chosen Maxeler dataflow machines. The following are some of the reasons for this choice. It could consist of both a conventional computer with a DataFlow Engine (DFE), a Maxeler card connected to the PCIe slot (e.g., Galava, Isca, Maia etc), or a set of 1U boxes that could be stacked into a 40U rack.

Since most programmers are used to programming control-flow type of processors, it is expected that the dataflow hardware will not be widely accepted before a paradigm shift occurs. However, such a paradigm shift in computer science would not be new. It has happened several times in the past, e.g., when moving from assembler language to high-level languages, and later, from procedural languages to object-oriented languages, etc. The importance of supporting high-level languages could be learned from [14].

One of the possible methodologies to bring dataflow hardware closer to the market is to attract as many educators and researchers as possible, who would be the driving force of future changes. However, many engineers are mostly concerned with their working environment. It is often much faster to start from a working application, rather than from scratch. Also in our case, we have chosen Maxeler dataflow machines as the underlying hardware technology. Some of the decisive reasons for this choice are as follows: (i) the Maxeler technology could act as a conventional computer with a DataFlow Engine (DFE), or (ii) as a single Maxeler card connected to the PCIe slot (e.g., Galava, Isca, Maia, etc.), or (iii) as a set of 1U boxes that could be stacked into 40U racks for maximum performance. (iv) Maxeler AppGallery [15, 16] provides dataflow hardware with the corresponding source code for enthusiastic dataflow programmers. Finally, (v) the Maxeler maintains trial servers that can be accessed around the world for an immediate start to developing dataflow applications, even before investing in the dataflow hardware.

8.3 Maxeler AppGallery

The main goal of the Maxeler Application Gallery project is to support efforts that make dataflow supercomputing applications available to the wider public and in order to increase the public awareness about the applicability of the dataflow paradigm. Although the dataflow paradigm has already existed for several decades, this project focuses on a general approach to accelerating critical loops by the automatic or semiautomatic formation of execution graphs and their mappings onto reconfigurable hardware. Today, the hardware of choice is based on FPGA, but the Maxeler Application Gallery is not strictly dependent on this technology.

The general Maxeler framework [17] is based on the Open Spatial Programming Language (OpenSPL) [18], and on a specific implementation for the development of AppGallery.Maxeler.com applications, which is based on the Maxeler Java (MaxJ) programming language. The MaxJ is in many aspects similar to the Java programming language. However, because the main purpose of the dataflow hardware is to accelerate the execution of algorithms, the MaxJ lacks many of the features of Java that are related to the GUI. On the other hand, it introduces new types of variables, which can be stored in an arrangement of wires. As a result, once the Java variable is stored in a MaxJ hardware variable, it cannot be restored as a Java variable, because hardware variables are determined, for example, by voltage levels, which in general vary around 0 or 5 V.

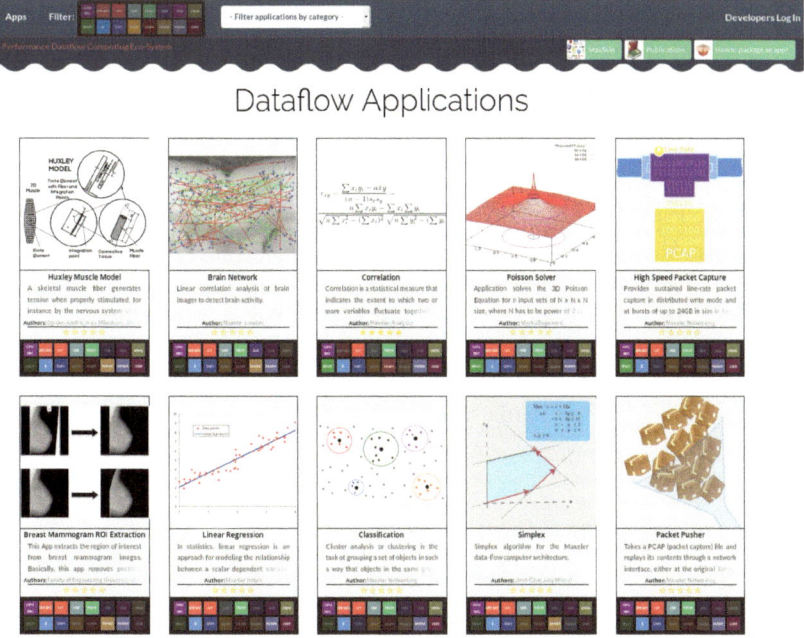

Fig. 8.1 AppGallery.Maxeler.com web site

Fig. 8.2 AppGallery.
Maxeler.com application
icons

The Maxeler Application Gallery is a technical infrastructure described with more details in [19]. Figure 8.1 illustrates the structure and contents of an App-Gallery.Maxeler.com panel with some of the commonly used applications in dataflow hardware. Each of the selected applications includes a high-level view of the algorithm, a picture presenting the essence of the algorithm, major highlights, and a short description.

Each of the applications includes the following elements, shown in Fig. 8.2: the CPU source code (marked by the icon CPU SRC), the dataflow hardware source code (marked by the icon DFE SRC), the Git repository (marked by the icon GIT), links to pdf files, including user and technical presentations (marked by the icon TECH), etc. Additionally, Maxeler-related implementation details, e.g., MAP, MAX3, etc., are also included.

8.4 Selected Examples of DataFlow Applications in Physics Simulations

There are many applications that fit well in dataflow hardware, e.g., the evaluation of polynomial and rational functions [20], conjugate gradient solvers [21], sorting networks [22], bitcoin mining [23], neural network algorithms [24, 25], processing big data [26], medicine [27], ontologies [28], etc. These applications are highly scalable on classic computers, but can also be implemented in an efficient way on dataflow hardware. DataFlow computing is also a good choice for the implementation of real-time biofeedback systems that require the massive processing of large data streams [29], and even for server applications [30, 31]. In this section, another class of applications is presented, which expresses a superior execution speedup if implemented on dataflow hardware.

8.4.1 N-Body Simulation

The N-body simulation is widely used when modeling the coordinates of bodies based on their mutual interactions. Use cases include astrophysics, molecular dynamics, plasma physics, plate tectonics, or chemical reaction network theory. The intensity of the interactions is a function of the bodies features. For example, a force Fij between the bodies i and j can be calculated as a product of their masses m1 and m2 divided by a squared distance r between two bodies and multiplied by an appropriate constant G, as shown in Fig. 8.3.

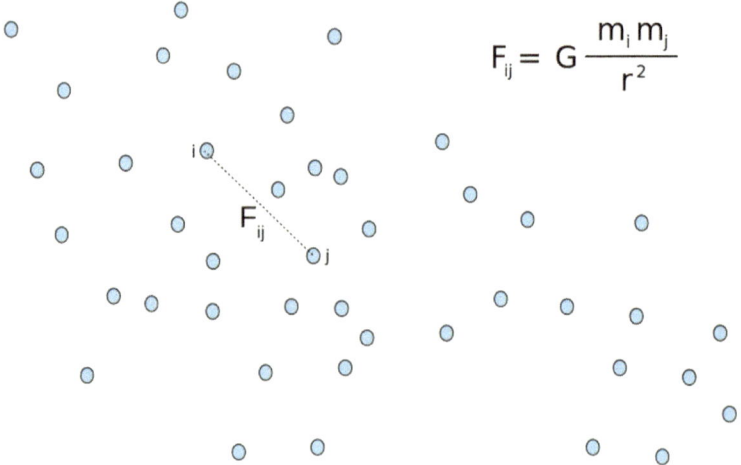

$$F_{ij} = G \frac{m_i m_j}{r^2}$$

Fig. 8.3 N-body simulation

Once the force is calculated, the acceleration of the bodies can be calculated by dividing the force by the body mass. The total acceleration of a body is calculated as the sum of the accelerations caused by all the other bodies of the system. Based on the acceleration and the numerical time step, the positions of all the bodies are updated. The calculation is repeated for the simulated time period, as shown in Algorithm 1, where the bodies are referred to as objects.

```
int time = 0;
while ( time < n\_time\_steps){
    for (int i = 0; i < n\_objects; i++){
        objects[i] = ResetForces(objects[i]);
        for (int j = 0; j < n\_objects; j++){
            if (i != j){
                objects[i] = AddForce(objects[i],
                                      objects[j]);
            }
        }
    }
    //updating position of objects
    //based on calculated forces
    for (int i = 0; i < n\_objects; i++){
        objects[i] = Update(objects[i], time\_step);
    }
    time += delta\_time;
}
```

Algorithm 1: N-body simulation algorithm

The number of calculations is proportional to the squared number of bodies, multiplied by the number of time steps. However, the scalability is limited, since the mutual influence of the bodies requires updated data for each time step, which requires a time-consuming all-to-all communication [32, 33].

The communication time could be reduced by bringing the data closer, i.e., increasing the number of calculations per chip die area. This makes the N-body simulation suitable for the dataflow paradigm. Since the same set of data, including coordinates, speeds, and accelerations, is calculated in each time step, it is justifiable to create a single dataflow instruction for all these tasks and to keep the data from the previous time step and the results from the current time step close to the place where the calculations are performed. In our case, they were kept in the main memory of the Maxeler dataflow card. In the case of a standard computer, even if the number of bytes per body is relatively low, for a large number of bodies, i.e., more than 109, it would not be possible to keep all of the body data in the cache memory, which could significantly deteriorate the execution time. Therefore, they could be kept in the main memory of the Maxeler dataflow card.

8.4.2   The Lattice–Boltzmann Method

The Lattice–Boltzmann method is commonly used in scientific computing for simulations of fluid dynamics. As the simulation algorithm can be designed in such a way that the ratio between the computing time and the communication time is high, the advantages of applying a heterogeneous dataflow paradigm might be significant [34]. The method is applied on a discretized problem domain and is local by definition. Therefore, an iterative kernel can update the solution array according to some fixed pattern, often referred to as a stencil. We should be aware that, before measuring the computation time, a very important issue is the time needed for the design of the dataflow hardware applications [35]. Since the Lattice–Boltzmann method is based on a straightforward programming construct, it is a good candidate for the semi-automatic transformation of the control-flow algorithm into the dataflow hardware [36–38]. Algorithm 2 depicts an essential part of the Lattice–Boltzmann code.

```
for (int iter=0; iter < maxIter; iter++){
    stream();
    apply\_BCs(); // apply boundary conditions
    collide();
}
```

Algorithm 2: Lattice–Boltzmann code

The function collide() is responsible for most of the calculations. Supposing that the stencil of the method is limited to neighboring sub-domains only, the execution of this function can be parallelized with dataflow hardware by letting the data flow through the neighboring columns of the discretized problem domain, as depicted in Fig. 8.4. The Lattice–Boltzmann kernel is responsible for updating the state of one elementary volume based on the surroundings. Therefore, it is fed by three columns during each clock cycle, and offsets are used to obtain the upper and lower neighboring elements. It is shown in [36–38] that the total execution time could be two orders of magnitude lower when using this approach.

8.4.3   Ray Casting

Virtual reality (VR) is one of the emerging fields in computer science. VR requires low and consistent latency (a few ms), while the processing is uniform, and therefore the amount of computation does not vary in the general case. This makes VR a potential candidate for an implementation in dataflow hardware. The most complex computational part of VR is the ray casting, which implements the tracing of every single ray from a pixel to every surface in a scene, as shown in Fig. 8.5.

One of the most important parts of the ray casting algorithm is to check where a ray intersects a plane. One of the possible implementations of this task is given in Algorithm 3. After tracing all of the rays, this is independent and can be done in parallel, the results are integrated through a summation of all the ray contributions.

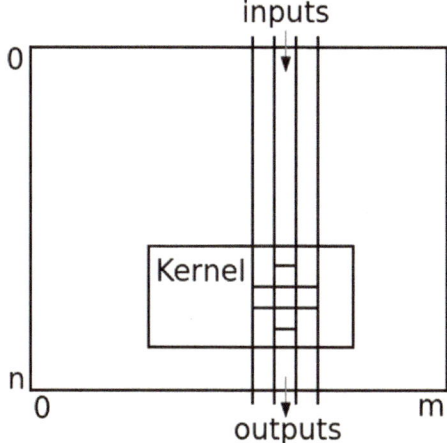

Fig. 8.4 Lattice-Boltzman kernel

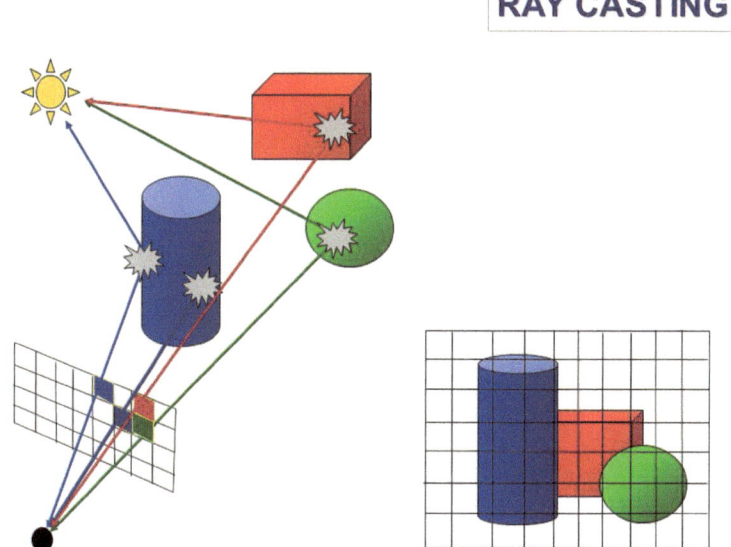

Fig. 8.5 Ray casting

```
const double EPSILON = 0.000001;
const double EPSILON = 0.000001;
bool RayIntersectsTriangle(Vector3D rayOrigin,
        Vector3D rayVector,
        Triangle* inTriangle,
        Vector3D\& outIntersectionPoint) {
    Vector3D vertex0 = inTriangle->vertex0;
    Vector3D vertex1 = inTriangle->vertex1;
    Vector3D vertex2 = inTriangle->vertex2;
    Vector3D edge1, edge2, h, s, q;
    double a, f, u, v;
    edge1 = vertex1 - vertex0;
    edge2 = vertex2 - vertex0;
    h = rayVector.crossProduct(edge2);
    a = edge1.dotProduct(h);
    if ((a > -EPSILON) \&\& (a < EPSILON))
        return false;
    f = 1/a;
    s = rayOrigin - vertex0;
    u = f * (s.dotProduct(h));
    if ((u < 0.0) || (u > 1.0))
        return false;
    q = s.crossProduct(edge1);
    v = f * rayVector.dotProduct(q);
    if ((v < 0.0) || (u + v > 1.0))
        return false;
    //computing t to find out
    //where the intersection point is on the line
    float t = f * edge2.dotProduct(q);
    if (t > EPSILON){ // ray intersection
        outIntersectionPoint = rayOrigin +
                                rayVector * t;
        return true;
    }
    else //there is a line intersection,
        //but not a ray intersection
        return false;
}
```

Algorithm 3: Ray casting program segment for the evaluation of the ray-plane inter-section

The painter's algorithm has a lower limit on latency, but the execution time changes significantly with the scene composition, so it is not useful for VR in the general case. DataFlow computing enables true-parallel execution, allowing operations to run in space with deterministic latencies. Friston et al. [39] implemented the dataflow implementation of a complete VR algorithm. They report that their system, with a lower average latency, can more faithfully draw the ideal VR system.

8.5 Conclusion

Reconfigurable dataflow hardware combined with the control-flow type of processing units can achieve a larger number of instructions per second than the solely control-flow type of processors, when running certain types of high-performance applications. A greater density of simpler processing units responsible for executing a single instruction and a lower frequency of the dataflow hardware result in a lower power consumption. The simulation of natural phenomena is often parallelized by decomposing the simulated domain into elementary sub-domains. This makes it possible for the dataflow hardware to be efficiently used, often in a heterogeneous computing platform, which provides a synergy between the classic control-flow approach and the emerging dataflow approach.

We demonstrated the dataflow implementations using solutions of a few selected applications in a physics simulation. All the examples are uniformly presented, which enables an easy design and performance comparison among the presented dataflow algorithms. Further work is needed to develop new and more sophisticated algorithms, e.g., in the area of signal processing and visualization that should enable a real-time analysis of big data sets.

References

1. Milutinovic V (1996) Surviving the design of a 200MHz RISC microprocessor. IEEE Computer Society Press, Washington DC, USA
2. Milutinovic V, Stenstrom P (1999) Special issue on distributed shared memory systems. Proc IEEE 87:399–404
3. Tartalja I, Milutinovic V (1997) The cache coherence problem in shared-memory multiprocessors: software solutions. IEEE Computer Society Press
4. Milutinovic V, Hurson A (2015) Dataflow processing, 1st edn. Academic, New York, pp 1–266
5. Hurson A, Milutinovic V (2015) Special issue on dataflow supercomputing. In: Advances in computers, vol 96
6. Mencer O, Gaydadjiev G, Flynn M (2012) OpenSPL: the Maxeler programming model for programming in space. Maxeler Technologies, UK
7. Feynman RP, Hey AJ, Allen RW (2000) Feynman lectures on computation. Perseus Books, Cambridge
8. Milutinovic V (1985) Trading latency and performance: a new algorithm for adaptive equalization. IEEE Trans Commun
9. Milutinovic V (1995) Splitting temporal and spatial computing: enabling a combinational dataflow in hardware. In: The ISCA ACM tutorial on advances in supercomputing, Santa Margherita Ligure, Italy
10. Trifunovic N, Milutinovic V, Salom J, Kos A (2015) Paradigm shift in big data supercomputing: dataflow vs. controlflow. J Big Data 2:4
11. Stojanovi S, Boji D, Bojovi M (2015) An overview of selected heterogeneous and reconfigurable architectures. In: Hurson A, Milutinovic V (eds) Dataflow processing, Advances in computers, vol 96. Academic Press, Waltham, 145 pp
12. Tomasevic M, Milutinovic V (1996) A simulation study of hardware-oriented DSM approaches. IEEE Parallel Distrib Technol 4(1)
13. Huang K, Liu Y, Korolija N, Carulli J, Makris Y (2015) Recycled IC detection based on statistical methods. IEEE Trans Comput-Aided Des Integr Circuits Syst 34(6):947–960

14. Milutinovic V (ed) (1988) High-level language computer architecture. Computer Science Press, New York, USA
15. Trifunovic N, Milutinovic V, Korolija N, Gaydadjiev G (2016) An AppGallery for dataflow computing. J Big Data 3(1)
16. Milutinovic V, Salom J, Veljovic D, Korolija N, Markovic D, Petrovic L (2017) Maxeler AppGallery revisited. Dataflow supercomputing essentials. Springer, Cham, pp 3–18
17. Flynn M, Mencer O, Milutinovic V, Rakocevic G, Stenstrom P, Valero M, Trobec R (2013) Moving from PetaFlops to PetaData. Commun ACM 39–43
18. Web site visited on 5 Feb 2018. http://www.OpenSPL.org
19. Trifunovic N et al (2016) The MaxGallery project, Advances in computers, vol 104. Springer, Berlin
20. Milutinovic V, Salom J, Veljovic D, Korolija N, Markovic D, Petrovic L (2017) Polynomial and rational functions. Dataflow supercomputing essentials. Springer, Cham, pp 69–105
21. Korolija N, Milutinovic V, Milosevic S (2007) Accelerating conjugate gradient solver: temporal versus spatial data. In: The IPSI BgD transactions on advanced research
22. Kos A, Rankovic V, Tomazic S (2015) Sorting networks on maxeler dataflow supercomputing systems. Adv Comput 96:139–186
23. Meden R, Kos A (2017) Rudarjenje bitcoinov s podatkovno pretokovnimi raunalniki Maxeler. Electrotech Rev 84(5):253–258
24. Ngom A, Stojmenovic I, Milutinovic V (2001) STRIP-a strip-based neural-network growth algorithm for learning multiple-valued functions. IEEE Trans Neural Netw 12:212–227
25. Milutinovic V (1989) Mapping of neural networks on the honeycomb architecture. Proc IEEE 77:1875–1878
26. Kos A, Toma S, Salom J, Trifunovic N, Valero M, Milutinovic V (2015) New benchmarking methodology and programming model for big data processing. Int J Distrib Sens Netw 11
27. Djordjevic S, Stancin S, Meglic A, Milutinovic V, Tomazic S (2011) Mc sensor A novel method for measurement of muscle tension. Sensors 11:9411–9425
28. Jakus G, Milutinovic V, Omerovic S, Tomazic S (2013) Concepts, ontologies, and knowledge representation. Springer, New York, pp 5–27
29. Umek A, Kos A (2016) The role of high performance computing and communication for real-time biofeedback in sport. Math Probl Eng
30. Knezevic P, Radnovic B, Nikolic N, Jovanovic T, Milanov D, Nikolic M, Milutinovic V et al (2000) The architecture of the Obelix-an improved internet search engine. In: Proceedings of the 33rd annual Hawaii international conference on IEEE system sciences, Hawaii, 7 pp
31. Kovacevic M, Diligenti M, Gori M, Milutinovic V (2004) Visual adjacency multigraphs-a novel approach for a web page classification. In: Proceedings of SAWM04 workshop
32. Milutinovi V, Furht B, Obradovi Z, Korolija N (2016) Advances in high performance computing and related issues. Math Probl Eng
33. Trobec R, Jerebic I, Jane D (1993) Parallel algorithm for molecular dynamics integration. Parallel Comput 19:1029–1039
34. Korolija N, Djukic T, Milutinovic V, Filipovic N (2013) Accelerating Lattice-Boltzman method using Maxeler dataflow approach. Trans Internet Res 9(2):5–10
35. Popovic J, Bojic D, Korolija N (2015) Analysis of task effort estimation accuracy based on use case point size. IET Softw 9(6):166–173
36. Korolija N, Popovi J, Cvetanovi M, Bojovi M (2017) Dataflow-based parallelization of control-flow algorithms. Creativity in computing and dataflow supercomputing. Advances in computers, vol 104. https://doi.org/10.1016/bs.adcom.2016.09.003. ISBN 9780128119556
37. Milutinovic V, Salom J, Veljovic D, Korolija N, Markovic D, Petrovic L (2017) Mini tutorial, dataflow supercomputing essentials. Springer, Cham, pp 133–147
38. Milutinovic V, Salom J, Veljovic D, Korolija N, Markovic D, Petrovic L (2017) Transforming applications from the control flow to the dataflow paradigm. Dataflow supercomputing essentials. Springer, Cham, pp 107–129
39. Friston S, Steed A, Tilbury S, Gaydadjiev G (2016) Construction and evaluation of an ultra low latency frameless renderer for VR. IEEE Trans Vis Comput Graph 22(4):1377–1386

Chapter 9
Bitcoin Mining Using Maxeler DataFlow Computers

Rok Meden and Anton Kos

Abstract Bitcoin, which is known as the world's first decentralized peer-to-peer payment network and cryptocurrency, introduced a decentralized mining process, where miners compete in confirming transactions in order to earn a certain amount of digital coins (bitcoins). Bitcoin mining is a repetitive and highly parallelizable process, and thus suitable for parallel computing. In this chapter, we present Maxeler dataflow paradigm as a form of parallel computing to process big data with low energy consumption and explain our dataflow implementation of the bitcoin mining algorithm for Maxeler MAX2B and MAX5C dataflow computers. With our dataflow design, we achieved up to 102 times faster and up to 256 times more energy-efficient bitcoin mining compared to standard multicore CPUs (Central Processing Units). While Maxeler dataflow computers are not able to compete against ASIC (Application-Specific Integrated Circuit) bitcoin mining rigs in terms of hash rate and energy efficiency, they are flexible and can be reprogrammed to do other tasks, while ASIC mining rigs are fixed (running only one specific algorithm) and usually become outdated in a few months.

9.1 Introduction

Cryptography became more widely available in the late 1980s and the very first cryptocurrencies appeared (b-money, digicash, hashcash, and many others). Those cryptocurrencies were usually backed by a national currency or a precious metal (gold, silver) and issued by clearing houses, just like in a traditional banking system. Although they worked well, they were centralized and therefore easy targets for hackers and worried governments.

R. Meden · A. Kos (✉)
Faculty of Electrical Engineering, University of Ljubljana, Trzaska cesta 25, 1000 Ljubljana, Slovenia
e-mail: anton.kos@fe.uni-lj.si

R. Meden
e-mail: rok.meden@gmail.com

© Springer Nature Switzerland AG 2019
V. Milutinovic and M. Kotlar (eds.), *Exploring the DataFlow Supercomputing Paradigm*, Computer Communications and Networks, https://doi.org/10.1007/978-3-030-13803-5_9

In October 2008, a whitepaper titled "Bitcoin: A Peer-to-Peer Electronic Cash System" [1] was published by pseudonymous author Satoshi Nakamoto. Bitcoin was presented as a completely decentralized digital payment system that did not rely on any central authority or control point to issue currency or to settle transactions. Bitcoin consists of [1, 2]:

- **a decentralized peer-to-peer network** (the Bitcoin protocol),
- **a public transaction ledger** (the blockchain),
- **a decentralized mathematical currency issuance** (distributed mining), and
- **a decentralized transaction verification system** (a transaction script).

In April 2011, Satoshi Nakamoto withdrew from the public and left the further development of the Bitcoin source code to a small group of people, which has expanded into a large community over time. Bitcoin is a free and open-source software that can be reviewed and developed by anyone.

The Bitcoin network is protected by volunteers called bitcoin miners who compete at confirming transactions in the Bitcoin network to earn rewards in bitcoins. Due to the limited supply of bitcoins, which consist of 21 million units in total, those rewards for miners diminish over time. Therefore, mining bitcoins in the Bitcoin network represents mining of a diminishing natural resource, such as gold.

Bitcoin mining rigs have been developed rapidly (in just a few years) due to the massive exploitation of parallelism of the bitcoin mining algorithm. At the early stage (2009), bitcoins were mined with classic computers and laptops (CPU, Central Processing Unit), and in the following years (2010–2011), they were mined with gaming graphics cards (GPU, Graphics Processing Unit) and customized FPGA (Field-Programmable Gate Array) boards. In 2013, the first ASIC (Application-Specific Integrated Circuit) bitcoin mining rigs were designed, manufactured, and used solely for mining bitcoins. Those proved to be the fastest and the most energy-efficient mining equipment, which made CPU, GPU, and FPGA based bitcoin mining rigs obsolete for the task. Most miners, who do not own ASIC mining rigs therefore, either stopped mining bitcoins due to a tremendous mining difficulty or moved onto mining less popular, but more computer-friendly alternative cryptocurrencies called altcoins that appeared soon after bitcoin (e.g., litecoin, dogecoin, ethereum).

Our primary goal was to evaluate the performance of Maxeler dataflow computers at bitcoin mining. During our research, we have followed the creativity method known as specialization [3]. The method starts from a well-established general approach (bitcoin mining algorithm that had been implemented in many programming languages for various hardware solutions), which is used to derive a specific knowledge for a specific domain or task (dataflow computing with Maxeler dataflow computers). Our expectations, based on experience with some other dataflow applications [4–9], were that dataflow implementation of bitcoin mining would outperform the CPU and GPU solutions, but would be inferior to ASIC solutions [10, 11]. The source code is available in the Appendix of this chapter and in Github repository [12].

This chapter first provides basic concepts of the bitcoin cryptocurrency and the bitcoin mining process. Next, a dataflow paradigm as a form of parallel computing to process big data is introduced, and our dataflow bitcoin miner implementation, which was tested on two Maxeler PCI-e accelerator cards, MAX2B, and MAX5C, is explained. Our dataflow design is then compared to various bitcoin mining rigs and followed by the discussion of results.

9.2 Bitcoin

The word "bitcoin" refers to several concepts; Bitcoin is open-source software which implements a Bitcoin protocol that is used by the decentralized peer-to-peer Bitcoin network. The bitcoin cryptocurrency allows fast and direct transactions between users over the Internet with digital coins called bitcoins which are mined by bitcoin miners and managed with bitcoin wallets.

9.2.1 Blockchain

The Bitcoin network uses a distributed public transaction ledger called blockchain, which contains information related to bitcoin transactions. Blockchain can be simply viewed as a chain of blocks that grows as more blocks with transactions are chained together (Fig. 9.1). Each block header contains a double-SHA256 hash of the previous block header to ensure the chronological order of blocks. Each block header also contains a hash called Merkle root that is a summary of all the transactions in the block and is used to verify transactions.

Security in the Bitcoin network is provided by bitcoin miners. Anyone with proper bitcoin mining rigs can participate in the bitcoin mining process. Miners, who successfully add valid blocks into the blockchain, have to wait for at least 100 confirmations (100 more blocks have to be added after their own block) to receive rewards in bitcoins on their bitcoin addresses. Those rewards consist of the initial value (which halves on every 210,000 blocks) and fees of confirmed transactions.

The very first transaction in the block is different from the rest of included transactions, because it is created by a miner or a mining pool. The so-called coinbase transaction has no inputs and only one output; a bitcoin address of the miner or the mining pool to which initial rewards and transaction fees are paid, if their block is accepted by the Bitcoin network consensus, added into the blockchain, and confirmed at least 100 times.

9.2.2 Bitcoin Wallets

The basic cryptocurrency unit is a bitcoin, also labeled as BTC or XBT. Bitcoins can be divided into the smallest currency units named satoshi; 1 BTC equals 10,000,000 satoshis.

Bitcoins are easy, but also risky to manage. Users have to take full responsibility for their assets in bitcoins as there is no central authority to help them in case of loss or fraud. A bitcoin wallet is either a software or hardware that maintains private and public keys and allows its users to send and receive bitcoins via bitcoin transactions (Fig. 9.2).

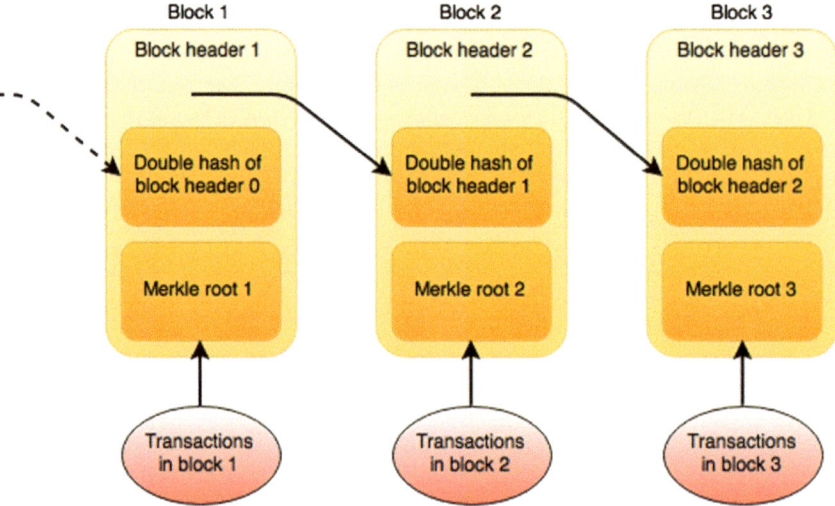

Fig. 9.1 A simplified view of the blockchain which grows in size as more blocks with transactions are chained together. Merkle root is a hash of all the transactions in the block and is used to validate transactions [13]

Fig. 9.2 Electrum is a lightweight software bitcoin wallet that does not store a full copy of the blockchain and is often recommended for beginners

9.2.3 Digital Keys and Bitcoin Addresses

A private key (denoted as k) is in most cases a random 256-bit number A public key (denoted as K) is calculated from the private key k using the asymmetric Elliptic Curve Digital Signature Algorithm (ECDSA) on the curve called *secp256k1*, as shown in Eq. (9.1).

$$K = \text{secp256k1}(k) \tag{9.1}$$

A bitcoin address (denoted as A) is calculated from the public key with two asymmetric functions, the Secure Hash Algorithm (SHA) and the RACE Integrity Primitives Evaluation Message Digest (RIPEMD), and is then generally *Base58Check*-encoded as shown in Eq. (9.2).

$$A = \text{Base58Check}(\text{RIPEMD160}(\text{SHA256}(K))) \tag{9.2}$$

Due to the one-way nature of asymmetric cryptography, calculating digital keys in the reversed direction is practically impossible (Fig. 9.3).

Only the recipient's public key or bitcoin address has to be known in order to receive bitcoins. But to manage bitcoins on a certain bitcoin address, the corresponding private key is needed. Anyone who has access to the private key can manage the correlated bitcoins, but if the private key is lost, bitcoins correlated to it cannot be managed and may remain unspent or lost forever.

Most bitcoin addresses consist of 34 random characters (letters and digits), where letters O, I, l, and a digit 0 are not present to avoid accidental mixing of visually analogous characters. The bitcoin address generally starts with a digit 1, such as *1PCxQ97Q6dAQ4D5J7gcRp18a9Azikh5Lxs*.

9.2.4 Bitcoin Transactions

A bitcoin transaction is technically not a transfer of bitcoins (bitcoins do not travel anywhere as they do not have physical form), but it is rather a transfer of ownership over bitcoins from the sender to the recipient. In addition, bitcoin wallets do not

Fig. 9.3 Asymmetric computation of digital keys and bitcoin addresses [2]

private key   public key   bitcoin address

actually hold any bitcoins, but they do maintain private and public keys that are used to access bitcoins on corresponding bitcoin addresses and to sign transactions. By digitally signing a transaction with the owner's private key, it is mathematically proven that only the owner of the private key could execute a transaction. Bitcoin transactions generally consist of three main pieces of data:

- **inputs**, which are outputs of previous transactions that gave the sender the ownership of bitcoins on their bitcoin address (Fig. 9.4);
- **amounts of bitcoins** that are being sent to the recipient; and
- **outputs**, which are the recipients' bitcoin addresses.

Transaction fee does not have any special input or output in the transaction. It is calculated as a difference between the summation of inputs and the summation of outputs and it cannot be negative, as shown in Eq. (9.3).

$$fee = \sum inputs - \sum outputs \geq 0 \qquad (9.3)$$

Although the Bitcoin network accepts a bitcoin transaction in a matter of seconds, confirming the transaction takes around 10 min on average. The recipient may spend received bitcoins immediately; however, it is more recommended to wait for several

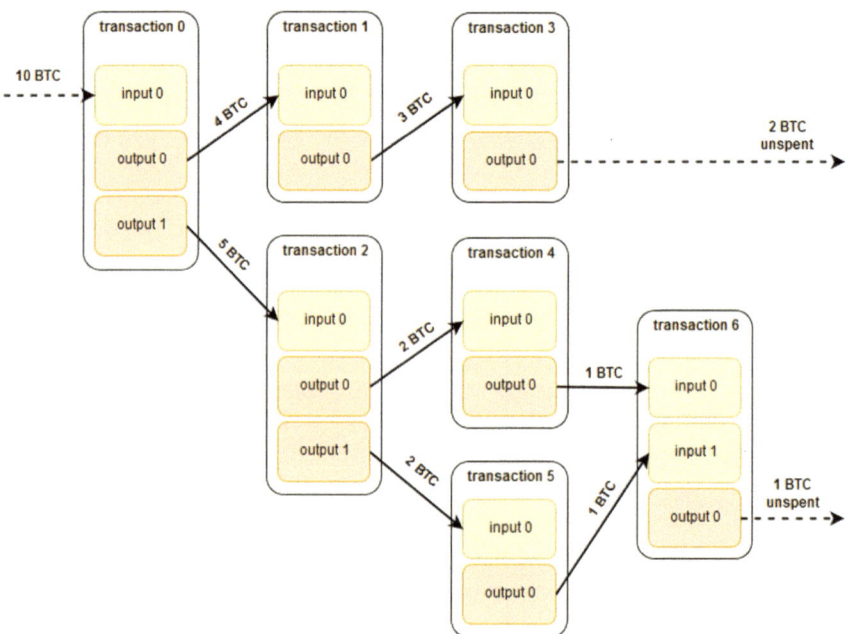

Fig. 9.4 One transaction's outputs are the next transaction's inputs. Transaction fee of 1 BTC is added to each transaction in the given example [13]

transaction confirmations. A confirmation is some sort of an agreement in the Bitcoin network about the ownership of bitcoins.

Packing a transaction into a block, which is then accepted by the Bitcoin network consensus and added to the blockchain, represents the first confirmation. Every 10 min on average, another block is added next to the current block, which extends the blockchain and it represents another confirmation (Fig. 9.5). With every added block in the blockchain, an agreement about the confirmation of transaction rises, and chances for cancelation of transaction are greatly reduced.

Bitcoins "appear" in the user's wallet when the user runs it even though he is not always connected to the Bitcoin network. This is because a bitcoin wallet does not actually accept and store bitcoins; the fact that the user received bitcoins is written in the updated blockchain. When the user runs a bitcoin wallet, the wallet downloads new blocks from the updated blockchain in the Bitcoin network and notices new transactions that it was not aware of before. The wallet then shows new transactions and newly received bitcoins to the user. Bitcoin wallets therefore do not have to be run all the time, but only when the user wants to execute transactions.

9.2.5   Bitcoin Mining

Bitcoin mining in the Bitcoin network is a heavy computational process, where miners in the Bitcoin network attempt to confirm bitcoin transactions by packing

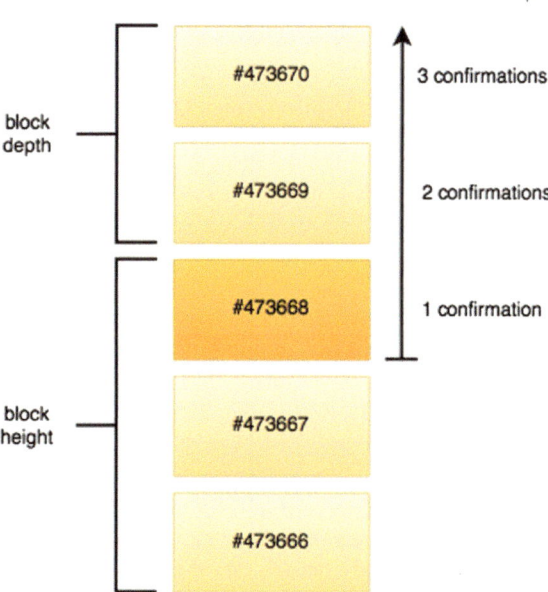

Fig. 9.5 Transactions are repeatedly confirmed as new blocks are chained into the blockchain [2]

them into blocks that satisfy the Bitcoin network consensus criteria. The process of mining bitcoins has an important role for the Bitcoin network [2]:

- It creates new bitcoins with each block, just like central bank prints new money. But unlike the traditional bank system, where new money is printed at will, issuing bitcoins is regular and predicted.
- It generates trust that transactions are only confirmed, if enough computational power was provided to generate a valid block with transactions. More blocks in the blockchain mean more provided computational power, which leads to more trust into the longest blockchain.

Bitcoin mining is often illustrated as gold mining, because it represents a diminishing supply of raw materials. However, the computational work done by miners is useful for the entire Bitcoin network, as it ensures its integrity and security. The mining process will be needed even after all 21 million bitcoins will be issued into circulation.

9.2.5.1 Reward for Miners

As stated before, reward for bitcoin miners consist of initial rewards and transaction fees. The initial reward for miners halves at every 210,000 blocks, which occurs in every 3–4 years (Fig. 9.6). At the early stage of Bitcoin (2009), the reward for miners was 50 BTC per block. In November 2012, after the first 210,000 blocks, it halved to 25 BTC per block and in July 2016, after 420,000 blocks in total, it halved again to 12.5 BTC per block. The amount of newly issued bitcoins is exponentially reduced until the minimal value (1 satoshi or 0.00000001 BTC) will be reached. After issuing all 21 million bitcoins, the reward for bitcoin miners will consist merely of transaction fees.

Limited supply and the lowering issuance of bitcoins create a fixed monetary supply which resists the inflation. Unlike standard monetary currencies, which can be printed indefinitely by central banks, bitcoin cannot be inflated at will.

9.2.5.2 Hash Rate, Network Difficulty, Target Threshold

Total computational power in the Bitcoin network has been continuously increasing ever since its existence (Table 9.1; Fig. 9.7). At the early stage (2009), bitcoins were mined with general purpose computers and laptops (CPUs). In years 2010 and 2011, it was discovered that gaming graphics cards (GPUs) are more suitable for bitcoin mining than CPUs because GPUs are mainly designed for executing repetitive tasks in parallel. GPUs appeared to be 50 times to 100 times faster in mining than CPUs. While FPGAs did not enjoy such an increase in hashing speed compared to GPUs, they were about five times more energy efficient. This improvement was enough to start bitcoin mining industry with bitcoin mining farms. In 2013, the first ASIC bitcoin mining rigs appeared, which led to the greatest increase of computational

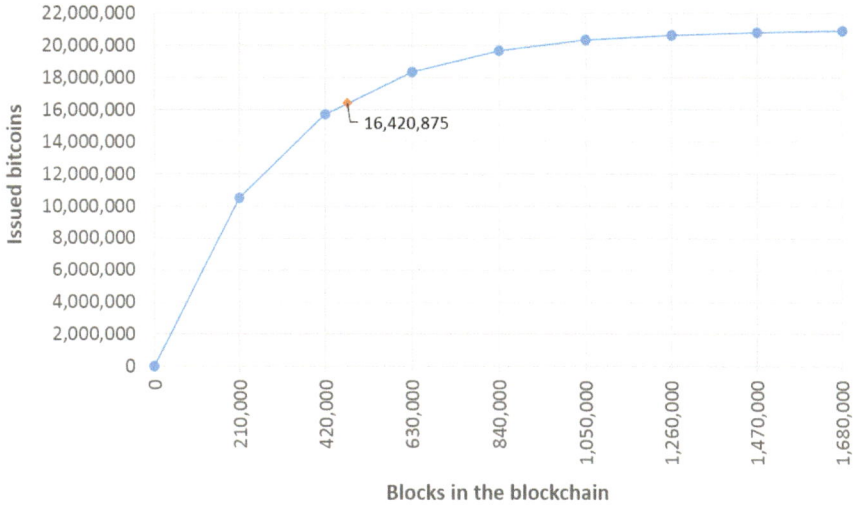

Fig. 9.6 Supply of issued bitcoins is slowly moving toward 21 million units; diamond marker represents the progress as of July 1st, 2017

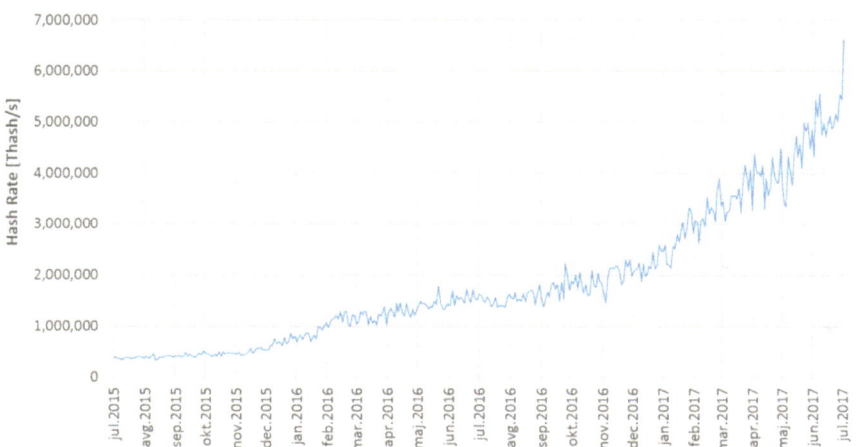

Fig. 9.7 Total hash rate in the Bitcoin network in the past 2 years (July 2015–July 2017) [14]

power; one ASIC bitcoin mining rig brought more computational power than the whole Bitcoin network in 2010. Nowadays, bitcoins can also be mined "in cloud"; miners can rent mining equipment in data centers and mine "on distance" for a certain price (e.g., 1.2 € per 10 Ghash/s per month).

The Bitcoin network difficulty and the target threshold as criteria to accept blocks into the blockchain change along with the contributed hash rate (Fig. 9.8); if total hash rate increases, so does the network difficulty by lowering the target threshold

Table 9.1 Total hash rate in the Bitcoin network in particular years [14]

Year	Contributed hash rate	Factor	Dominating hardware
2009	5 Mhash/s–9 Mhash/s	2	CPU
2010	9 Mhash/s–121 Ghash/s	13	GPU
2011	121 Ghash/s–9 Thash/s	74	GPU, FPGA
2012	9 Thash/s–22 Thash/s	2	GPU, FPGA
2013	22 Thash/s–12 Phash/s	545	ASIC
2014	12 Phash/s–328 Phash/s	27	ASIC
2015	328 Phash/s–759 Phash/s	2	ASIC
2016	759 Phash/s–2.3 Ehash/s	3	ASIC
2017 (Jan–Jul)	2.3 Ehash/s–5.9 Ehash/s	3	ASIC

to slow down block generation and vice versa, because it is planned to generate one block per 10 min on average.

The target threshold is a 256-bit number that starts with a certain number of zeros and serves as a criterion to compare against a double-SHA256 hash of a block header; the double-SHA256 hash of the block header has to be lower than the given target threshold to satisfy the criteria. The network difficulty is reciprocal to the target threshold, and it tells how much more computational power is needed in order to find a double-SHA256 hash of block header compared to the easiest possible target threshold to satisfy the current criteria, as shown in Eq. (9.4):

$$difficulty = \frac{target_1}{target_{network}} \tag{9.4}$$

where $target_1$ is the largest possible target threshold that represents the lowest possible difficulty (difficulty 1),

00000000ffff0000000000000000000000000000000000000000000000000000
00000000000,

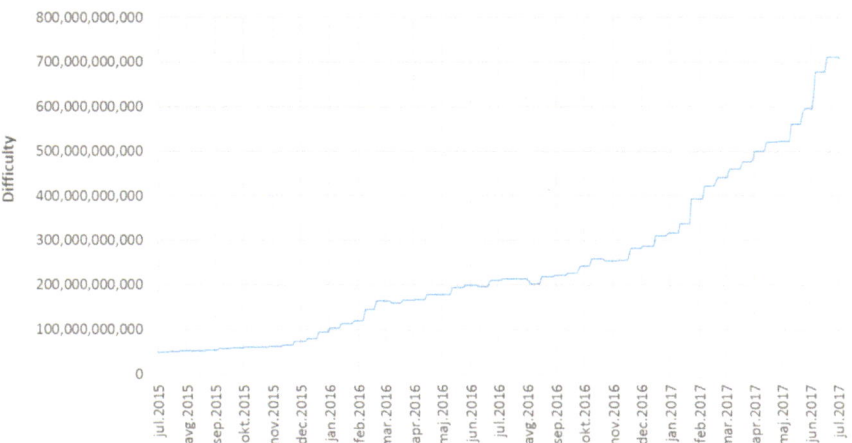

Fig. 9.8 The Bitcoin network difficulty in the past 2 years (July 2015–July 2017) [15]

and $target_{network}$ is the current target threshold in the Bitcoin network which is much lower than $target_1$. Mining pools often use non-truncated target threshold, for example,

```
00000000ffffffffffffffffffffffffffffffffffffffffffffffffffffffff
ffffffffffff.
```

The network difficulty and the target threshold are recalculated on every 2016 blocks (14 days on average) to ensure the predicted speed of generating next 2016 blocks in the following 14 days (10 min per block on average). If blocks are generated too fast, the target threshold of the Bitcoin network is lowered (the network difficulty is increased) to slow down the generation of valid blocks, and vice versa. Adjusting the network difficulty and the target threshold is limited to factors 4 and ¼ to avoid large-scale changes in the Bitcoin network.

9.2.5.3 Bitcoin Mining Process

In order to mine bitcoins, an Internet connection to the Bitcoin network and appropriate mining equipment (software and hardware) are needed. Miners can mine bitcoins individually (although not recommended), and therefore, a full mining node in the Bitcoin network has to be established (either Bitcoin Core or bitcoind daemon, see Fig. 9.9). A full mining node is set on a dedicated computer or a server with a large hard disk capacity (the blockchain file size has already exceeded 120 GB), at least 2 GB RAM, and a broadband Internet connection with upload speed of at least 50 kB/s.

Miners prefer to join mining pools to cooperate with other miners at generating valid blocks (Fig. 9.10). In this case, miners do not need to establish and maintain a full mining node. They also have smaller but more frequent payouts (shared rewards),

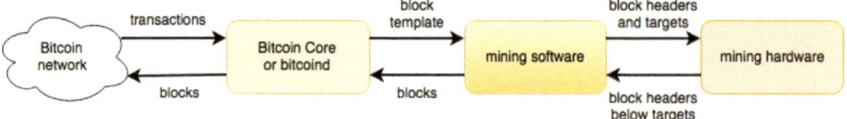

Fig. 9.9 Individual mining approach [13]

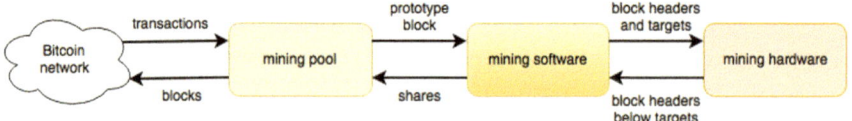

Fig. 9.10 Pool mining approach [13]

which makes rewards for miners a steady income; if a valid block is generated and accepted by the Bitcoin network consensus, all participants in the winning mining pool share the reward in bitcoins.

Miners can connect to mining pools through one of three available mining protocols: GetWork, GetBlockTemplate, and Stratum [13]. The first and easiest to use is Getwork, where mining pool constructs block header directly for mining equipment to process; only nonce has to be changed and the block header has to be hashed twice. But this protocol has been deprecated since 2013, because ASIC mining rigs with hash rate above 4 Ghash/s would have to generate hundreds of requests for work, which would result in congestion of mining pool's Internet bandwidth.

Only GetBlockTemplate and Stratum are used nowadays, because those two contribute to saving of Internet bandwidth; mining rigs only receive block templates and fill missing data on their own to generate so-called shares. Shares are solutions of block headers that do not necessarily meet the Bitcoin network difficulty, but are good enough to meet the lowered mining pool's difficulty.

Mining pools estimate cooperating miners' contributed computational power (hash rate) by checking frequency ("how often") and difficulty ("how hard") of their sent shares. The contributed computational power of the largest bitcoin mining pools on July 1st, 2017, is shown in Fig. 9.11.

Both, individual miners and mining pools collect unconfirmed transactions in the Bitcoin network and store them into their local transaction pools. They also generate and add their coinbase transactions with no inputs and their bitcoin address as output (denoted as T_A in the following example). From collected transactions, a hash tree (Merkle tree) is built and a Merkle root (which is a part of the block header) is calculated (Fig. 9.12).

Then, double-SHA256 hashes (denoted as H_A, H_B, H_C) of collected transactions (denoted as T_A, T_B, T_C) are calculated as shown in Eq. (9.5), and those represent the leaves of the Merkle tree.

$$H_A = SHA256(SHA256(T_A))$$

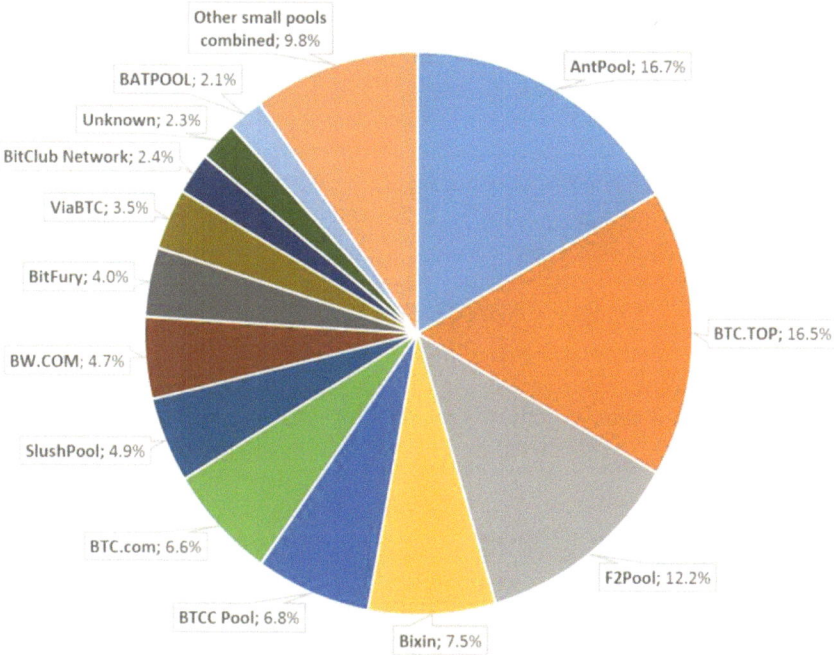

Fig. 9.11 Contributed hash rate of largest mining pools in the Bitcoin network on July 1st, 2017 [16]

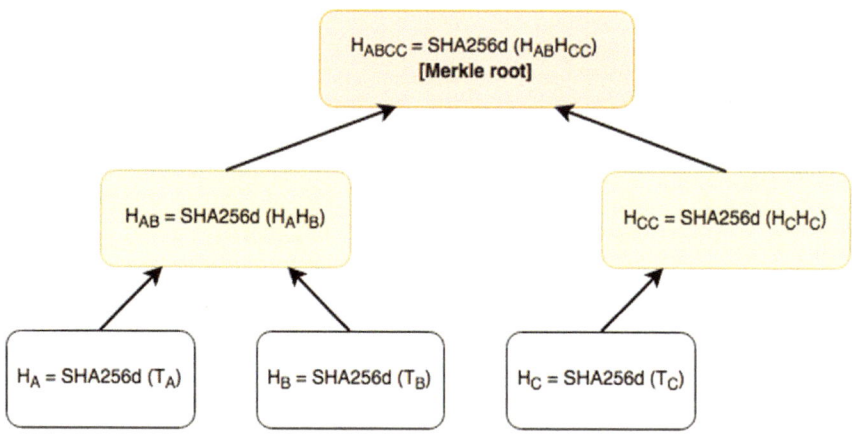

Fig. 9.12 Building a Merkle tree and calculating a Merkle root from collected transactions by applying double-SHA256 hashing algorithm [2]

$$H_B = SHA256(SHA256(T_B))$$
$$H_C = SHA256(SHA256(T_C)) \qquad (9.5)$$

Those 256-bit hashes are then concatenated together by two and two into 512-bit numbers (just like a and b are concatenated into ab) and hashed twice again, as shown in Eq. (9.6). It can happen that only one leaf of the Merkle tree is left at the end and there is no other leaf to be paired with. If this is the case, the last remaining leaf is duplicated and concatenated with itself.

$$H_{AB} = SHA256(SHA256(H_A H_B))$$
$$H_{CC} = SHA256(SHA256(H_C H_C)) \qquad (9.6)$$

This procedure continues until only one final 256-bit hash remains; this is a Merkle root (denoted as H_{ABCC}) which represents a summary of all included transactions, as shown in Eq. (9.7).

$$H_{ABCC} = SHA256(SHA256(H_{AB} H_{CC})) \qquad (9.7)$$

A Merkle tree is a very efficient data structure, because at most $2 * \log_2 (N)$ calculations are needed to verify the presence of any transaction (Table 9.2), where N is the amount of transactions in a block.

After calculating the Merkle root of collected transactions, a block header is generated; this is a 640 bits (or 80 bytes) long data structure that consists of 6 fields (Tables 9.3 and 9.4).

Block header is generally written in a single line:
`0100000081cd02ab7e569e8bcd9317e2fe99f2de44d49ab2b8851`
`a4a308000000000000e320b6c2fffc8d750423db8b1eb942ae710e`
`951ed797f7affc8892b0f1fc122bc7f5d74df2b9441a`**00000000**.

The only variable part of the block header is the last 32 bits which represent an arbitrary number "nonce" (short for number-once) that can take any of 2^{32} possible values; it starts at 0 (**00000000**) and ends at $2^{32} - 1$ (**ffffffff**). Nonce has to be

Table 9.2 Comparison between the size of Merkle path and block size based on a number of transactions in a block

Transactions	Merkle path	Estimated block size	Size of Merkle path (B)	Size factor
16	4	4 kB	128	32
512	9	128 kB	288	455
1024	10	256 kB	320	819
2048	11	512 kB	352	1489
4096	12	1 MB	384	2731

Adapted from [2]

Table 9.3 Structure of block header [2]

Field	Description	Size (bits)
Version	Block version number	32
Previous block hash	256-bit hash of the previous block header	256
Merkle root	256-bit hash which is the summary of all transactions in a block	256
Timestamp	Current timestamp as seconds since 1. 1. 1970, 00:00 UTC	32
Difficulty target	Current target in compact format	32
Nonce	A counter used for proof-of-work algorithm (SHA256d)	32

Table 9.4 Example of block header in hex

Field	Value (hex)
Version	01000000
Previous block hash	81cd02ab7e569e8bcd9317e2fe99f2de44d49ab2b8851ba4 a308000000000000
Merkle root	e320b6c2fffc8d750423db8b1eb942ae710e951ed797f7af fc8892b0f1fc122b
Timestamp	c7f5d74d
Difficulty target	f2b9441a
Nonce	00000000

changed (incremented by 1) and the modified block header has to be hashed twice repeatedly until a hash with several leading zeros is found.

Let's take an example of the previously given block header with nonce **42a14695**, thus yielding a block header

`0100000081cd02ab7e569e8bcd9317e2fe99f2de44d49ab2b8851b`
`a4a308000000000000e320b6c2fffc8d750423db8b1eb942ae710e`
`951ed797f7affc8892b0f1fc122bc7f5d74df2b9441a`**42a14695**.

Its double hash is

`1dbd981fe6985776b644b173a4d0385ddc1aa2a829688d1e000000`
`0000000000`,

and after swapping its byte order, we get

`00000000000000001e8d6829a8a21adc5d38d0a473b144b6765798`
`e61f98bd1d`.

This hash has now leading zeros and it can be compared against the target threshold, for example,

`00000000000044b9f200000000000000000000000000000000000000`
`0000000000`.

In this case, the double-SHA256 hash of the given block header is smaller than the target threshold, therefore a nonce **42a14695** is one of few possible solutions to the problem. When such nonce is found, a mining equipment either generates a block with transactions and broadcast it to the Bitcoin network (in case of individual mining) or generates and sends a share including the nonce solution to the chosen mining pool in the Bitcoin network (in case of pooled mining).

A block is a collection of valid transactions that are waiting to be confirmed. It consists of the block header (its double hash has to start with a certain number of zeros) and a list of transactions that represent the majority of block (Table 9.5).

The block header is 80 B long, but the average transaction is at least 250 B and the average block contains at least 500 transactions. The currently allowed maximum block size is 1 MB and therefore miners may confirm at most 4096 transactions in 10 min (or 6.7 transactions per second). Due to this limitation, most miners prioritize transactions with the highest transaction fees.

Once a block with transactions is generated, it is broadcasted to the Bitcoin network. If the broadcasted block is accepted by the Bitcoin network consensus and is added into the blockchain, all transactions in the block are confirmed, and the miner or the mining pool may be rewarded with the initial reward and transaction fees (100 additional blocks have to be chained after the accepted block). The whole Bitcoin network then immediately starts searching for the next valid block.

A proof-of-work in each individual block depends on its previous block, which ensures the chronological order of all blocks in the blockchain. The chances to cancel past transactions are lowered exponentially by adding new blocks into the blockchain; to cancel a transaction in a certain block, a whole computational work for a certain block and all the following blocks would have to be redone. Miners therefore cannot cheat the Bitcoin network by increasing their own rewards or confirming invalid transactions, because the other mining nodes in the Bitcoin network would immediately reject a block with invalid data. This makes the Bitcoin network secure even if some miners are not trustworthy.

Because the blockchain is a distributed and decentralized data structure, its locally stored copies in individual nodes are not always the same. The nodes always assume that only the longest blockchain (the one with most blocks) is correct.

If two different nodes broadcast different valid blocks at nearly the same time, some nodes receive the first block first, and some other nodes receive the second

Table 9.5 Block structure [2]

Size	Field	Description
4 B	Block size	The size of block in bytes
80 B	Block header	The block header which consists of six fields
1–9 B	Number of transactions	The number of transactions that are packed inside a block
Variable	Transactions	Transactions that are packed inside a block

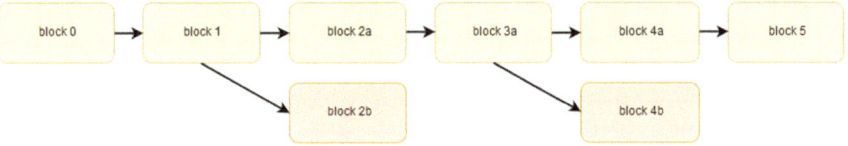

Fig. 9.13 Short blockchain forking is common and does not harm the Bitcoin network [13]

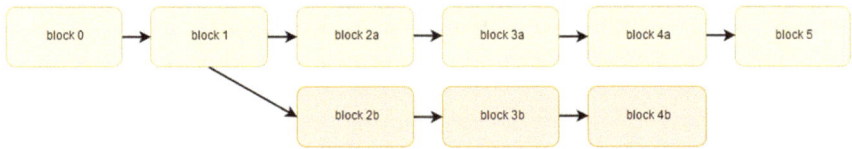

Fig. 9.14 Long blockchain forking is rare, and it can be troublesome for the Bitcoin network [13]

block first. All nodes work with the block they received earlier. Sooner or later, one of the blockchains becomes longer and nodes that worked on a shorter blockchain immediately switch to the longer blockchain, thus making the shorter blockchain obsolete (or orphaned); see Figs. 9.13 and 9.14.

9.2.5.4 Mining Profitability

Most miners in the Bitcoin network mine bitcoins to sell them afterward. The mining process itself is quite expensive; miners have to purchase and maintain their mining rigs, they have to pay for the electricity, etc. Return of their investment into their mining rigs mainly depends on the current network difficulty, price of bitcoins, and electricity costs. Therefore, the following parameters have to be considered when purchasing a mining rig:

- **Price**: Faster mining rigs are generally more expensive.
- **Hash rate**: How many double hashes per second can a mining rig compute and compare against the target threshold; more is better.
- **Power consumption**: How much electric energy a mining rig consumes during computation; less is better.
- **Energy efficiency**: Mining rigs consume a lot of electric energy, and therefore it is recommended to select a rig that has high ratio between hash rate and electric power, as shown in Eq. (9.8):

$$eff\,[\text{hash/s/W}] = \frac{hr[\text{hash/s}]}{P[\text{W}]} \tag{9.8}$$

where eff is energy efficiency, hr is hash rate, and P is electric power of a mining rig during mining. Let's compare two of the most popular ASIC bitcoin mining rigs [17, 18]: ASIC mining rig Antminer S7 mines with a hash rate of 4.73 Thash/s at

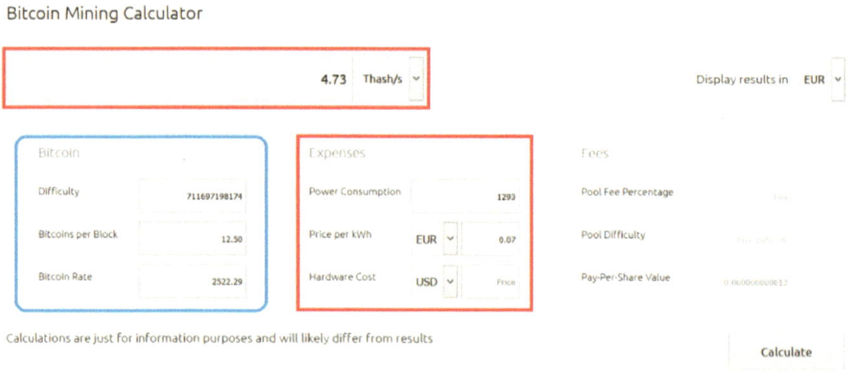

Bitcoin Mining Calculator

Fig. 9.15 Bitcoin mining profitability calculator BTCServ [19] used for calculating a mining profitability with ASIC AntMiner S7 bitcoin mining rig on July 1st, 2017

1293 W, and its energy efficiency is 3.65 Ghash/s/W. ASIC mining rig SP20 Jackson mines with a hash rate of 1.7 Thash/s at 1100 W, and thus its energy efficiency is 1.54 Ghash/s/W. ASIC Antminer S7 is about 2.4× more energy efficient than ASIC SP20 Jackson and is therefore a better option for serious miners.

The so-called bitcoin mining profitability calculators are mainly used to determine whether mining bitcoins with certain mining rigs is profitable or not (e.g., BTCServ [19]). The information about mining rigs, (such as price of mining equipment, energy consumption, and price of electricity) have to be entered manually (Fig. 9.15, red rectangles), while the variable factors of the Bitcoin network, (such as the network difficulty, current reward in bitcoins, price of bitcoins) are inserted automatically by the calculator itself (Fig. 9.15, blue oval).

Profitability calculators show predicted earning or loss in time intervals (e.g., hour, day, week, month, year) and additional information about mining rigs (such as hardware break even, theoretical time to generate a block and energy efficiency), as shown in Fig. 9.16.

9.3 Multiscale DataFlow Computing

CPU frequency stopped increasing several years ago due to physical limitations of shrinking transistors, the increased power density, and heat emissions. Another large issue with the traditional von Neumann computer architecture and control-flow approach is the processor-memory bottleneck (also known as von Neumann bottleneck), where the memory bus limits overall computer system performance; modern CPU spends most of the time idling while waiting for data to be fetched from memory, which results in a considerably slowed data movement between CPU and memory.

Fig. 9.16 BTCServ calculator [19] predicted the average earning of 65,909 satoshis (1.41 €) per day with ASIC mining rig AntMiner S7 on July 1st, 2017

While there are attempts and approaches to overcome this bottleneck (such as caching, multithreading, shared memory) [20–27], parallel computing is also being extensively researched as a new approach to process big data. Maxeler Technologies introduced an alternative approach to parallel computing; multiscale dataflow computing with Maxeler dataflow computers [28–34], where data simply flows from the start (input streams) through a field of simple arithmetic units (calculation through dataflow graphs) to the end (output streams). This approach contrasts the previously mentioned traditional CPU architecture with multiple level caches and shows great speedups in processing big data, such as digital image processing [8] and neural networks [9, 35].

DataFlow applications are developed within MaxIDE, a programming environment based on Eclipse (the open-source platform to program in Java language), where a modified Java language, *MaxJ*, is used.

9.3.1 DataFlow Engines

CPU of a host workstation runs uncommon operations and programs that are generally written in *C*, *C++*, *Fortran*, *Python*, and many other languages, while a DataFlow engine (DFE) runs *.max* configuration files and executes dataflow computing.

CPU application uploads a *.max* configuration file onto DFE. After calling an automatically generated SLiC (Simple Live CPU) interface, data is sent to DFE and back to CPU via PCI-e or Infiniband connection while dataflow computation is

Fig. 9.17 DataFlow engine architecture [28]

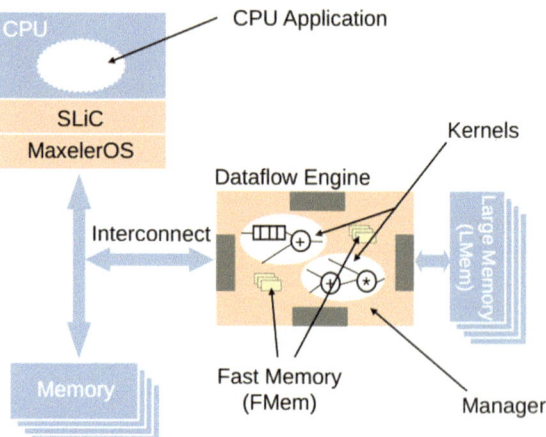

running (Fig. 9.17). DFE is programmed with only one Manager and at least one Kernel (can be multiple Kernels). The Manager controls data transfer while Kernels run the actual dataflow computation.

MaxelerOS is an operating system that runs within Linux operating system and DFEs themselves. The main jobs of MaxelerOS are data transfer management and dynamic optimization of applications at runtime.

DFE consists of an integrated FPGA circuit and few types of memories; Fast Memory (FMem) which can store several megabytes of data with a memory bandwidth of several TB/s and Large Memory (LMem) which can store several gigabytes of data with a memory bandwidth of several GB/s.

DataFlow applications are usually accelerated for several magnitudes compared to traditional control-flow applications, because memory and computation are both laid in space rather than in time and data can be held in memory close to computation units.

9.3.2 DataFlow Versus Control-Flow

One analogy for moving from control-flow to dataflow model is replacing individual artisans with manufacturing model. In a factory, each worker has a simple task and all workers operate in lines. Just like the manufacturing approach, a dataflow approach is a method to speed up the computation [28–32].

The source code of a control-flow program is first translated into a list of instructions for CPU, which is then loaded into memory. CPU then repeatedly accesses memory for reading and writing data. The control-flow programming model is sequential and the computing performance is therefore affected by the latency of data access in memory and the processor's frequency (Fig. 9.18).

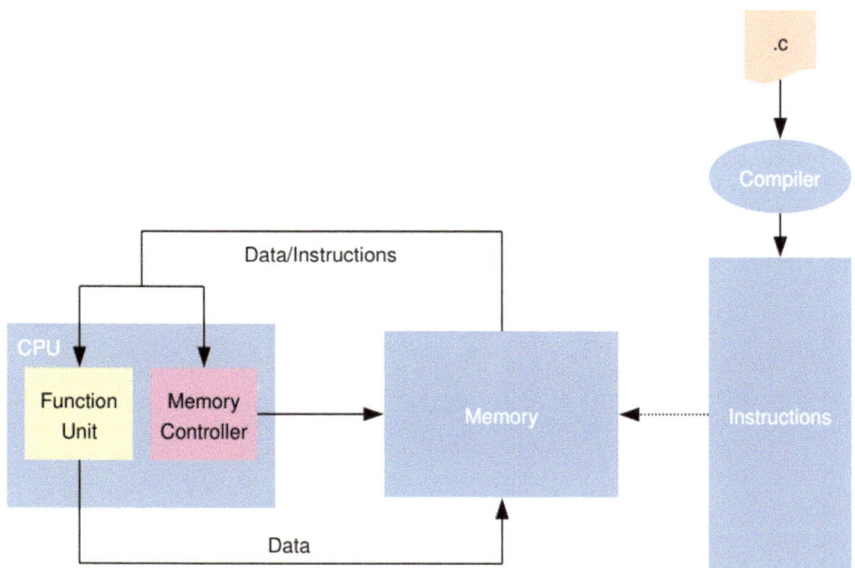

Fig. 9.18 Control-flow computation is sequential and laid in time [28]

On the other side, a DFE structure itself represents the computation, and therefore no instructions are needed; the instructions are replaced with arithmetic units laid in space and together they represent a certain structure of an operation (dataflow graph) [28, 34]. Because there are no instructions, there is no need for decoding logic, caching instructions, or predictions; DFE resources are completely dedicated to the computation (Fig. 9.19). DFE simply processes large data sets while CPU runs uncommon operations and controls the communication with DFE.

DataFlow computation runs in units of time called ticks unlike control-flow computation which runs in clock cycles. In general cases with one input and one output stream, Kernel executes one step of the computation and consumes one input value per tick. Output elements are not produced immediately, because values flow down through dataflow graph one step per tick and have to reach the end of dataflow graph (output). After D ticks, where D is depth of dataflow graph, one output value is produced per tick.

Every single arithmetic unit in DFE executes only one operation (i.e., addition, subtraction, multiplication, division, comparison, bit shift). A single DFE can contain thousands of them due to their simplicity. Unlike a sequential control-flow computing, where operations are executed on the same functional units at different times, a dataflow computing is laid in space (on a chip). DataFlow graphs (Fig. 9.20) may contain several types of nodes:

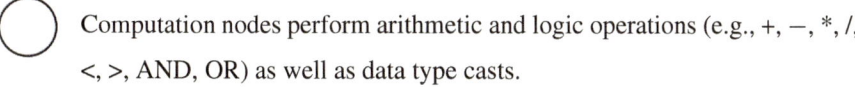 Computation nodes perform arithmetic and logic operations (e.g., $+$, $-$, $*$, $/$, $<$, $>$, AND, OR) as well as data type casts.

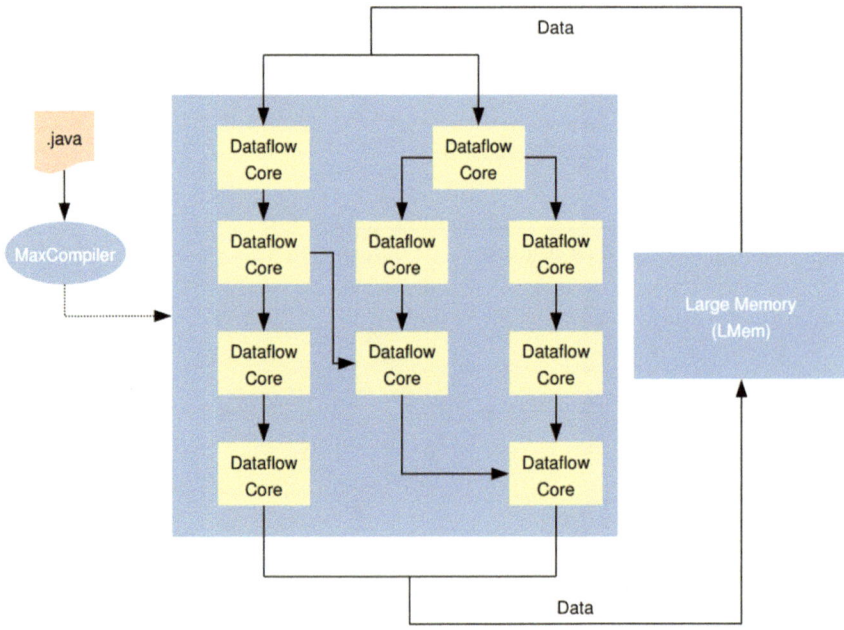

Fig. 9.19 DataFlow computation is parallel and laid in space [28]

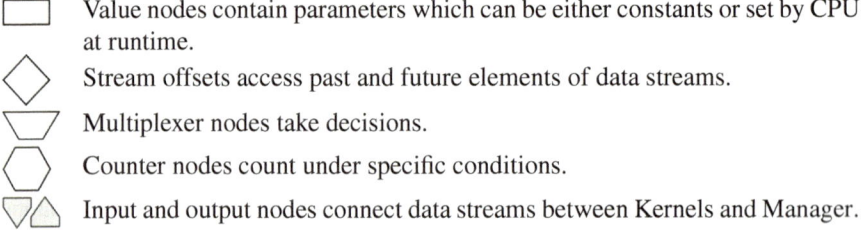 Value nodes contain parameters which can be either constants or set by CPU at runtime.

Stream offsets access past and future elements of data streams.

Multiplexer nodes take decisions.

Counter nodes count under specific conditions.

Input and output nodes connect data streams between Kernels and Manager.

The main advantage of dataflow computation over control-flow computation is parallelization; the given graph can be, for example, copied into multiple instances, which leads to even greater acceleration of a dataflow application (Fig. 9.21).

9.4 DataFlow Implementation

Bitcoin mining algorithm, where a deprecated mining protocol Getwork is used to simplify implementation and explanation, consists of looped actions:

1. fetch work via Getwork mining protocol and process block header;
2. change nonce (32-bit variable number) in the block header;
3. apply double-SHA256 hashing algorithm [36] to the modified block header;

Fig. 9.20 A dataflow graph which represents a simple function, $f(x) = x^2 + x$. Graph includes an input data stream x, which flows through arithmetic units (a multiplier and an adder), and the results flow into an output data stream y [28]

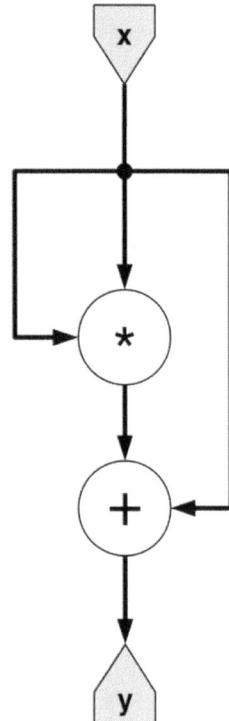

4. swap byte order (endianness) of the calculated double hash and compare it against the target threshold;
5. if double hash is smaller than the target threshold, store and use the "matching nonce" to generate a share or a block; and
6. repeat steps 2–5 until all 2^{32} possible nonce values have been exhausted, then return to step 1.

This algorithm can be parallelized into multiple dataflow graphs (analogous to threads in control-flow programming model), as shown in Fig. 9.22. The target threshold is a common 256-bit number and is therefore shared amongst pipelines, while each pipeline stores its own copy of the block header and generates its own unique nonce. Each pipeline then applies its own nonce to its own copy of the block header and hashes it twice, swaps the byte order of the calculated double hash and compares it against the target threshold. If double hash is smaller than the target threshold, the "matching nonce" is saved into mapped memory (part of FMem that is accessible by both CPU and DFE), and the host CPU program uses it to generate shares which are then sent to the mining node. With this approach, multiple nonces and hashes can be tried per kernel tick, which multiplies hash rate of dataflow bitcoin miner application.

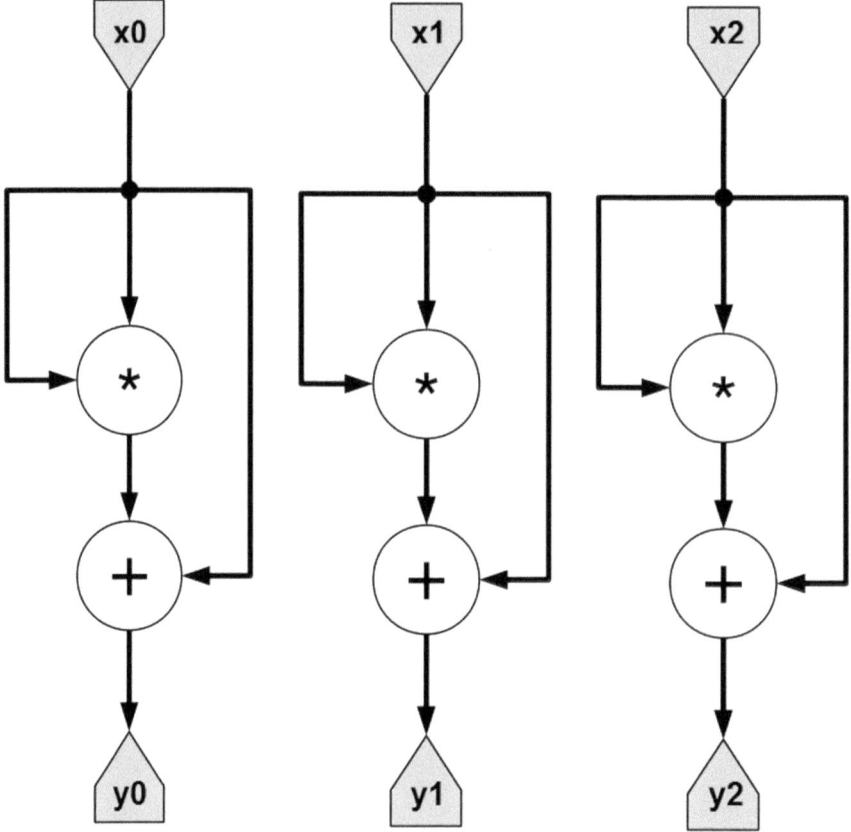

Fig. 9.21 Multiple instances of dataflow graph. Adapted from [28]

9.4.1 Used Hardware

Bitcoin miner application was implemented and tested on two Maxeler PCI-e accelerator cards, MAX2B and MAX5C; see Tables 9.6 and 9.7. Maxeler MAX2B PCI-e accelerator card is shown in Fig. 9.23.

9.4.2 Used Software

Our dataflow bitcoin miner application was written within MaxIDE 2013.3 for MAX2B board and MaxIDE 2015.2 for MAX5C board; see Fig. 9.24. MaxIDE is a working environment for writing dataflow applications and is a part of Max-Compiler which builds simulation and hardware designs for Maxeler DFEs. It is

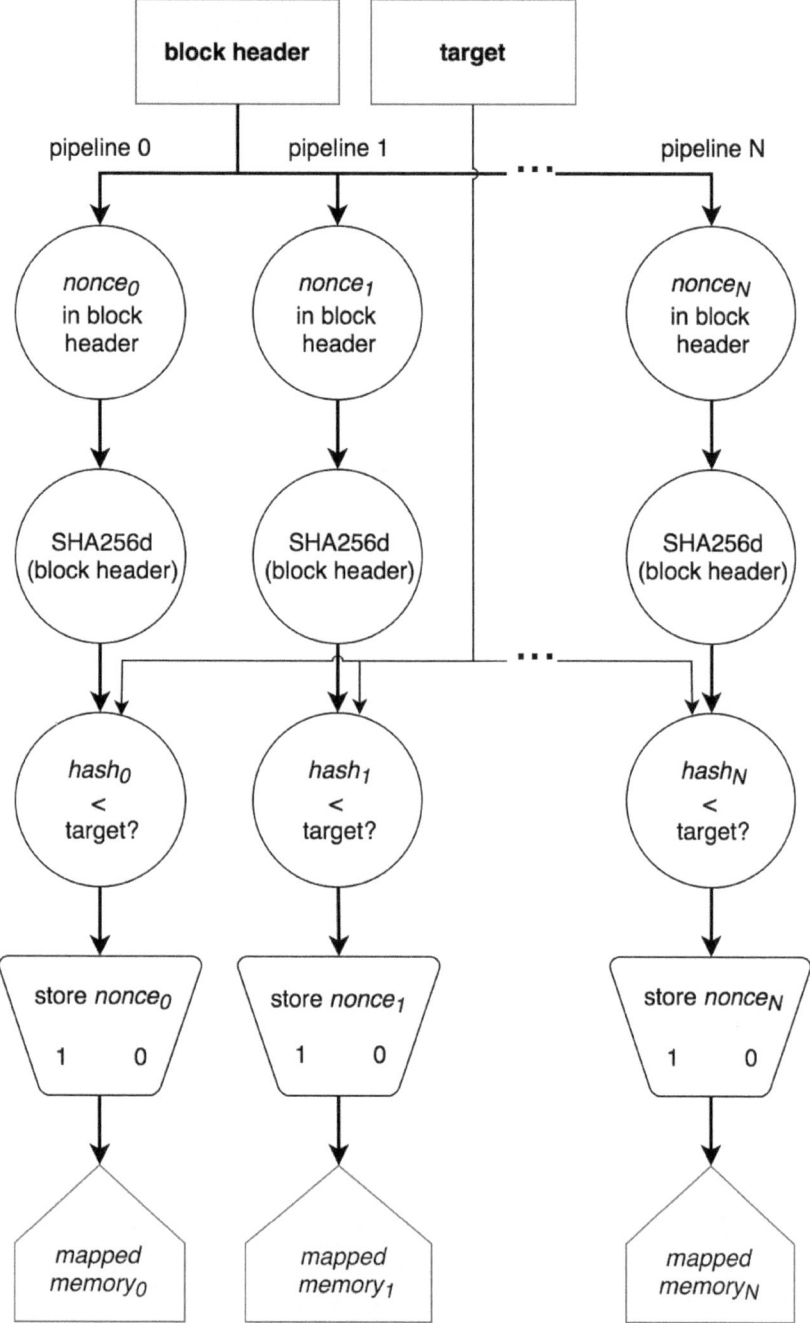

Fig. 9.22 Abstracted dataflow bitcoin miner implementation with N pipelines to process N nonces and hashes per Kernel tick

Table 9.6 Properties of MAX2B and MAX5C PCI-e accelerator cards and their host workstations

	MAX2B	MAX5C
FPGA	Virtex-5 XC5VLX330T	Altera Stratix V 5SGSMD8N2F45C2
LMem	12 GB DDR2	48 GB DDR3
Logic utilization	207,360	262,400
LUTs	207,360	/
Primary FFs	207,360	524,800
Secondary FFs	/	524,800
Multipliers	192 (25 × 18)	3926 (18 × 18)
DSP blocks	192	1963
Block memory	648 (BRAM18)	2567 (M20K)
Host CPU	Intel Core 2 Quad Q9400 @ 2.66 GHz	Intel Core i7-6700K @ 4.00 GHz
Host OS	Linux CentOS 6.5	Linux CentOS 6.9
PCI express	x4	x8
MaxIDE version	2013.3	2015.2

Table 9.7 Consumed chip resources by DFE hardware designs of dataflow bitcoin miner application

	MAX2B	MAX5C
Pipelines	3	7
Stable frequency (MHz)	95	210
Expected hash rate	285 Mhash/s	1470 Mhash/s
Actual hash rate	282 Mhash/s	1430 Mhash/s
Logic utilization	199,295/207,360 (96.11%)	225,083/262,400 (85.78%)
LUTs	178,919/207,360 (86.28%)	/
Primary FFs	169,889/207,360 (81.93%)	387,413/524,800 (73.82%)
Secondary FFs	/	49,456/524,800 (9.42%)
Multipliers	0/192 (0.00%)	0/3926 (0.00%)
DSP blocks	0/192 (0.00%)	0/1963 (0.00%)
Block memory	68/648 (10.49%)	1268/2567 (49.40%)

based on Eclipse and it uses extended version of Java language, *MaxJ*. Source code files have an extension *.maxj* to distinguish them from pure Java language.

MaxCompiler creates a *.max* configuration file from Kernels and a Manager, and builds an executable binary dataflow application for DFE which can be run in MaxCompiler or via terminal command. Kernels contain dataflow graphs for dataflow computing and Manager controls data transfer between CPUs and DFEs. DataFlow computing is only efficient if there is a lot more data than there are steps

Fig. 9.23 Maxeler MAX2B PCI-e accelerator card

in a dataflow graph; it appears this is not an issue for our dataflow bitcoin miner application, as it has to process data indefinitely.

9.4.3 DataFlow Application

Our dataflow bitcoin miner application consists of host CPU source code, written in *C* language:

- **BitcoinMinerCpuCode.c** (main CPU host code; see Appendix 1);
- **libcurl.a** (open-source CURL C library to handle communication via HTTP);
- **libjson-c.a** (open-source JSON C library to handle JSON objects);

and dataflow source code, written in *MaxJ* language:

- **BitcoinMinerEngineParameters.maxj**
  (parameters which affect hardware design, see Appendix 2);
- **BitcoinMinerKernel.maxj**
  (dataflow hashing algorithm implementation, see Appendix 3);
- **BitcoinMinerManager.maxj**
  (manager settings to control-flow of data, see Appendix 4).

The architecture of our working dataflow bitcoin miner application is shown in Fig. 9.25.

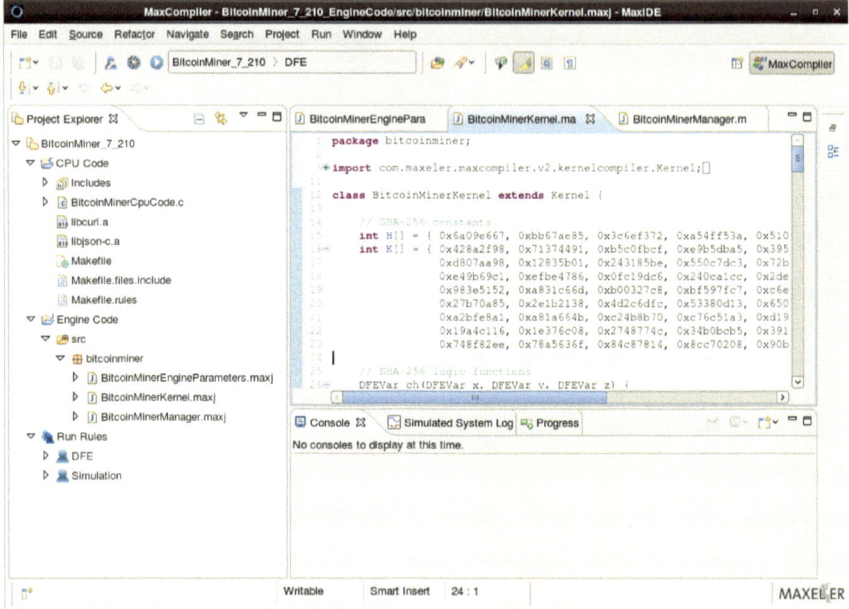

Fig. 9.24 MaxIDE 2015.2 workstation

9.4.3.1 Accept User Inputs

DataFlow bitcoin miner application, which is a binary executable file named *BitcoinMiner*, accepts same user inputs like any other cryptocurrency mining program: mining pool's URL address, username, and password, which are entered via terminal command, for example,

```
./BitcoinMiner -o http://46.4.148.123:9332 -u
1PCxQ97Q6
dAQ4D5J7gcRp18a9Azikh5Lxs -p anything.
```

Application uses Getwork mining protocol to fetch work from the chosen mining node, and it can communicate directly if mining node supports Getwork protocol (Fig. 9.26).

Nowadays, most mining pools strongly prefer Stratum mining protocol over deprecated Getwork. Communication between the application that uses Getwork and the mining nodes that use Stratum can be bridged with the bridging program called "stratum-mining-proxy" [37], which acts as proxy to relay messages (Fig. 9.27).

9.4.3.2 Fetch Work for Mining

Application uses an external library *libcurl* that handles communication with the chosen mining pool via HTTP requests and responses, where user authentication

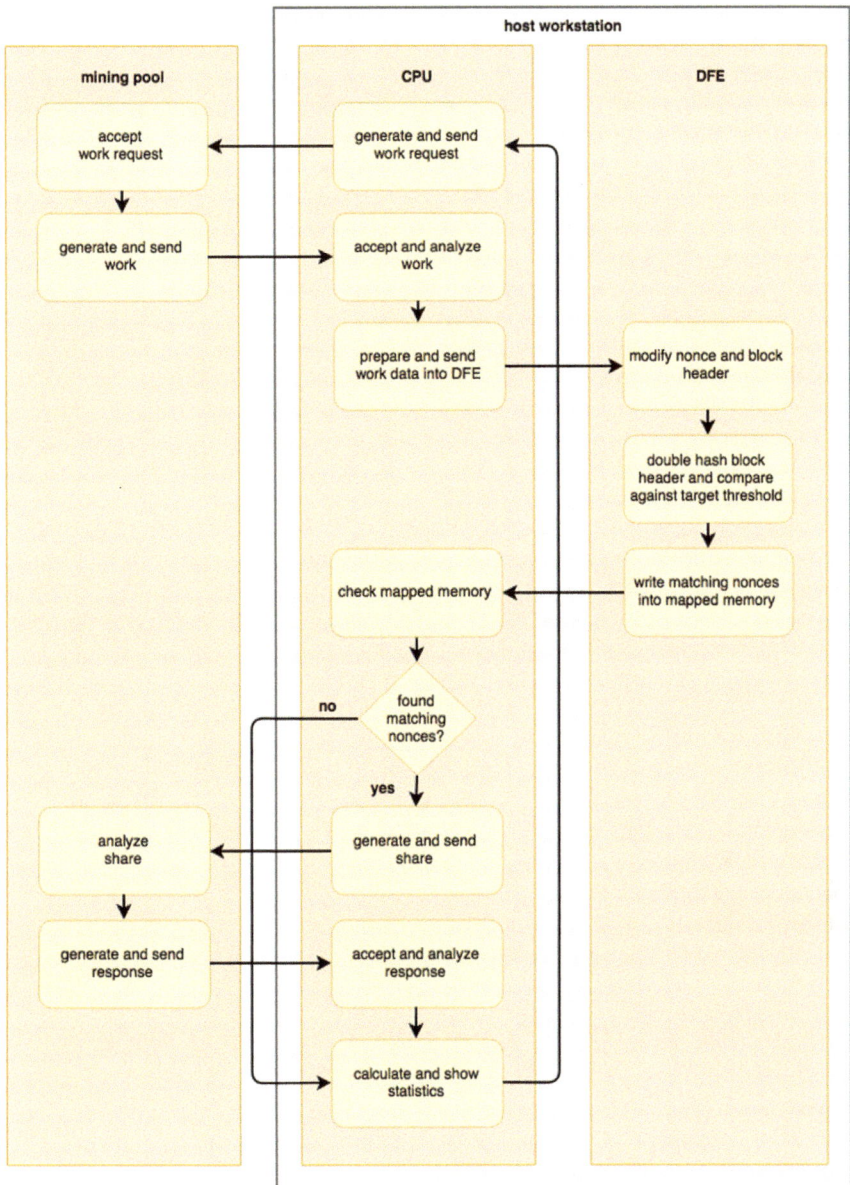

Fig. 9.25 Architecture of dataflow bitcoin miner application

Fig. 9.26 DataFlow bitcoin miner application and the mining node can communicate directly if they both support Getwork mining protocol

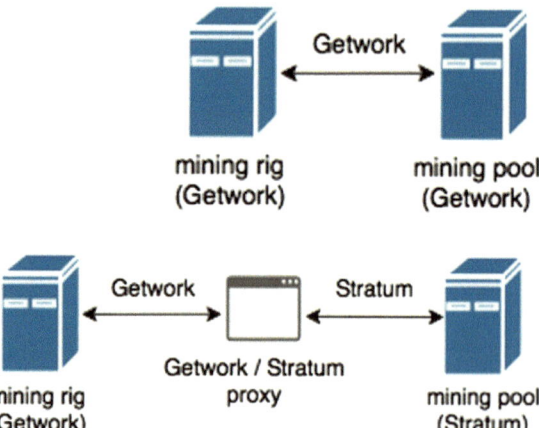

mining rig
(Getwork)

mining pool
(Getwork)

Fig. 9.27 Bridging program can be used as proxy between the mining equipment with Getwork protocol and the mining pool with Stratum protocol

Getwork / Stratum
proxy

mining rig
(Getwork)

mining pool
(Stratum)

is necessary to distinguish between miners in the Bitcoin network. Application first combines the provided username and password with a colon, in this example `1PCxQ97Q6dAQ4D5J7gcRp18a9Azikh5Lxs:anything`. This string is then base64-encoded for a basic HTTP authentication scheme while generating and sending an HTTP request to the chosen mining pool:

POST/HTTP/1.1
Host: 95.128.48.209:9332
Authorization: Basic MVBDeFE5N1E2ZEFRNEQ1SjdnY1JwMThhOUF6aWtoN
Ux4czphbnl0aGluZw==
User-Agent: libcurl-agent/1.0
Accept: application/json
Content-Type: application/json
Content-Length: 44
`{"method": "getwork", "params": [], "id": 1}`

The chosen mining pool receives a HTTP request to fetch work from the application and sends back a HTTP response which contains work for mining in JSON format, for example:

HTTP/1.1 200 OK
Content-Length: 664
X-Roll-Ntime: expire = 100
X-Long-Polling: /long-polling
Server: TwistedWeb/13.2.0
X-Is-P2pool: true
X-Stratum: stratum + tcp://95.128.48.209:9332
Date: Wed, 24 Aug 2016 12:28:19 GMT
Content-Type: application/json
`{"error": null, "jsonrpc": "2.0", "id": 1, "result": {"`

hash1": "000000000000000000000000000000000000000000000000
0000000000000000000000008000000000000000000000000000000000
0000000000000000000010000", **"target"**: **"ffffffffffffff
ffffffffffffffffffffffffffffffffffffffffffff00000000"**, "s
ubmitold": true, "identifier": "250871", **"data"**: **"20000
000f1a37b7986db0a38832ee208f1e0f526365bff1d04e5d0660000
000000000000cc2ff8b29bd51dac0abf6dbc41fb91f1f4bedd5975e
0f56d70b48d714279974457cda0291804fb08000000000000008000
000000000000000000000000000000000000000000000000000000000
000000000000000000000080020000"**, **"midstate"**: **"dfbb6ee1
9eb5beb0e26644035c018997d64a41acc0469a7ad405427a2133ce1
b"**}}

9.4.3.3 Process Work for Mining

Application processes the received HTTP response with help of both included libraries, *libcurl* and *libjson-c*. Three strings of digits (marked bold) have to be parsed into JSON objects and then converted into arrays of 32-bit unsigned integers (*uint*): *midstate*, *target*, and *data*. Because "midstate" hash is already included in the fetched work, calculating it by mining equipment itself is optional. Example of C implementation to calculate "midstate" hash is given in Appendix 5.

The following explanation of calculating "midstate" hash is derived from [36] and has been modified for this specific example. SHA-256 hashing algorithm uses 8 initial H constants as shown in Eq. (9.9).

$$H_0^{(0)} = 6a09e667$$
$$H_1^{(0)} = bb67ae85$$
$$H_2^{(0)} = 3c6ef372$$
$$H_3^{(0)} = a54ff53a$$
$$H_4^{(0)} = 510e527f$$
$$H_5^{(0)} = 9b05688c$$
$$H_6^{(0)} = 1f83d9ab$$
$$H_7^{(0)} = 5be0cd19 \tag{9.9}$$

SHA-256 hashing algorithm also uses 64 K constants:

```
428a2f98, 71374491, b5c0fbcf, e9b5dba5, 3956c25b,
59f111f1,923f82a4, ab1c5ed5, d807aa98, 12835b01,
243185be, 550c7dc3,72be5d74, 80deb1fe, 9bdc06a7,
c19bf174, e49b69c1, efbe4786,0fc19dc6, 240ca1cc,
2de92c6f, 4a7484aa, 5cb0a9dc, 76f988da,983e5152,
```

a831c66d, b00327c8, bf597fc7, c6e00bf3,
d5a79147,06ca6351, 14292967, 27b70a85, 2e1b2138,
4d2c6dfc, 53380d13,650a7354, 766a0abb, 81c2c92e,
92722c85, a2bfe8a1, a81a664b,c24b8b70, c76c51a3,
d192e819, d6990624, f40e3585, 106aa070,19a4c116,
1e376c08, 2748774c, 34b0bcb5, 391c0cb3,
4ed8aa4a,5b9cca4f, 682e6ff3, 748f82ee, 78a5636f,
84c87814, 8cc70208,90befffa, a4506ceb, bef9a3f7,
c67178f2.

Furthermore, six logical functions are used, as shown in Eq. (9.10). Those functions operate on 32-bit unsigned integers and produce 32-bit unsigned integers, which are also referred to as "words":

$$Ch(x, y, z) = (x \wedge y) \oplus (\neg x \wedge z),$$
$$Maj(x, y, z) = (x \wedge y) \oplus (x \wedge z) \oplus (y \wedge z),$$
$$\Sigma_0(x) = ROTR^2(x) \oplus ROTR^{13}(x) \oplus ROTR^{22}(x),$$
$$\Sigma_1(x) = ROTR^6(x) \oplus ROTR^{11}(x) \oplus ROTR^{25}(x),$$
$$\sigma_0(x) = ROTR^7(x) \oplus ROTR^{18}(x) \oplus SHR^3(x),$$
$$\sigma_1(x) = ROTR^{17}(x) \oplus ROTR^{19}(x) \oplus SHR^{10}(x) \qquad (9.10)$$

where $\wedge$ is bitwise AND, $\oplus$ is bitwise XOR, and is bitwise NOT. $ROTR^n(x)$ is right rotation by n bits and $SHR^n(x)$ is right shift by n bits, as shown in Eq. (9.11).

$$SHR^n(x) = x \gg n; \quad n \geq 0$$
$$ROTR^n(x) = (x \gg n) \vee (x \ll 32 - n); \quad 32 > n \geq 0 \qquad (9.11)$$

SHA-256 hashing algorithm pads a message into a multiple of 512 bits, then parses it into 512-bit message blocks, and processes them one at a time. The 640 bits long block header is therefore padded into a 1024-bit message by adding a bit 1 at the end of the block header, followed by 319 zeros; k is the number of zeros that is determined as the smallest nonnegative solution for Eq. (9.12):

$$l + 1 + k = 448 \bmod 512 \qquad (9.12)$$

where l is the message length in bits (in this case 640) and the last 64 bits of the block header represent the message length in hex (640 bits equal 280 in hex). The padded 1024-bit block header is then

00000020797ba3f1380adb8608e22e8326f5e0f11dff5b3666d0e5
040000000000000000b2f82fccac1dd59bbc6dbf0af191fb4159dd
bef46df5e075718db4704497794229a0cd5708fb041800000000**80**
00000000000000000000000000000000000000000000000000000000
0000000000000000000000000000000000000000280.

As stated before, this information is given as string *data* from the fetched work, but it is byte-swapped in little-endian on 32-bit basis:

```
20000000f1a37b7986db0a38832ee208f1e0f526365bff1d04e5d0
660000000000000000cc2ff8b29bd51dac0abf6dbc41fb91f1f4be
dd5975e0f56d70b48d714279974457cda0291804fb080000000000
0000800000000000000000000000000000000000000000000000000
0000000000000000000000000000000000080020000.
```

The padded 1024-bit block header is then split into two 512-bit message blocks, the first message block (denoted as $M^{(0)}$) being

```
00000020797ba3f1380adb8608e22e8326f5e0f11dff5b3666d0e5
040000000000000000b2f82fccac1dd59bbc6dbf0af191fb4159dd
bef46df5e075718db470
```

and the second message block (denoted as $M^{(1)}$) being

```
4497794229a0cd5708fb04180000000008000000000000000000000
0000000000000000000000000000000000000000000000000000000
00000000000000000280.
```

The first 512-bit message block, $M^{(0)}$, is split into 16 words:

```
00000020, 797ba3f1, 380adb86, 08e22e83, 26f5e0f1, 1dff5b36,
66d0e504, 00000000, 00000000, b2f82fcc, ac1dd59b, bc6dbf0a,
f191fb41, 59ddbef4, 6df5e075, 718db470.
```

Message is then scheduled into 64 words which are denoted as W_j, as shown in Eq. (9.13).

$$W_j = M_j^{(0)}; 0 \le j \le 15$$
$$W_j = \sigma_1(W_{j-2}) + W_{j-7} + \sigma_0(W_{j-15}) + W_{j-16}; 16 \le j \le 63 \qquad (9.13)$$

Eight working variables (a, b, c, d, e, f, g, h) are set to their initial values that are taken from Eq. (9.9), as shown in Eq. (9.14).

$$a = H_0^{(0)} = \text{6a09e667}$$
$$b = H_1^{(0)} = \text{bb67ae85}$$
$$c = H_2^{(0)} = \text{3c6ef372}$$
$$d = H_3^{(0)} = \text{a54ff53a}$$
$$e = H_4^{(0)} = \text{510e527f}$$
$$f = H_5^{(0)} = \text{9b05688c}$$
$$g = H_6^{(0)} = \text{1f83d9ab}$$
$$h = H_7^{(0)} = \text{5be0cd19} \qquad (9.14)$$

SHA-256 compression function is applied to shuffle bits, as shown in Eq. (9.15). Two temporal words (denoted as T_1 and T_2) are added at this point.

$$\text{for } j = 0 \text{ to } 63\{$$
$$T_1 = h + \sum\nolimits_1 (e) + Ch(e, f, g) + K_j + W_j$$
$$T_2 = \sum\nolimits_0 (a) + Maj(a, b, c)$$
$$h = g$$
$$g = f$$
$$f = e$$
$$e = d + T_1 \tag{9.15}$$
$$d = c$$
$$c = b$$
$$b = a$$
$$a = T_1 + T_2$$
$$\}$$

The last part is calculating intermediate hash values, as shown in Eq. (9.16).

$$H_0^{(1)} = H_0^{(0)} + a,$$
$$H_1^{(1)} = H_1^{(0)} + b,$$
$$H_2^{(1)} = H_2^{(0)} + c,$$
$$H_3^{(1)} = H_3^{(0)} + d,$$
$$H_4^{(1)} = H_4^{(0)} + e,$$
$$H_5^{(1)} = H_5^{(0)} + f$$
$$H_6^{(1)} = H_6^{(0)} + g,$$
$$H_7^{(1)} = H_7^{(0)} + h \tag{9.16}$$

For the given block header example, the calculated intermediate hash values are `e16ebbdf`, `b0beb59e`, `034466e2`, `9789015c`, `ac414ad6`, `7a9a46c0`, `7a4205d4`, `1bce3321`, respectively, and together they represent a 256-bit intermediate hash $H^{(1)}$, that is

`e16ebbdfb0beb59e034466e29789015cac414ad67a9a46c07a4205`
`d41bce3321`.

This intermediate hash does not change when modifying block header with all 2^{32} possible nonce values, and therefore it only has to be calculated once and not 2^{32} times. As stated before, calculating "midstate" hash is optional and can be skipped, because it is already precalculated by the mining node and provided from the fetched work as string *midstate*:

`dfbb6ee19eb5beb0e26644035c018997d64a41acc0469a7ad40542`
`7a2133ce1b`,

which is written in little-endian format and has to be byte-swapped into big-endian format on 32-bit basis into

`e16ebbdfb0beb59e034466e29789015cac414ad67a9a46c07a4205`
`d41bce3321`.

This "midstate" hash is split and packed into an array of eight words (`e16ebbdf`, `b0beb59e`, `034466e2`, `9789015c`, `ac414ad6`, `7a9a46c0`, `7a4205d4`, `1bce3321`) that is named *midstate*.

Thanks to the pre-calculated "midstate" hash, the first 512-bit message block can be skipped and only the second 512-bit message block has to be processed:

`4497794229a0cd5708fb0418000000000``80000000000000000000000000`
`000000000000000000000000000000000000000000000000000000000000000000`
`00000000000000000280`.

It appears only the first four 32-bit words of the block header (marked bold) are variable, and the rest of the block header does not change (regardless of retrieved block headers). The first three numbers, in this example **44977942**, **29a0cd57** and **08fb0418**, are extracted from a string *data* that is a part of the fetched work and the fourth number is a nonce that starts at **00000000**. Therefore, those first three numbers are packed into an array of three elements that is named *data*. A string *target*, in this example

`ffffffffffffffffffffffffffffffffffffffffffffffffffffffffffffffff`
`ff00000000`,

has to be byte-swapped on 256-bit basis (or in other words, from right to left) to represent a 256-bit target threshold

`00000000ffffffffffffffffffffffffffffffffffffffffffffffffffffffff`
`ffffffffff`.

The 256-bit target threshold can be packed into an array of eight 32-bit numbers (`00000000`, `ffffffff`, `ffffffff`, `ffffffff`, `ffffffff`, `ffffffff`, `ffffffff`, `ffffffff`), named *target*. However, the given target threshold is not mandatory, as miners can apply their own target thresholds.

At this point we have three arrays of numbers to send into DFE:

- *midstate* (`e16ebbdf`, `b0beb59e`, `034466e2`, `9789015c`, `ac414ad6`, `7a9a46c0`, `7a4205d4`, `1bce3321`);
- *data* (`44977942`, `29a0cd57`, `08fb0418`);
- *target* (`00000000`, `ffffffff`, `ffffffff`, `ffffffff`, `ffffffff`, `ffffffff`, `ffffffff`, `ffffffff`).

Those three arrays are sent to the available DFE via SLiC (Simple Live CPU) interface as scalar inputs (single values), for example

```
BitcoinMiner(N, block, midstate, target, output,
mappedRam[0], mappedRam[0], mappedRam[1],
mappedRam[1],
…, mappedRam[M], mappedRam[M]);
```

where *N* is size of output data stream, *output* is actual output data stream to run ticks in kernel, and *mappedRam[M]* pairs are mapped memory inputs (to reset "matching nonces") and outputs (to store "matching nonces") for each pipeline in Kernel.

9.4.3.4 Compute on DFE

A loop for trying out all 2^{32} possible nonce values would be implemented in control-flow program as follows:

```
for nonce = 0 to ffffffff {
    … // further computation is placed within this loop
}
```

But in dataflow computing, this loop does not to be implemented explicitly; unique nonce values are generated and used per each kernel tick (or output data stream element).

From array *data*, which contains the first three numbers of the second 512-bit message block $M^{(1)}$, (in this example 44977942, 29a0cd57 and 08fb0418), the second message block, $M^{(1)}$, is reconstructed as

4497794229a0cd5708fb0418xxxxxxxx800000000000000000000000
000000000000000000000000000000000000000000000000000000000000
0000000000000000000280

where **xxxxxxxx** is variable nonce value (marked as $M_3^{(1)}$) and the message block is therefore scheduled into 64 words as shown in Eq. (9.17).

$$W_j = M_j^{(1)}, \quad \text{where } M_3^{(1)} = nonce; \quad 0 \le j \le 15$$
$$W_j = \sigma_1(W_{j-2}) + W_{j-7} + \sigma_0(W_{j-15}) + W_{j-16}; \quad 16 \le j \le 63 \tag{9.17}$$

For further explanation, we will use the expanded block header (message block $M^{(1)}$) with a nonce value **fad80419** as an example, thus yielding a modified block header

000000000000000000000000000000000000000000000000000000000000
000000000000000000000000000000000000000000000000000000000000
0000000000000000000280.

Because this second message block $M^{(1)}$ is successor of the first message block $M^{(0)}$, intermediate hash values of the first message block $M^{(0)}$, that are denoted as $H_i^{(1)}$, are used to initialize the working variables (a, b, c, d, e, f, g, h) instead of initial H constants, as shown in Eq. (9.18). In this example, these hash values are as follows: e16ebbdf, b0beb59e, 034466e2, 9789015c, ac414ad6, 7a9a46c0, 7a4205d4, 1bce3321.

$$a = H_0^{(1)} = midstate_0$$
$$b = H_1^{(1)} = midstate_1$$
$$c = H_2^{(1)} = midstate_2$$
$$d = H_3^{(1)} = midstate_3$$
$$e = H_4^{(1)} = midstate_4$$
$$f = H_5^{(1)} = midstate_5$$

$$g = H_6^{(1)} = midstate_6$$
$$h = H_7^{(1)} = midstate_7 \qquad (9.18)$$

SHA-256 compression function is applied, as shown in Eq. (9.19) which is identical to Eq. (9.15).

$$
\begin{aligned}
&\text{for } j = 0 \text{ to } 63\{ \\
&\quad T_1 = h + \sum\nolimits_1 (e) + Ch(e, f, g) + K_j + W_j \\
&\quad T_2 = \sum\nolimits_0 (a) + Maj(a, b, c) \\
&\quad h = g \\
&\quad g = f \\
&\quad f = e \\
&\quad e = d + T_1 \\
&\quad d = c \\
&\quad c = b \\
&\quad b = a \\
&\quad a = T_1 + T_2 \\
&\}
\end{aligned}
\qquad (9.19)
$$

The complete hash of the block header is calculated, as shown in Eq. (9.20).

$$
\begin{aligned}
H_0^{(2)} &= H_0^{(1)} + a, \\
H_1^{(2)} &= H_1^{(1)} + b, \\
H_2^{(2)} &= H_2^{(1)} + c, \\
H_3^{(2)} &= H_3^{(1)} + d, \\
H_4^{(2)} &= H_4^{(1)} + e, \\
H_5^{(2)} &= H_5^{(1)} + f \\
H_6^{(2)} &= H_6^{(1)} + g, \\
H_7^{(2)} &= H_7^{(1)} + h
\end{aligned}
\qquad (9.20)
$$

Hash values (denoted as $H_i^{(2)}$) for the given block header with a nonce value **fad80419** are 93dfffee, 9469bafb, 8baa79d5, e0f4a981, caf77d87, cf0937fd, b3880d3f, and 0ec5b2f3, respectively. Together they represent a 256-bit hash

93dfffee9469bafb8baa79d5e0f4a981caf77d87cf0937fdb3880d
3f0ec5b2f3.

But these hash values do not have to be concatenated at this point yet, as this hash has to be hashed once more and is thus padded by SHA-256 standard into a 512-bit message block; a bit 1 is added at the end of the hash, and according to Eq. (9.12), 191 zeros follow. The last 64 bits are now `0000000000000100` unlike in the previous hashing round (`0000000000000280`), because the message length l is now 256 bits (`100` in hex) and not 640 bits (`280` in hex) like before. Padded hash is then the following message block, $M^{(0)}$:

`93dfffee9469bafb8baa79d5e0f4a981caf77d87cf0937fdb3880d`
`3f0ec5b2f3800000000000000000000000000000000000000000000`
`0000000000000000000100`.

The message schedule W, which consists of 64 words, is calculated as shown in Eq. (9.21) which is identical to Eq. (9.13).

$$W_j = M_j^{(1)}; \quad 0 \le j \le 15$$
$$W_j = \sigma_1(W_{j-2}) + W_{j-7} + \sigma_0(W_{j-15}) + W_{j-16}; \quad 16 \le j \le 63 \qquad (9.21)$$

Working variables (a, b, c, d, e, f, g, h) are initialized with initial H values, as shown in Eq. (9.22) which is identical to Eq. (9.14), because it is a start of a new (second) hashing round. In this second hashing round, only one message block, $M^{(0)}$, has to be processed.

$$a = H_0^{(0)} = \texttt{6a09e667}$$
$$b = H_1^{(0)} = \texttt{bb67ae85}$$
$$c = H_2^{(0)} = \texttt{3c6ef372}$$
$$d = H_3^{(0)} = \texttt{a54ff53a}$$
$$e = H_4^{(0)} = \texttt{510e527f}$$
$$f = H_5^{(0)} = \texttt{9b05688c}$$
$$g = H_6^{(0)} = \texttt{1f83d9ab}$$
$$h = H_7^{(0)} = \texttt{5be0cd19} \qquad (9.22)$$

SHA-256 compression function is applied, as shown in Eq. (9.23) which is identical to Eqs. (9.15) and (9.19).

$$\text{for } j = 0 \text{ to } 63 \ \{$$
$$T_1 = h + \textstyle\sum_1 (e) + Ch(e, f, g) + K_j + W_j$$
$$T_2 = \textstyle\sum_0 (a) + Maj(a, b, c)$$
$$h = g$$
$$g = f$$
$$f = e$$
$$e = d + T_1 \qquad\qquad (9.23)$$
$$d = c$$
$$c = b$$
$$b = a$$
$$a = T_1 + T_2$$
$$\}$$

Hash values are calculated as shown in Eq. (9.24), which is identical to Eq. (9.16).

$$H_0^{(1)} = H_0^{(0)} + a,$$
$$H_1^{(1)} = H_1^{(0)} + b,$$
$$H_2^{(1)} = H_2^{(0)} + c,$$
$$H_3^{(1)} = H_3^{(0)} + d,$$
$$H_4^{(1)} = H_4^{(0)} + e,$$
$$H_5^{(1)} = H_5^{(0)} + f$$
$$H_6^{(1)} = H_6^{(0)} + g,$$
$$H_7^{(1)} = H_7^{(0)} + h \qquad\qquad (9.24)$$

In this example, those hash values are as follows: `07723479`, `ea735579`, `93659b5d`, `ec9a7d47`, `2c149aa2`, `4d630e1b`, `65477bb1`, `00000000`. When these hash values are concatenated together, they represent a double 256-bit hash (denoted as $H^{(1)}$) in big-endian:

`07723479ea73557993659b5dec9a7d472c149aa24d630e1b65477b`
`b100000000.`

But to compare it against the target threshold, the double hash has to be byte-swapped into little-endian on 256-bit basis, or in other words, from right to left:

`00000000b17b47651b0e634da29a142c477d9aec5d9b6593795573`
`ea79347207.`

This byte-swapped hash is then compared against the target threshold, in this example

`00000000ffffffffffffffffffffffffffffffffffffffffffffffff`
`ffffffffffff.`

Because the double-SHA-256 hash of the given block header with a nonce value **fad80419** is smaller than target threshold, this nonce is stored in mapped memory. In our dataflow implementation, this is done by writing a pair of match indicator (number 1 represents Boolean `true`) and nonce into mapped memory that is accessed by both CPU and DFE.

Mapped memory instances were set for each pipeline and it was determined to store up to eight values per pipeline; four pairs of match indicator and nonces, as there is no need for more than that. The writing addresses are controlled by an advanced counter *addressCount* (see Appendix 3) that is increased by 2 per each match (in each pipeline) and goes up to 8. If a matching nonce 00000003 would be found, then a match indicator, 1, would be written into mapped memory address 0 and a nonce 00000003 would be written into address 1. The counter *addressCount* would then increase to 2. If the next matching nonce 00000006 would be found in the same pipeline, then a match indicator, 1, would be written into mapped memory address 2 and nonce 00000006 would be written into address 3. The counter *addressCount* would then increase to 4 and so on.

The only thing left to do in the Kernel is setting a dummy output stream to actually run ticks in Kernel, which can be generated in Kernel itself.

Implementations of described bitcoin mining algorithm are given in Appendix 3 (dataflow, *MaxJ* language) and in Appendix 5 (control-flow, *C* language).

9.4.3.5   Create and Send Share

Host CPU program of dataflow bitcoin miner application, *BitcoinMinerCpuCode.c*, checks mapped memory for pairs of match indicators and nonces. When a match indicator 1 is found, nonce has to be byte-swapped from big-endian back to little-endian; in this example, nonce **fad80419** is byte-swapped into **1904d8fa** and is used to create a share; the string *data* that was given in the fetched work,

```
20000000f1a37b7986db0a38832ee208f1e0f526365bff1d04e5d0
660000000000000000cc2ff8b29bd51dac0abf6dbc41fb91f1f4be
dd5975e0f56d70b48d714279974457cda0291804f0000000000000
b0808000000000000000000000000000000000000000000000000
0000000000000000000000000000000080020000,
```

is modified into

```
20000000f1a37b7986db0a38832ee208f1e0f526365bff1d04e5d0
660000000000000000cc2ff8b29bd51dac0abf6dbc41fb91f1f4be
dd5975e0f56d70b48d714279974457cda0291804f1904d8fa00000
b0808000000000000000000000000000000000000000000000000
0000000000000000000000000000000080020000
```

and added into a HTTP request which is then sent to the mining pool:

POST/HTTP/1.1
Host: 95.128.48.209:9332
Authorization: Basic MUNjblZwZXhBUkJMR2ljS3ByOUNmbnF6ZDJwUFlNc3d
xQTpsb2xMT0xsb2w=
User-Agent: libcurl-agent/1.0
Accept: application/json

Content-Type: application/json
Content-Length: 302

{"method":"getwork", "params":["20000000f1a37b7986db0a3
8832ee208f1e0f526365bff1d04e5d06600000000000000000cc2ff8
b29bd51dac0abf6dbc41fb91f1f4bedd5975e0f56d70b48d7142799
74457cda0291804fb08**1904d8fa**000000800000000000000000000000
0000000000000000000000000000000000000000000000000000000000
0000080020000"], "id":1}

9.4.3.6   Analysis of HTTP Response

The mining pool may accept or reject the sent share, and may optionally state the reason for rejection:

HTTP/1.1 200 OK
Content-Length: 58
X-Roll-Ntime: expire $= 100$
X-Long-Polling: /long-polling
Server: TwistedWeb/13.2.0
X-Is-P2pool: true
X-Stratum: stratum + tcp://95.128.48.209:9332
Date: Wed, 24 Aug 2016 12:28:19 GMT
Content-Type: application/json
{"error": null, "jsonrpc": "2.0", "id": 1, **"result":**
true}

If the content of HTTP response contains a word **false**, then the mining pool refused the sent share (most likely the fetched work became obsolete and/or the share was sent too late). Otherwise, if the content of HTTP response contains a word **true** (as in the given example above), it means the mining pool accepted the sent share.

9.4.3.7   Mining Statistics

Our application shows real-time bitcoin mining statistics, which is implemented in the host CPU program, *BitcoinMinerCpuCode.c*; it calculates hash rate, amount of accepted shares, ratio between accepted and all sent shares, and delay between sending shares and receiving responses from the mining pool. It also reports errors, such as bad user inputs, connection errors, and other failed actions (Fig. 9.28).

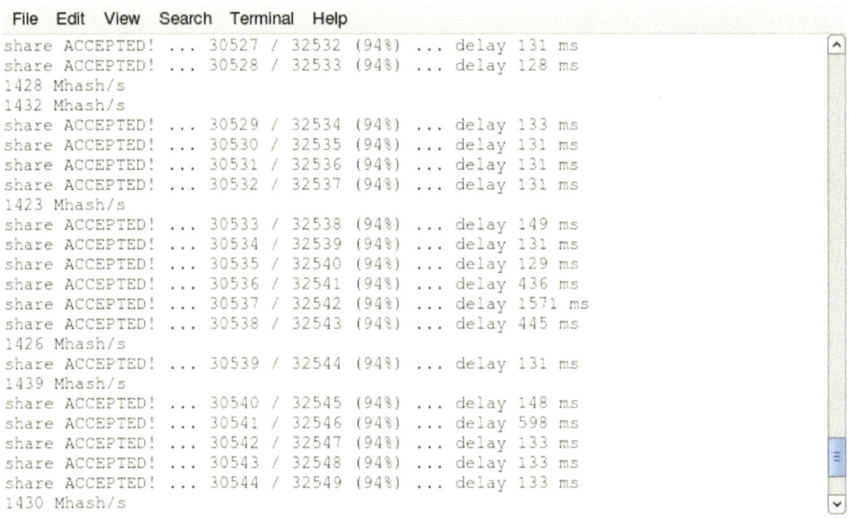

```
File  Edit  View  Search  Terminal  Help
share ACCEPTED! ... 30527 / 32532 (94%) ... delay 131 ms
share ACCEPTED! ... 30528 / 32533 (94%) ... delay 128 ms
1428 Mhash/s
1432 Mhash/s
share ACCEPTED! ... 30529 / 32534 (94%) ... delay 133 ms
share ACCEPTED! ... 30530 / 32535 (94%) ... delay 131 ms
share ACCEPTED! ... 30531 / 32536 (94%) ... delay 131 ms
share ACCEPTED! ... 30532 / 32537 (94%) ... delay 131 ms
1423 Mhash/s
share ACCEPTED! ... 30533 / 32538 (94%) ... delay 149 ms
share ACCEPTED! ... 30534 / 32539 (94%) ... delay 131 ms
share ACCEPTED! ... 30535 / 32540 (94%) ... delay 129 ms
share ACCEPTED! ... 30536 / 32541 (94%) ... delay 436 ms
share ACCEPTED! ... 30537 / 32542 (94%) ... delay 1571 ms
share ACCEPTED! ... 30538 / 32543 (94%) ... delay 445 ms
1426 Mhash/s
share ACCEPTED! ... 30539 / 32544 (94%) ... delay 131 ms
1439 Mhash/s
share ACCEPTED! ... 30540 / 32545 (94%) ... delay 148 ms
share ACCEPTED! ... 30541 / 32546 (94%) ... delay 598 ms
share ACCEPTED! ... 30542 / 32547 (94%) ... delay 133 ms
share ACCEPTED! ... 30543 / 32548 (94%) ... delay 133 ms
share ACCEPTED! ... 30544 / 32549 (94%) ... delay 133 ms
1430 Mhash/s
```

Fig. 9.28 MAX5C bitcoin miner dataflow application shows basic mining statistics in the console; share status, ratio between accepted and all sent shares, and delay between sending share and receiving response

Table 9.8 Benchmarked bitcoin mining equipment

Hardware	Software
CPU Intel 2 Core Quad Q9400 @ 2.66 GHz	cpuminer-minerd 2.4.5 (open-source)
CPU Intel i7-6700K @ 4.00 GHz	cpuminer-minerd 2.4.5 (open-source)
GPU AMD Radeon 6950 HD	cgminer 3.7.2 (open-source)
DFE MAX2B	Our dataflow implementation
DFE MAX5C	Our dataflow implementation

9.5 Tests and Results

We benchmarked various bitcoin mining equipment in three categories, named A, B, and C, respectively. Used bitcoin mining rigs are shown in Table 9.8.

The results show that Maxeler DFEs are faster and more energy efficient than CPUs and GPUs, but inferior to specialized ASIC bitcoin mining rigs.

9.5.1 Test A

It is known that DFEs do not achieve the top computation speed immediately, but it is generally increasing as more data are processed at once and it stops at a certain point. Therefore we ran our dataflow application that produced output streams with

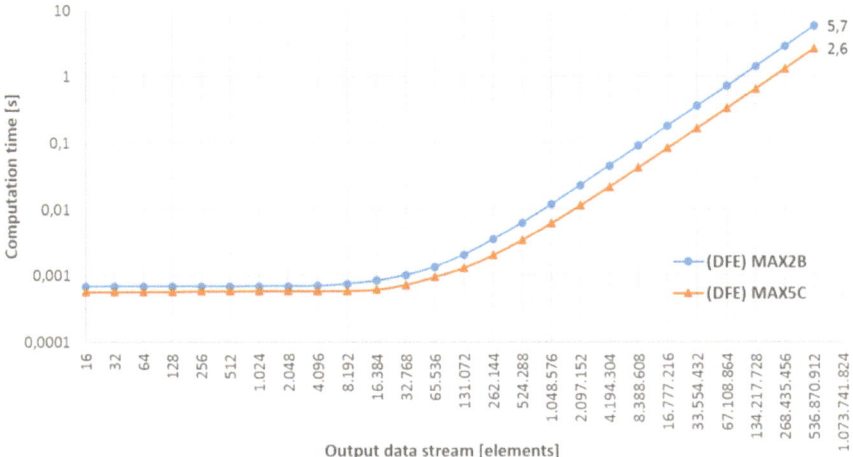

Fig. 9.29 Measured computation time of dataflow bitcoin miner application with different sizes of output stream. Both axes are in logarithmic scale

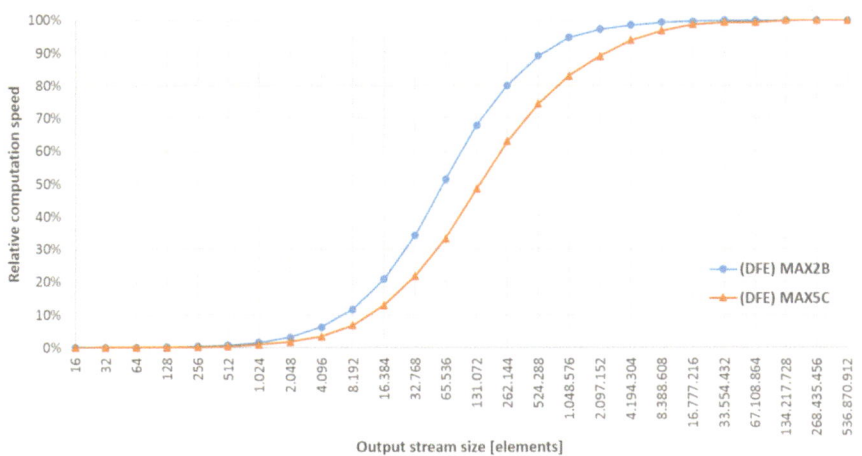

Fig. 9.30 Relative computation speed based on output data stream size

different sizes, measured the required computation time (Fig. 9.29) and calculated relative computation speed based on output data stream size (Fig. 9.30).

While the computing time of CPUs generally increases linearly with data size [38], the computing time of DFEs is rather "curved". DFEs require a certain minimum amount of time to start and stop streaming data; input stream data have to be transferred into DFE, the data has to run through kernel graphs which consist of several hundreds of arithmetic units, and the output stream data have to be transferred back to CPU. This amount of time remains the same, because it does not depend on the amount of output data stream elements. In our case, MAX2B board needed at least 685 μs and MAX5C board needed at least 572 μs to finish computation and streaming data (Fig. 9.29). Therefore significant speedups can be achieved only with very large data sets (Fig. 9.30).

Both MAX2B and MAX5C boards achieved top computation speed at streaming at least $2^{27} = 134{,}217{,}728$ output elements. In addition, dataflow application for MAX2B board was realized with 3 pipelines and dataflow application for MAX5C board was realized with 7 pipelines; so MAX2B board computed at full speed when it processed at least 402,653,184 nonce values and MAX5C board computed at full speed when it processed at least 939,524,096 nonce values.

The computation speed was not an issue for our dataflow implementation, as all $2^{32} = 4{,}294{,}967{,}296$ nonce values have to be processed per single block header.

9.5.2 Test B

First, we measured the average hash rates of mining hardware components during bitcoin mining (Fig. 9.31) and compared them against each other (Table 9.9). No instruments were needed because hash rates are reported by the mining software.

Next, we retrieved the electric powers of hardware components during mining (Fig. 9.32) and compared them against each other (Table 9.10). Electric power of MAX2B board was calculated as a difference between electric power of the host workstation with fully running MAX2B board and electric power of the same host workstation without MAX2B board, both measured with energy consumption meter VOLTCRAFT Energy Logger 4000. Electric power of MAX5C board was retrieved with a terminal command that gives detailed reports of specific Maxeler

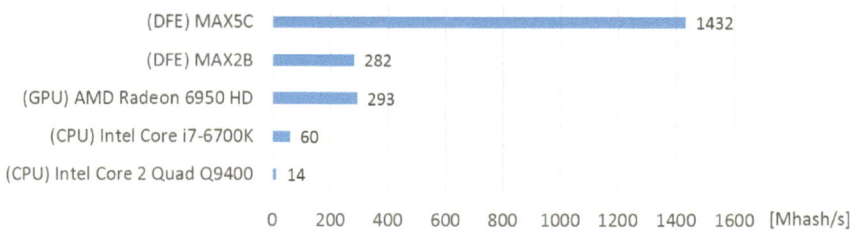

Fig. 9.31 Measured hash rates of mining hardware components during bitcoin mining (more is better)

Table 9.9 Hash rate factors between mining hardware components

	MAX5C	MAX2B
vs. MAX2B	5	/
vs. GPU	5	1
vs. CPU (i7-6700K)	24	5
vs. CPU (Q9400)	102	20

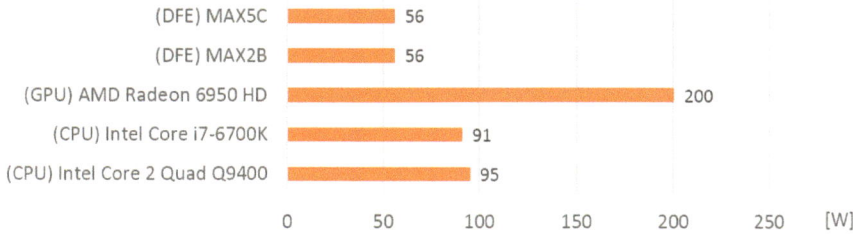

Fig. 9.32 Measured (DFEs) and theoretical (CPUs and GPU) electric powers of mining hardware components during bitcoin mining (less is better)

Table 9.10 Electric power factors between mining hardware components

	MAX5C	MAX2B
vs. MAX2B	1	/
vs. GPU	0.28	0.28
vs. CPU (i7-6700K)	0.62	0.62
vs. CPU (Q9400)	0.59	0.59

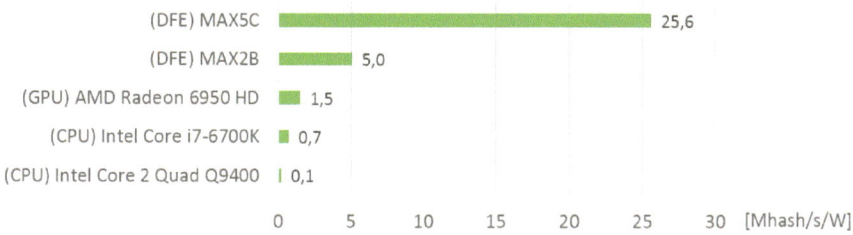

Fig. 9.33 Calculated energy efficiency of mining hardware components during bitcoin mining (more is better)

Table 9.11 Energy efficiency factors between mining hardware components

	MAX5C	MAX2B
vs. MAX2B	5	/
vs. GPU	17	3
vs. CPU (i7-6700K)	37	7
vs. CPU (Q9400)	256	50

cards, `maxtop -v`. CPUs and GPU were given theoretical values of thermal design powers (TDP) that are specified in hardware specifications.

As stated before, the most important parameter of mining rigs is energy efficiency, which can be calculated as shown in Eq. (9.8). It has many different units available, but the most used and understandable is hash/s/W, where higher hash rate per unit of electric power is better (Fig. 9.33). Comparison of hardware in energy efficiency is shown in Table 9.11.

Table 9.12 Comparison between Maxeler MAX5C board and ASIC Antminer S7 bitcoin mining rig

	MAX5C	Antminer S7	Factor
Hash rate	1.43 Ghash/s	4.73 Thash/s	3303
Power consumption	56 W	1300 W	23
Energy efficiency	25.6 Mhash/s/W	3.64 Ghash/s/W	142

Factors are calculated as a ratio between ASIC Antminer S7 and MAX5C board properties

MAX5C board appeared to be superior compared to the rest of tested mining hardware in all three categories: hash rate, electric power, and energy efficiency. However, comparison of MAX5C board against one of the most popular ASIC bitcoin mining rigs, Antminer S7, shows a considerable difference between the two (Table 9.12).

9.5.3 Test C

First, we calculated mining profitability of MAX2B and MAX5C boards with a mining profitability calculator BTCServ [19], where electricity price 0.07 € / kWh was taken into account (Figs. 9.34 and 9.35).

Fig. 9.34 A daily loss at bitcoin mining with MAX2B board, calculated with BTCServ calculator [19]

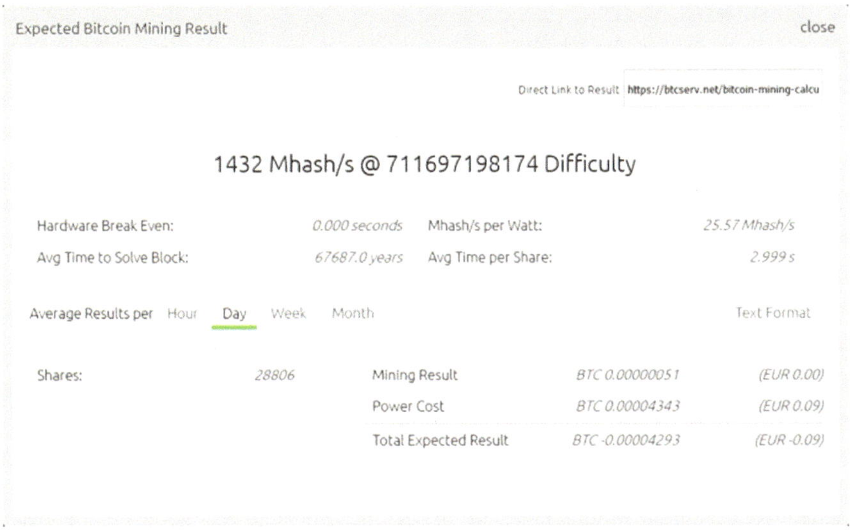

Expected Bitcoin Mining Result
close
Direct Link to Result https://btcserv.net/bitcoin-mining-calcu
1432 Mhash/s @ 711697198174 Difficulty
Hardware Break Even: 0.000 seconds
Mhash/s per Watt: 25.57 Mhash/s
Avg Time to Solve Block: 67687.0 years
Avg Time per Share: 2.999 s
Average Results per Hour Day Week Month
Text Format
Shares: 28806
Mining Result BTC 0.00000051 (EUR 0.00)
Power Cost BTC 0.00004343 (EUR 0.09)
Total Expected Result BTC -0.00004293 (EUR -0.09)

Fig. 9.35 A daily loss at bitcoin mining with MAX5C board, calculated with BTCServ calculator [19]

9.5.3.1 Mining Pool: P2Pool

The currently active mining nodes of decentralized mining pool P2Pool that has implemented a paying scheme PPLNS (Pay Per Last N Shares) can be seen on a few websites [39]. P2Pool is probably the only remaining mining pool that still supports the long deprecated Getwork mining protocol. P2Pool does not require registration on a website unlike most other mining pools; the username is a miner's bitcoin address (e.g., *1PCxQ97Q6dAQ4D5J7gcRp18a9Azikh5Lxs*), and password does not matter (e.g., *anything*). Measured hash rates by P2Pool node are shown in Figs. 9.36 and 9.37.

Miners
1CcnVpexARBLGicKpr9Cfnqzd2pPYMswqA
400MH/s
300MH/s
200MH/s
100MH/s
0.0H/s
Fri 30
03 AM
06 AM
09 AM
12 PM
03 PM
06 PM
09 PM
Total
Mean: 275MH/s
Area: 23.7TH
Dead
Mean: 61.6MH/s
Area: 5.31TH

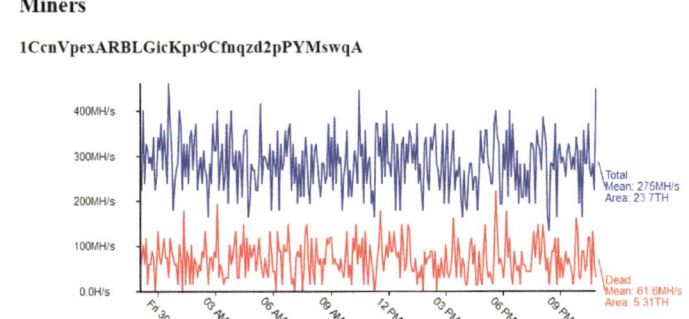

Fig. 9.36 Estimated hash rate of MAX2B board by P2Pool node in 24 h

Miners

1PCxQ97Q6dAQ4D5J7gcRp18a9Azikh5Lxs

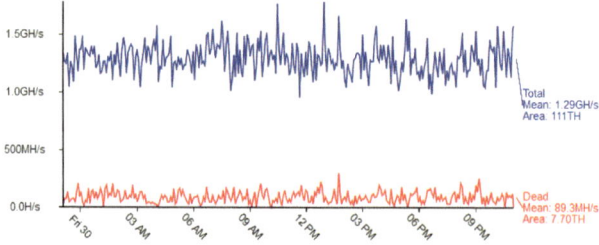

Fig. 9.37 Estimated hash rate of MAX5C board by P2Pool node in 24 h

P2Pool rejects a certain amount of shares produced by bitcoin mining rigs, also known as Dead on Arrival (DoA); in other words, it measures "dead hash rate". The fetched work from the mining nodes often becomes obsolete very soon and therefore the mining rigs with generally low hash rate (in this case MAX2B board) will have relatively higher DoA compared to their total hash rate.

As mentioned before, P2Pool shares rewards using a paying scheme PPLNS where the only miners that provide shares with high difficulty get paid. Unfortunately for us, our dataflow bitcoin miner applications did not find any shares with such high difficulty (around 300,000) to be paid small amounts in bitcoins for the effort; the highest share difficulty found by our bitcoin miners was ca. 3000.

9.5.3.2 Mining Pool: Slush Pool

The mining pool SlushPool requires an account registration on its website (https://slushpool.com/home/). Once registered, the so-called workers for the account have to be created. We created rmeden.MAX2B and rmeden.MAIA workers for our bitcoin mining with MAX2B and MAX5C boards; note the difference in naming convention between the worker and the board, as MAIA was later renamed into MAX5C.

SlushPool does not support Getwork, but it is possible to connect our implemented bitcoin miner that uses Getwork with a mining pool that supports Stratum with the bridging program (stratum-mining-proxy, available on https://github.com/slush0/stratum-mining-proxy); see Fig. 9.38. The implemented bitcoin miner therefore needs an IP address or an URL address of a computer that runs the bridging program and not the URL address of the mining pool as usual. In addition, the bridging program has to be configured to connect to the desired mining pool.

With our dataflow bitcoin miner applications, we would not earn enough to cover the expenses; we were able to mine 18 satoshis per day with MAX2B board and 76 satoshis per day with MAX5C board, which is practically less than 0,01 €. But to be paid in bitcoins by the mining pool, we would have to reach the minimum

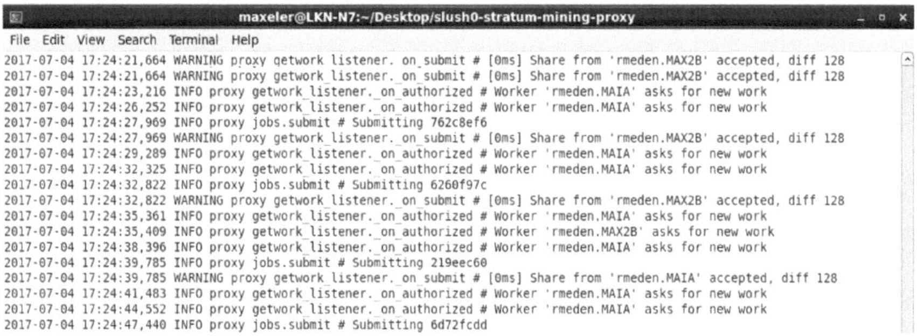

```
[maxeler@LKN-N7:~/Desktop/slush0-stratum-mining-proxy]                    _  □  ×
File  Edit  View  Search  Terminal  Help
2017-07-04 17:24:21,664 WARNING proxy getwork listener. on_submit # [0ms] Share from 'rmeden.MAX2B' accepted, diff 128
2017-07-04 17:24:21,664 WARNING proxy getwork listener. on_submit # [0ms] Share from 'rmeden.MAX2B' accepted, diff 128
2017-07-04 17:24:23,216 INFO proxy getwork listener. on_authorized # Worker 'rmeden.MAIA' asks for new work
2017-07-04 17:24:26,252 INFO proxy getwork listener. on_authorized # Worker 'rmeden.MAIA' asks for new work
2017-07-04 17:24:27,969 INFO proxy jobs.submit # Submitting 762c8ef6
2017-07-04 17:24:27,969 WARNING proxy getwork listener. on_submit # [0ms] Share from 'rmeden.MAX2B' accepted, diff 128
2017-07-04 17:24:29,289 INFO proxy getwork listener. on_authorized # Worker 'rmeden.MAIA' asks for new work
2017-07-04 17:24:32,325 INFO proxy getwork listener. on_authorized # Worker 'rmeden.MAIA' asks for new work
2017-07-04 17:24:32,822 INFO proxy jobs.submit # Submitting 6260f97c
2017-07-04 17:24:32,822 WARNING proxy getwork listener. on_submit # [0ms] Share from 'rmeden.MAX2B' accepted, diff 128
2017-07-04 17:24:35,361 INFO proxy getwork listener. on_authorized # Worker 'rmeden.MAIA' asks for new work
2017-07-04 17:24:35,409 INFO proxy getwork listener. on_authorized # Worker 'rmeden.MAX2B' asks for new work
2017-07-04 17:24:38,396 INFO proxy getwork listener. on_authorized # Worker 'rmeden.MAIA' asks for new work
2017-07-04 17:24:39,785 INFO proxy jobs.submit # Submitting 219eec60
2017-07-04 17:24:39,785 WARNING proxy getwork listener. on_submit # [0ms] Share from 'rmeden.MAIA' accepted, diff 128
2017-07-04 17:24:41,483 INFO proxy getwork listener. on_authorized # Worker 'rmeden.MAIA' asks for new work
2017-07-04 17:24:44,552 INFO proxy getwork listener. on_authorized # Worker 'rmeden.MAIA' asks for new work
2017-07-04 17:24:47,440 INFO proxy jobs.submit # Submitting 6d72fcdd
```

Fig. 9.38 Stratum-mining-proxy program running on MAX2B host workstation with Linux CentOS 6.5 (Python 2.7 is required)

limit (currently at 0.001 BTC or 100,000 satoshis), which would take approximately 21 years with MAX2B board or 5 years with MAX5C board.

9.6 Conclusion

We mentioned the main weakness of the first cryptocurrencies and introduced the first decentralized cryptocurrency bitcoin from the user and technical perspectives. We also described bitcoin mining process and implemented dataflow mining algorithm on Maxeler PCI-e boards, MAX2B, and MAX5C. We tested our dataflow implementations and discussed the results.

MAX5C board appears to be superior compared to the rest of the tested mining hardware solutions, while MAX2B board did not show much improvement. However, nowadays, it is not recommended to use Maxeler dataflow computers for real bitcoin mining as it would result in a loss due to electricity price.

This is because of dominating ASIC bitcoin mining rigs which made bitcoin mining with any general purpose mining hardware (based on CPU, GPU, FPGA) obsolete; the mining difficulty was increased to the levels, where only miners with ASIC mining rigs can make profit with bitcoin mining. Most other miners eventually gave up on mining bitcoins and moved onto mining less popular but more CPU-friendly alternative cryptocurrencies called altcoins, such as litecoin, dogecoin, and ethereum.

By using ASIC mining rigs to mine bitcoins, a centralization of computational power to protect the Bitcoin network occurred, which contradicts the very essence of Bitcoin—the decentralization. Bitcoin mining has become an industry of bitcoin mining farms with several hundreds of ASIC bitcoin mining rigs. It is estimated that at least 75% of all computational power bitcoin mining is owned by Chinese miners due to their cheap mass production of ASIC mining rigs and cheap electricity. Additional problem for miners is the initial reward halving, which occurs on every

210,000 mined blocks, or in other words, in roughly every 3–4 years. This lowers the already low miners' profitability or even turns it into a loss.

It was first believed that altcoins were ASIC-resistant; especially altcoins based on Scrypt proof-of-work (e.g., litecoin and dogecoin). But when enough interest was shown into mining those altcoins, ASIC mining rigs for Scrypt-based cryptocurrencies were produced. Soon after that, the altcoin ethereum became quite popular for mining, because no ASIC mining rigs for mining ether (currency unit of ethereum) were developed yet; mining rigs with several GPUs are heavily used for mining instead. This is a practical proof that ASIC mining rigs can be designed and manufactured for any cryptocurrency, if enough interest is shown, and if their proof-of-work algorithm does not change. ASIC-resistant cryptocurrency is thus only a nicer way to say that cryptocurrency is not yet worth investing into developing ASIC mining rigs to mine it.

FPGA and Maxeler dataflow computers will never be able to compete against ASIC circuits in terms of speed and energy efficiency. However, ASIC bitcoin mining rigs usually become obsolete in a matter of months while FPGA s and Maxeler dataflow computers can be reprogrammed to do other tasks, such as mining other cryptocurrency or run big data analysis.

Acknowledgements The authors acknowledge the financial support from the Slovenian Research Agency (research core funding No. P2-0246). Special thanks goes to Belgrade team at Maxeler Technologies for providing us access to Maxeler MAX5C board for testing purposes.

Appendix 1: BitcoinMinerCpuCode.c

```
// system library headers
#include <math.h>
#include <string.h>
#include <pthread.h>
#include <unistd.h>

// external library headers
#include "curl/curl.h"
#include "json-c/json.h"

// MaxCompiler headers
#include "Maxfiles.h"
#include "MaxSLiCInterface.h"

int errorCount = 0; // counter of connection errors
int acceptCount = 0; // counter of accepted shares
int shareCount = 0; // counter of all shares
```

```
// data structure - HTTP headers
struct curl_slist *headers = NULL;

// data structure - content of HTTP responses
struct responseStruct {
  char *content; // content string
  size_t size; // content size
};

// function - write content of HTTP response into memory;
// adapted from "https://curl.haxx.se/libcurl/c/getinmemory.html"
static size_t getInMemory(void *contents, size_t size, size_t nmemb, void *userp) {

  size_t realsize = size * nmemb;
  struct responseStruct *response = (struct responseStruct *) userp;
  response->content = realloc(response->content, response->size + realsize + 1);

  if (response->content == NULL) {
    fprintf(stderr, "ERROR: Not enough memory\n");
    return 0;
  }

  memcpy(&(response->content[response->size]), contents, realsize);
  response->size += realsize;
  response->content[response->size] = 0;
  return realsize;
}
// function - swap byte order (endianness) of 32-bit words;
// e.g. 0x12345678 into 0x78563412
uint swap(uint x) {
  uint x0 = (x >> 0) & 0xff;
  uint x1 = (x >> 8) & 0xff;
  uint x2 = (x >> 16) & 0xff;
  uint x3 = (x >> 24) & 0xff;
  return x0 << 24 | x1 << 16 | x2 << 8 | x3 << 0;
}

// improvised global semaphore
bool semaphore = true;

// data structure - thread data
struct threadStruct {
  char *url; // URL address of bitcoin mining node
  char *userpwd; // username and password
  char *string; // "data" from work as string
  char *nonce; // swapped nonce as string
};
```

```
// thread function - create and send shares with matching nonces
void *sendShare(void *arg) {

  // copy strings into memory
  struct threadStruct *threadData = (struct threadStruct *) arg;
  char *string = strdup(threadData->string);
  char *nonce = strdup(threadData->nonce);

  // let the main thread continue after copying strings into memory
  semaphore = true;

  // initialize CURL handle and CURL status code
  CURL *CURL;
  CURLcode CURLcode;
  CURL = curl_easy_init();

  // if CURL handle initialization succeeds, continue
  if (CURL) {

    // write a matching nonce into a string
    for (int i = 0; i < 8; i++) {
      string[153 + i] = nonce[i];
    }

    // expand string into a share
    char share[303];
    const char *share1 = "{\"method\": \"getwork\", \"params\": [";
    const char *share2 = "], \"id\": 1}";

    strcpy(share, share1);
    strcat(share, string);
    strcat(share, share2);

    // allocate memory for the content of HTTP response
    struct responseStruct response;
    response.content = malloc(1);
    response.size = 0;

    // create a HTTP request to send share
    curl_easy_setopt(CURL, CURLOPT_CUSTOMREQUEST, "POST");
    curl_easy_setopt(CURL, CURLOPT_URL, threadData->url);
    curl_easy_setopt(CURL, CURLOPT_USERPWD, threadData->userpwd);
    curl_easy_setopt(CURL, CURLOPT_USERAGENT, "libcurl-agent/1.0");
    curl_easy_setopt(CURL, CURLOPT_NOSIGNAL, 1);
    curl_easy_setopt(CURL, CURLOPT_TIMEOUT, 3);
    curl_easy_setopt(CURL, CURLOPT_HTTPHEADER, headers);
    curl_easy_setopt(CURL, CURLOPT_POSTFIELDS, share);
    curl_easy_setopt(CURL, CURLOPT_WRITEFUNCTION, getInMemory);
    curl_easy_setopt(CURL, CURLOPT_WRITEDATA, (void *) &response);
```

```c
// get the starting time snapshot of submitting share
struct timeval start, stop, diff;
gettimeofday(&start, NULL);

// send share to the mining node
CURLcode = curl_easy_perform(CURL);
curl_easy_cleanup(CURL);

// if sending share succeeds, continue
if (CURLcode == CURLE_OK) {

  // parse the content of HTTP response
  struct json_object *json, *jsonResult;
  json = json_tokener_parse(response.content);
  json_object_object_get_ex(json, "result", &jsonResult);

  // check if share was accepted ...
  if (strcmp(json_object_to_json_string(jsonResult), "true") == 0) {
    acceptCount++;
    shareCount++;
    printf("share ACCEPTED! ... ");
  }

  // ... or rejected
  else if (strcmp(json_object_to_json_string(jsonResult), "false") == 0) {
    shareCount++;
    printf("share REJECTED! ... ");
  }

  // print out share statistics
  printf("%d / %d (%.0Lf%%) ... ", acceptCount, shareCount,
    (long double) acceptCount / (long double) shareCount * 100);

  // calculate and print out a delay between submitting share and receiving response
  gettimeofday(&stop, NULL);
  timersub(&stop, &start, &diff);
  long double delay = (long double) diff.tv_sec + (long double) diff.tv_usec / 1000000;
  printf("delay %.0Lf ms\n", delay * 1000);

  // free memory
  json_object_put(json);

}
  // if sending share fails, print out error
  else fprintf(stderr, "Share not sent\n");

  // free memory
  free(response.content);

}

// if CURL handle initialization fails, print out error
else {
  fprintf(stderr, "ERROR: Trouble with CURL library\n");
}

// free memory
free(string);
free(nonce);

// end of thread function
return NULL;

}
```

```c
// function - display instructions to run program
void use() {
  puts("Required parameters:");
  puts("   -o: a bitcoin mining node URL address");
  puts("   -u: a bitcoin miner's username");
  puts("   -p: a bitcoin miner's password");
  puts("Optional parameters:");
  puts("   -h: display this help message and exit");
  puts("Command example:");
  puts("./BitcoinMiner -o http://12.34.56.78:9332 -u username -p password");
}

// main function - accept user inputs;
// -o mining node URL address -u username -p password || -h help (optional)
int main (int argc, char **argv) {

  char *url = NULL;
  char *user = NULL;
  char *pwd = NULL;
  int c;

  // check for user inputs
  while ((c = getopt (argc, argv, "ho:u:p:")) != -1) {
    switch (c) {
      case 'h':
        use();
        return 0;
      case 'o':
        url = optarg;
        break;
      case 'u':
        user = optarg;
        break;
      case 'p':
        pwd = optarg;
        break;
      default:
        use();
        abort();
    }
  }

  // if user inputs are missing, warn user and terminate program
  if (url == NULL || user == NULL || pwd == NULL) {
    use();
    return 0;
  }
```

```
// otherwise continue program
int burstSize = 16; // CPU stream must be a multiple of 16 bytes
int intervals = 16; // number of hashing intervals per block header
int pipelines = BitcoinMiner_pipelines; // number of pipelines in kernel

// round up the size of dummy CPU output stream
uint64_t size = pow(2, 32) / intervals / pipelines;
while (size % burstSize != 0) {
  ++size;
}

// allocate memory for dummy CPU output stream
uint8_t *output = malloc(size * sizeof(uint8_t));

// allocate memory for mapped RAM (pairs of matches and nonces)
uint64_t mappedRam[pipelines][8];
for (int p = 0; p < pipelines; p++) {
  for (int r = 0; r < 8; r++) {
    mappedRam[p][r] = 0;
  }
}

// add headers of HTTP requests to an existing data structure
headers = curl_slist_append(headers, "Accept: application/json");
headers = curl_slist_append(headers, "Content-Type: application/json");

// initialize thread identifier
pthread_t pth;

// concatenate username and password into one string
char userpwd[256];
char *colon = ":";
strcpy(userpwd, user);
strcat(userpwd, colon);
strcat(userpwd, pwd);

// add user inputs to thread data
struct threadStruct threadData;
threadData.url = url;
threadData.userpwd = userpwd;

// initialize CURL handle and CURL status code
CURL *CURL;
CURLcode CURLcode;

// data structure - content of HTTP response
struct responseStruct response;

// a part of HTTP request to fetch work for mining
const char *getwork = "{\"method\": \"getwork\", \"params\": [], \"id\": 1}";

// data structure - json objects for parsing contents of HTTP responses
struct json_object *json, *jsonResult, *jsonData, *jsonMidstate, *jsonTarget;

// strings to parse work for bitcoin miner
char *dataString = "";
char *midsString = "";
char *targString = "";

// arrays to prepare work for bitcoin miner
char temp[9];
uint data[3];
uint midstate[8];
uint target[8];
```

```
// data structure - time snapshots
struct timeval start, stop, diff;

// nonce-related variables
uint base;
uint nonce;
char nonceString[8];

// statistic variables - elapsed time and hash rate
long double hashTime;
long double hashRate;

// announce the start of bitcoin miner
puts("");
puts("Starting Bitcoin Miner ...");
puts("");

// program will run in infinite loop
while(1) {

  // initialize CURL handle
  CURL = curl_easy_init();

  // if CURL handle initialization succeeds, continue
  if (CURL) {

    // allocate memory for the content of HTTP response
    response.content = malloc(1);
    response.size = 0;

    // create a HTTP request to fetch work for bitcoin miner
    curl_easy_setopt(CURL, CURLOPT_CUSTOMREQUEST, "POST");
    curl_easy_setopt(CURL, CURLOPT_URL, url);
    curl_easy_setopt(CURL, CURLOPT_USERPWD, userpwd);
    curl_easy_setopt(CURL, CURLOPT_USERAGENT, "libcurl-agent/1.0");
    curl_easy_setopt(CURL, CURLOPT_NOSIGNAL, 1);
    curl_easy_setopt(CURL, CURLOPT_TIMEOUT, 3);
    curl_easy_setopt(CURL, CURLOPT_HTTPHEADER, headers);
    curl_easy_setopt(CURL, CURLOPT_POSTFIELDS, getwork);
    curl_easy_setopt(CURL, CURLOPT_WRITEFUNCTION, getInMemory);
    curl_easy_setopt(CURL, CURLOPT_WRITEDATA, (void *) &response);

    // send HTTP request to the mining node
    CURLcode = curl_easy_perform(CURL);
    curl_easy_cleanup(CURL);

    // if fetching work succeeds, continue
    if (CURLcode == CURLE_OK) {

      // reset counter of connection errors
      errorCount = 0;
```

```
// parse the content of HTTP response (work for bitcoin miner)
json = json_tokener_parse(response.content);
json_object_object_get_ex(json, "result", &jsonResult);
json_object_object_get_ex(jsonResult, "data", &jsonData);
json_object_object_get_ex(jsonResult, "midstate", &jsonMidstate);
json_object_object_get_ex(jsonResult, "target", &jsonTarget);

dataString = (char *) json_object_to_json_string(jsonData);
midsString = (char *) json_object_to_json_string(jsonMidstate);
targString = (char *) json_object_to_json_string(jsonTarget);

// convert the content of HTTP response into numbers
for (int i = 0; i < 3; i++) {
  strncpy(temp, dataString + 129 + i * 8, 8);
  temp[8] = '\0'; // null-terminator must be added for a proper conversion
  data[i] = swap((uint) strtoul(temp, 0, 16));
}

for (int i = 0; i < 8; i++) {
  strncpy(temp, midsString + 1 + i * 8, 8);
  temp[8] = '\0'; // null-terminator must be added for a proper conversion
  midstate[i] = swap((uint) strtoul(temp, 0, 16));
}

for (int i = 0; i < 8; i++) {
  strncpy(temp, targString + 1 + i * 8, 8);
  temp[8] = '\0'; // null-terminator must be added for a proper conversion
  target[7-i] = swap((uint) strtoul(temp, 0, 16));
}

// get the starting time snapshot of DFE run
gettimeofday(&start, NULL);

// DFE will run for several hashing intervals
for (int i = 0; i < intervals; i++) {

  // nonce base updates after each DFE run
  base = i * pipelines * (uint) size;

  // SLiC interface
  BitcoinMiner(size, base, data, midstate, target, output,
    mappedRam[0], mappedRam[0], mappedRam[1], mappedRam[1], mappedRam[2],
    mappedRam[2]);

  // check for matching nonces after each DFE hashing interval
    for (int p = 0; p < pipelines; p++) {
      for (int r = 0; r < 8; r += 2) {

        // if match indicator is positive ...
        if (mappedRam[p][r] == 1) {

          // wait for any existing threads to finish copying strings into memory
          while(!semaphore) { }
          semaphore = false;

          // prepare strings for a thread function
          threadData.string = dataString;
          nonce = (uint) mappedRam[p][r+1];
          sprintf(nonceString, "%08x", swap(nonce));
          threadData.nonce = nonceString;

          // create and detach a thread to handle sending share independently
          pthread_create(&pth, NULL, sendShare, (void *) &threadData);
          pthread_detach(pth);
```

```
      // reset contents of current mapped RAM instance
      mappedRam[p][r] = 0;
      mappedRam[p][r+1] = 0;

    }

   }

  }

}
// get the ending time snapshot of DFE run
gettimeofday(&stop, NULL);

// calculate and print out DFE hash rate
timersub(&stop, &start, &diff);
hashTime = (long double) diff.tv_sec + (long double) diff.tv_usec / 1000000;
hashRate = pow(2, 32) / hashTime / 1000000;
printf("%.0Lf Mhash/s\n", hashRate);

// wait for any existing threads to finish copying strings into memory
while(!semaphore) { }

// free memory
json_object_put(json);

}
// if fetching work fails, print out instructions for user ...
else if (CURLcode == CURLE_URL_MALFORMAT ||
 CURLcode == CURLE_COULDNT_RESOLVE_HOST) {
 puts("Please specify a valid bitcoin mining node's URL address (-o http://ip:port)");
 return 0;
}
// ... or print out errors
else {
    errorCount++;
    if (errorCount == 1) fprintf(stderr, "Work not fetched\n");
    else if (errorCount == 100) fprintf(stderr, "Trouble with the chosen bitcoin mining
node\n");
    }

   // free memory
   free(response.content);

  }

  // if CURL handle initialization fails, print out error
  else {
   fprintf(stderr, "ERROR: Trouble with CURL library\n");
  }

 }
 // end of infinite loop, shall never be reached
 free(output);
 return 0;

}
```

Appendix 2: BitcoinMinerEngineParameters.maxj

```java
package bitcoinminer;

import com.maxeler.maxcompiler.v2.build.EngineParameters;

public class BitcoinMinerEngineParameters extends EngineParameters {

    public BitcoinMinerEngineParameters(String[] argv) {
        super(argv);
    }

    private static final String frequency = "frequency";
    private static final String pipelines = "pipelines";
    private static final String useGlobalClockBuffer = "useGlobalClockBuffer";

    public int getFrequency() {
        return getParam(frequency);
    }

    public int getPipelines() {
        return getParam(pipelines);
    }

    public Boolean getUseGlobalClockBuffer() {
        return getParam(useGlobalClockBuffer);
    }

    @Override
    protected void declarations() {
        declareParam(frequency, DataType.INT, 95);
        declareParam(pipelines, DataType.INT, 3);
        declareParam(useGlobalClockBuffer, DataType.BOOL, true);
    }

    @Override
    protected void validate() {

    }

    @Override
    public String getBuildName() {
        return getMaxFileName() + "_" + getDFEModel() + "_" + getTarget();
    }

}
```

Appendix 3: BitcoinMinerKernel.maxj

```
package bitcoinminer;

import com.maxeler.maxcompiler.v2.kernelcompiler.Kernel;
import com.maxeler.maxcompiler.v2.kernelcompiler.KernelParameters;
import com.maxeler.maxcompiler.v2.kernelcompiler.stdlib.core.Count;
import com.maxeler.maxcompiler.v2.kernelcompiler.stdlib.core.Count.Counter;
import com.maxeler.maxcompiler.v2.kernelcompiler.stdlib.core.Mem.RamWriteMode;
import com.maxeler.maxcompiler.v2.kernelcompiler.stdlib.memory.Memory;
import com.maxeler.maxcompiler.v2.kernelcompiler.types.base.DFEVar;
import com.maxeler.maxcompiler.v2.utils.Bits;

class BitcoinMinerKernel extends Kernel {

    // SHA-256 constants
    int H[] = { 0x6a09e667, 0xbb67ae85, 0x3c6ef372, 0xa54ff53a, 0x510e527f, 0x9b05688c,
        0x1f83d9ab, 0x5be0cd19 };
    int K[] = { 0x428a2f98, 0x71374491, 0xb5c0fbcf, 0xe9b5dba5, 0x3956c25b, 0x59f111f1,
        0x923f82a4, 0xab1c5ed5, 0xd807aa98, 0x12835b01, 0x243185be, 0x550c7dc3,
        0x72be5d74, 0x80deb1fe, 0x9bdc06a7, 0xc19bf174, 0xe49b69c1, 0xefbe4786,
        0x0fc19dc6, 0x240ca1cc, 0x2de92c6f, 0x4a7484aa, 0x5cb0a9dc, 0x76f988da,
        0x983e5152, 0xa831c66d, 0xb00327c8, 0xbf597fc7, 0xc6e00bf3, 0xd5a79147,
        0x06ca6351, 0x14292967, 0x27b70a85, 0x2e1b2138, 0x4d2c6dfc, 0x53380d13,
        0x650a7354, 0x766a0abb, 0x81c2c92e, 0x92722c85, 0xa2bfe8a1, 0xa81a664b,
        0xc24b8b70, 0xc76c51a3, 0xd192e819, 0xd6990624, 0xf40e3585, 0x106aa070,
        0x19a4c116, 0x1e376c08, 0x2748774c, 0x34b0bcb5, 0x391c0cb3, 0x4ed8aa4a,
        0x5b9cca4f, 0x682e6ff3, 0x748f82ee, 0x78a5636f, 0x84c87814, 0x8cc70208,
        0x90befffa, 0xa4506ceb, 0xbef9a3f7, 0xc67178f2 };

    // function to swap byte order of 32-bit words; e.g. 0x12345678 into 0x78563412
    DFEVar swap(DFEVar x) {
        DFEVar x0 = x.shiftRight(0) & 0xff;
        DFEVar x1 = x.shiftRight(8) & 0xff;
        DFEVar x2 = x.shiftRight(16) & 0xff;
        DFEVar x3 = x.shiftRight(24) & 0xff;
        return x0.shiftLeft(24) | x1.shiftLeft(16) | x2.shiftLeft(8) | x3.shiftLeft(0);
    }

    // function to workaround addition with modulo 2^32
    DFEVar modulo(DFEVar x) {
        return x.slice(0, 32).cast(dfeUInt(32));
    }
```

```
// function to cast the type of constants (INT32 to UINT32) using bits layout
DFEVar uint32(int n) {
   Bits bits = new Bits(32);
   bits.setBits(n);
   return constant.var(dfeUInt(32), bits);
}

// SHA-256 logic functions
DFEVar ch(DFEVar x, DFEVar y, DFEVar z) {
   optimization.pushPipeliningFactor(0);
   DFEVar ch = (x & y) ^ (~x & z);
   optimization.popPipeliningFactor();
   return ch;
}

DFEVar maj(DFEVar x, DFEVar y, DFEVar z) {
   optimization.pushPipeliningFactor(0);
   DFEVar maj = (x & y) ^ (x & z) ^ (y & z);
   optimization.popPipeliningFactor();
   return maj;
}

DFEVar Sigma0(DFEVar x) {
   optimization.pushPipeliningFactor(0);
   DFEVar Sigma0 = x.rotateRight(2) ^ x.rotateRight(13) ^ x.rotateRight(22);
   optimization.popPipeliningFactor();
   return Sigma0;
}

DFEVar Sigma1(DFEVar x) {
   optimization.pushPipeliningFactor(0);
   DFEVar Sigma1 = x.rotateRight(6) ^ x.rotateRight(11) ^ x.rotateRight(25);
   optimization.popPipeliningFactor();
   return Sigma1;
}

DFEVar sigma0(DFEVar x) {
   optimization.pushPipeliningFactor(0);
   DFEVar sigma0 = x.rotateRight(7) ^ x.rotateRight(18) ^ x.shiftRight(3);
   optimization.popPipeliningFactor();
   return sigma0;
}

DFEVar sigma1(DFEVar x) {
   optimization.pushPipeliningFactor(0);
   DFEVar sigma1 = x.rotateRight(17) ^ x.rotateRight(19) ^ x.shiftRight(10);
   optimization.popPipeliningFactor();
   return sigma1;
}

// bitcoin miner kernel
protected BitcoinMinerKernel
(KernelParameters kernelParams, BitcoinMinerEngineParameters params) {
   super(kernelParams);
```

```
// dummy output variable
DFEVar output = constant.var(dfeUInt(8), 0);

// work for bitcoin miner (scalar inputs)
DFEVar[] data = new DFEVar[3];
data[0] = io.scalarInput("data0", dfeUInt(32));
data[1] = io.scalarInput("data1", dfeUInt(32));
data[2] = io.scalarInput("data2", dfeUInt(32));

DFEVar[] midstate = new DFEVar[8];
midstate[0] = io.scalarInput("midstate0", dfeUInt(32));
midstate[1] = io.scalarInput("midstate1", dfeUInt(32));
midstate[2] = io.scalarInput("midstate2", dfeUInt(32));
midstate[3] = io.scalarInput("midstate3", dfeUInt(32));
midstate[4] = io.scalarInput("midstate4", dfeUInt(32));
midstate[5] = io.scalarInput("midstate5", dfeUInt(32));
midstate[6] = io.scalarInput("midstate6", dfeUInt(32));
midstate[7] = io.scalarInput("midstate7", dfeUInt(32));

DFEVar[] target = new DFEVar[8];
target[0] = io.scalarInput("target0", dfeUInt(32));
target[1] = io.scalarInput("target1", dfeUInt(32));
target[2] = io.scalarInput("target2", dfeUInt(32));
target[3] = io.scalarInput("target3", dfeUInt(32));
target[4] = io.scalarInput("target4", dfeUInt(32));
target[5] = io.scalarInput("target5", dfeUInt(32));
target[6] = io.scalarInput("target6", dfeUInt(32));
target[7] = io.scalarInput("target7", dfeUInt(32));

// concatenate target components into a 256-bit target threshold
DFEVar targetCompare = target[0].cat(target[1]).cat(target[2]).cat(target[3])
    .cat(target[4]).cat(target[5]).cat(target[6]).cat(target[7]).cast(dfeUInt(256));

// count ticks in kernel
DFEVar ticks = control.count.simpleCounter(32);

// parallelize hashing
for (int p = 0; p < params.getPipelines(); p++) {

    // group nodes for pipelines
    pushGroup("Group" + p);

    // nonce candidates for each pipeline
    DFEVar nonce = ticks * params.getPipelines() + p;

    // SHA-256, round 1
    DFEVar[] W = new DFEVar[64];
    for (int i = 0; i <= 2; i++) {
        W[i] = data[i];
    }

    W[3] = nonce;
    W[4] = uint32(0x80000000);
```

```
for (int i = 5; i <= 14; i++) {
   W[i] = uint32(0x00000000);
}

W[15] = uint32(0x00000280);

for (int j = 16; j <= 63; j++) {
   W[j] = (sigma1(W[j-2]) + W[j-7] + sigma0(W[j-15]) + W[j-16]);
}

DFEVar a = midstate[0];
DFEVar b = midstate[1];
DFEVar c = midstate[2];
DFEVar d = midstate[3];
DFEVar e = midstate[4];
DFEVar f = midstate[5];
DFEVar g = midstate[6];
DFEVar h = midstate[7];

for (int j = 0; j < 64; j++) {
optimization.pushEnableBitGrowth(true);
DFEVar hh = modulo(h + uint32(K[j]) + W[j]);
optimization.pushPipeliningFactor(0);
DFEVar T1 = modulo(hh + Sigma1(e) + ch(e, f, g));
DFEVar T2 = modulo(Sigma0(a) + maj(a, b, c));
optimization.popPipeliningFactor();
optimization.popEnableBitGrowth();
h = g;
g = f;
f = e;
e = d + T1;
d = c;
c = b;
b = a;
a = T1 + T2;
}
DFEVar[] hash = new DFEVar[8];
hash[0] = midstate[0] + a;
hash[1] = midstate[1] + b;
hash[2] = midstate[2] + c;
hash[3] = midstate[3] + d;
hash[4] = midstate[4] + e;
hash[5] = midstate[5] + f;
hash[6] = midstate[6] + g;
hash[7] = midstate[7] + h;

// SHA-256, round 2
for (int j = 0; j <= 7; j++) {
   W[j] = hash[j];
}

W[8] = uint32(0x80000000);

for (int j = 9; j <= 14; j++) {
   W[j] = uint32(0x00000000);
}

W[15] = uint32(0x00000100);

for (int j = 16; j <= 63; j++) {
   W[j] = (sigma1(W[j-2]) + W[j-7] + sigma0(W[j-15]) + W[j-16]);
}
```

```
a = uint32(H[0]);
b = uint32(H[1]);
c = uint32(H[2]);
d = uint32(H[3]);
e = uint32(H[4]);
f = uint32(H[5]);
g = uint32(H[6]);
h = uint32(H[7]);

for (int j = 0; j < 64; j++) {
    optimization.pushEnableBitGrowth(true);
    DFEVar hh = modulo(h + uint32(K[j]) + W[j]);
    optimization.pushPipeliningFactor(0);
    DFEVar T1 = modulo(hh + Sigma1(e) + ch(e, f, g));
    DFEVar T2 = modulo(Sigma0(a) + maj(a, b, c));
    optimization.popPipeliningFactor();
    optimization.popEnableBitGrowth();
    h = g;
    g = f;
    f = e;
    e = d + T1;
    d = c;
    c = b;
    b = a;
    a = T1 + T2;
}

    hash[0] = uint32(H[0]) + a;
    hash[1] = uint32(H[1]) + b;
    hash[2] = uint32(H[2]) + c;
    hash[3] = uint32(H[3]) + d;
    hash[4] = uint32(H[4]) + e;
    hash[5] = uint32(H[5]) + f;
    hash[6] = uint32(H[6]) + g;
    hash[7] = uint32(H[7]) + h;

// swap byte order and concatenate double hash components
// into a 256-bit double hash
DFEVar hashCompare = swap(hash[7]).cat(swap(hash[6])).cat(swap(hash[5]))
    .cat(swap(hash[4])).cat(swap(hash[3])).cat(swap(hash[2])).cat(swap(hash[1]))
    .cat(swap(hash[0])).cast(dfeUInt(256));

// compare 256-bit double hash against the 256-bit target threshold
DFEVar match = hashCompare < targetCompare ? constant.var(dfeUInt(1), 1) : 0;

// mapped memory is set to hold up to 8 values; 4 pairs of matches and nonces
Memory<DFEVar> mappedRam = mem.alloc(dfeUInt(32), 8);
mappedRam.mapToCPU("mappedRam" + p);

// mapped memory writing addresses are controlled by an advanced counter
Count.Params addressParams =
    control.count.makeParams(3).withInc(2).withMax(8).withEnable(match);
Counter addressCount = control.count.makeCounter(addressParams);
DFEVar address = addressCount.getCount();
```

```
      // write pairs of matches and nonces into mapped memory
      mappedRam.port
          (address, match.cast(dfeUInt(32)), match, RamWriteMode.WRITE_FIRST);
      mappedRam.port(address + 1, nonce, match, RamWriteMode.WRITE_FIRST);

      // dummy output variable interacting with pipelined results
      output += match.cast(dfeUInt(8));

      // grouping nodes for pipelines ends here
      popGroup();
    }

    // output stream
    io.output("output", output, dfeUInt(8));
  }

}
```

Appendix 4: BitcoinMinerManager.maxj

```
package bitcoinminer;

import com.maxeler.maxcompiler.v2.kernelcompiler.Kernel;
import com.maxeler.maxcompiler.v2.kernelcompiler.KernelConfiguration;
import com.maxeler.maxcompiler.v2.managers.BuildConfig;
import com.maxeler.maxcompiler.v2.managers.engine_interfaces.CPUTypes;
import com.maxeler.maxcompiler.v2.managers.engine_interfaces.EngineInterface;
import com.maxeler.maxcompiler.v2.managers.engine_interfaces.InterfaceParam;
import com.maxeler.maxcompiler.v2.managers.engine_interfaces.InterfaceParamArray;
import com.maxeler.maxcompiler.v2.managers.standard.Manager;
import com.maxeler.maxcompiler.v2.managers.standard.Manager.IOType;

public class BitcoinMinerManager {

  public static void main(String[] args) {
    BitcoinMinerEngineParameters params = new BitcoinMinerEngineParameters(args);
    Manager manager = new Manager(params);

    KernelConfiguration kernelConfig = manager.getCurrentKernelConfig();
    kernelConfig.optimization.setUseGlobalClockBuffer(params.getUseGlobalClockBuffer());
    Kernel kernel = new BitcoinMinerKernel
    (manager.makeKernelParameters("BitcoinMinerKernel"), params);

    manager.setKernel(kernel);
    manager.setIO(IOType.ALL_CPU);
    manager.setClockFrequency(params.getFrequency());
    manager.createSLiCinterface(interfaceDefault());
    manager.addMaxFileConstant("pipelines", params.getPipelines());
```

```
      buildConfig(manager, params);
      manager.build();
   }

   private static EngineInterface interfaceDefault() {
      EngineInterface ei = new EngineInterface();

      InterfaceParam N = ei.addParam("N", CPUTypes.UINT64);
      InterfaceParamArray data = ei.addParamArray("data", CPUTypes.UINT32);
      InterfaceParamArray midstate = ei.addParamArray("midstate", CPUTypes.UINT32);
      InterfaceParamArray target = ei.addParamArray("target", CPUTypes.UINT32);
      ei.setScalar("BitcoinMinerKernel", "data0", data[0]);
      ei.setScalar("BitcoinMinerKernel", "data1", data[1]);
      ei.setScalar("BitcoinMinerKernel", "data2", data[2]);
      ei.setScalar("BitcoinMinerKernel", "midstate0", midstate[0]);
      ei.setScalar("BitcoinMinerKernel", "midstate1", midstate[1]);
      ei.setScalar("BitcoinMinerKernel", "midstate2", midstate[2]);
      ei.setScalar("BitcoinMinerKernel", "midstate3", midstate[3]);
      ei.setScalar("BitcoinMinerKernel", "midstate4", midstate[4]);
      ei.setScalar("BitcoinMinerKernel", "midstate5", midstate[5]);
      ei.setScalar("BitcoinMinerKernel", "midstate6", midstate[6]);
      ei.setScalar("BitcoinMinerKernel", "midstate7", midstate[7]);
      ei.setScalar("BitcoinMinerKernel", "target0", target[0]);
      ei.setScalar("BitcoinMinerKernel", "target1", target[1]);
      ei.setScalar("BitcoinMinerKernel", "target2", target[2]);
      ei.setScalar("BitcoinMinerKernel", "target3", target[3]);
      ei.setScalar("BitcoinMinerKernel", "target4", target[4]);
      ei.setScalar("BitcoinMinerKernel", "target5", target[5]);
      ei.setScalar("BitcoinMinerKernel", "target6", target[6]);
      ei.setScalar("BitcoinMinerKernel", "target7", target[7]);

      ei.setTicks("BitcoinMinerKernel", N);
      ei.setStream("output", CPUTypes.UINT8, N * CPUTypes.UINT8.sizeInBytes());

      return ei;
   }

   private static void buildConfig
   (Manager manager, BitcoinMinerEngineParameters params) {
      BuildConfig buildConfig = manager.getBuildConfig();
      buildConfig.setMPPRCostTableSearchRange
        (params.getMPPRStartCT(), params.getMPPREndCT());
      buildConfig.setMPPRParallelism(params.getMPPRThreads());
      buildConfig.setMPPRRetryNearMissesThreshold(params.getMPPRRetryThreshold());
   }

}
```

Appendix 5: SHA256-CPU.c

```c
#include <math.h>
#include <stdio.h>
#include <stdlib.h>

// SHA-256 constants
unsigned int H[8] = { 0x6a09e667, 0xbb67ae85, 0x3c6ef372, 0xa54ff53a, 0x510e527f,
    0x9b05688c, 0x1f83d9ab, 0x5be0cd19 };
unsigned int K[64] = { 0x428a2f98, 0x71374491, 0xb5c0fbcf, 0xe9b5dba5, 0x3956c25b,
    0x59f111f1, 0x923f82a4, 0xab1c5ed5, 0xd807aa98, 0x12835b01, 0x243185be,
    0x550c7dc3, 0x72be5d74, 0x80deb1fe, 0x9bdc06a7, 0xc19bf174, 0xe49b69c1,
    0xefbe4786, 0x0fc19dc6, 0x240ca1cc, 0x2de92c6f, 0x4a7484aa, 0x5cb0a9dc,
    0x76f988da, 0x983e5152, 0xa831c66d, 0xb00327c8, 0xbf597fc7, 0xc6e00bf3,
    0xd5a79147, 0x06ca6351, 0x14292967, 0x27b70a85, 0x2e1b2138, 0x4d2c6dfc,
    0x53380d13, 0x650a7354, 0x766a0abb, 0x81c2c92e, 0x92722c85, 0xa2bfe8a1,
    0xa81a664b, 0xc24b8b70, 0xc76c51a3, 0xd192e819, 0xd6990624, 0xf40e3585,
    0x106aa070, 0x19a4c116, 0x1e376c08, 0x2748774c, 0x34b0bcb5, 0x391c0cb3,
    0x4ed8aa4a, 0x5b9cca4f, 0x682e6ff3, 0x748f82ee, 0x78a5636f, 0x84c87814,
    0x8cc70208, 0x90befffa, 0xa4506ceb, 0xbef9a3f7, 0xc67178f2 };

// SHA-256 logic functions
#define shift(x, n) (x >> n)
#define rotate(x, n) ((x >> n) | (x << (32-n)))
#define Ch(x, y, z) ((x & y) ^ (~x & z))
#define Maj(x, y, z) ((x & y) ^ (x & z) ^ (y & z))
#define Sigma0(x) (rotate(x, 2) ^ rotate(x, 13) ^ rotate(x, 22))
#define Sigma1(x) (rotate(x, 6) ^ rotate(x, 11) ^ rotate(x, 25))
#define sigma0(x) (rotate(x, 7) ^ rotate(x, 18) ^ shift(x, 3))
#define sigma1(x) (rotate(x, 17) ^ rotate(x, 19) ^ shift(x, 10))

// main function
int main(void) {

    // define working variables and arrays
    unsigned int a, b, c, d, e, f, g, h, T1, T2;
    unsigned int midstate[8];
    unsigned int hash[8];
    unsigned int W[64];

// padded block header example which is split into two message blocks, M1 and M2
unsigned int M1[] = { 0x00000020, 0x797ba3f1, 0x380adb86, 0x08e22e83, 0x26f5e0f1,
    0x1dff5b36, 0x66d0e504, 0x00000000, 0x00000000, 0xb2f82fcc, 0xac1dd59b,
    0xbc6dbf0a, 0xf191fb41, 0x59ddbef4, 0x6df5e075, 0x718db470 };
unsigned int M2[] = { 0x44977942, 0x29a0cd57, 0x08fb0418, 0x00000000, 0x80000000,
    0x00000000, 0x00000000, 0x00000000, 0x00000000, 0x00000000, 0x00000000,
    0x00000000, 0x00000000, 0x00000000, 0x00000000, 0x00000280 };

// process first message block, M1, to calculate "midstate" (only once)
// schedule message (first message block)
for (int j = 0; j < 16; j++) {
    W[j] = M1[j];
}

for (int j = 16; j < 64; j++) {
    W[j] = sigma1(W[j-2]) + W[j-7] + sigma0(W[j-15]) + W[j-16];
}
```

```
// declare and set working variables to initial H values
a = H[0];
b = H[1];
c = H[2];
d = H[3];
e = H[4];
f = H[5];
g = H[6];
h = H[7];
```

```
// apply SHA-256 compression function
for (int j = 0; j < 64; j++) {
    T1 = h + Sigma1(e) + Ch(e, f, g) + K[j] + W[j];
    T2 = Sigma0(a) + Maj(a, b, c);
    h = g;
    g = f;
    f = e;
    e = d + T1;
    d = c;
    c = b;
    b = a;
    a = T1 + T2;
}
```

```
// calculate intermediate hash known as "midstate"
midstate[0] = H[0] + a;
midstate[1] = H[1] + b;
midstate[2] = H[2] + c;
midstate[3] = H[3] + d;
midstate[4] = H[4] + e;
midstate[5] = H[5] + f;
midstate[6] = H[6] + g;
midstate[7] = H[7] + h;
```

```
// process second message block, M2, where nonce is being changed
for (unsigned int nonce = 0; nonce <= 0xffffffff; nonce++) {

    M2[3] = nonce;

    for (int j = 0; j < 16; j++) {
        W[j] = M2[j];
    }

    for (int j = 16; j < 64; j++) {
        W[j] = sigma1(W[j-2]) + W[j-7] + sigma0(W[j-15]) + W[j-16];
    }

    a = midstate[0];
    b = midstate[1];
    c = midstate[2];
    d = midstate[3];
    e = midstate[4];
    f = midstate[5];
    g = midstate[6];
    h = midstate[7];
```

```
for (int j = 0; j < 64; j++) {
T1 = h + Sigma1(e) + Ch(e, f, g) + K[j] + W[j];
T2 = Sigma0(a) + Maj(a, b, c);
h = g;
g = f;
f = e;
e = d + T1;
d = c;
c = b;
b = a;
a = T1 + T2;
}

hash[0] = midstate[0] + a;
hash[1] = midstate[1] + b;
hash[2] = midstate[2] + c;
hash[3] = midstate[3] + d;
hash[4] = midstate[4] + e;
hash[5] = midstate[5] + f;
hash[6] = midstate[6] + g;
hash[7] = midstate[7] + h;
```

```
// first SHA-256 hashing is complete,
// hash has to be padded for second SHA-256 hashing
unsigned int M3[] = { hash[0], hash[1], hash[2], hash[3], hash[4], hash[5], hash[6],
   hash[7], 0x80000000, 0, 0, 0, 0, 0, 0, 0x100 };

for (int j = 0; j < 16; j++) {
   W[j] = M3[j];
}

for (int j = 16; j < 64; j++) {
   W[j] = sigma1(W[j-2]) + W[j-7] + sigma0(W[j-15]) + W[j-16];
}

a = H[0];
b = H[1];
c = H[2];
d = H[3];
e = H[4];
f = H[5];
g = H[6];
h = H[7];

for (int j = 0; j < 64; j++) {
   T1 = h + Sigma1(e) + Ch(e, f, g) + K[j] + W[j];
   T2 = Sigma0(a) + Maj(a, b, c);
   h = g;
   g = f;
   f = e;
   e = d + T1;
   d = c;
   c = b;
   b = a;
   a = T1 + T2;
}
```

```
hash[0] = H[0] + a;
hash[1] = H[1] + b;
hash[2] = H[2] + c;
hash[3] = H[3] + d;
hash[4] = H[4] + e;
hash[5] = H[5] + f;
hash[6] = H[6] + g;
hash[7] = H[7] + h;

if (hash[7] == 0) {
    // do something nice with matching nonce
    printf("matching nonce: %08x\n", nonce);
}

}
return 0;

}
```

References

Nakamoto S (2008) Bitcoin: a peer-to-peer electronic cash system. Available https://bitcoin.org/bitcoin.pdf. Accessed 1 July 2017 (Online)

Antonopoulos A (2015) Mastering bitcoin: unlocking digital cryptocurrencies. O'Reilly Media, Sebastopol

Blagojević V et al (2016) A systematic approach to generation of new ideas for PhD research in computing. Adv Comput 104:1–19 Elsevier

Apps|Maxeler AppGallery. Available http://appgallery.maxeler.com/. Accessed 1 July 2017 (Online)

Trifunović N, Milutinović V, Korolija N, Gaydadjiev G (2016) An AppGallery for dataflow computing. J Big Data 3(1):4

Ranković V, Kos A, Tomažič S, Milutinović V (2013) Performance of the bitonic mergesort network on a dataflow computer. In: 2013 21st telecommunications forum telfor (TELFOR), Belgrade, 2013, pp 849–852

Kos A, Ranković V, Tomažič S (2015) Sorting networks on maxeler dataflow supercomputing systems. Adv Comput 96:139–186

Milanković I, Mijailović N, Peulić A, Filipović N (2018) Application of data flow engines in biomedical images processing. IPSI BgD Trans Adv Res (TAR), 14(1). ISSN 1820 – 4511

Kotlar M, Milutinović V (2017) Implementing neural networks using the dataflow paradigm. IPSI BgD Trans Adv Res (TAR) 13(1). ISSN 1820 – 4511

Reese L (2018) Comparing hardware for artificial intelligence: FPGAs vs. GPUs vs. ASICs | embedded Intel® Solutions, Eecatalog.com, 2018. Available http://eecatalog.com/intel/2018/07/24/comparing-hardware-for-artificial-intelligence-fpgas-vs-gpus-vs-asics/. Accessed 1 August 2018 (Online)

Amara A, Amiel F, Ea T (2006) FPGA vs. ASIC for low power applications. Microelectron J 37:669–677

Meden R (2017) Maxeler bitcoin miner [source code]. https://github.com/medenko/Maxeler-Bitcoin-Miner

Developer Guide—Bitcoin (2009) Available https://bitcoin.org/en/developer-guide. Accessed 1 July 2017 (Online)

Hash Rate—Blockchain. Available https://blockchain.info/en/charts/hash-rate?timespan=2years. Accessed 1 July 2017 (Online)

Difficulty—Blockchain. Available https://blockchain.info/en/charts/difficulty?timespan=2years. Accessed 1 July 2017 (Online)

Bitcoin Hashrate Distribution—Blockchain.info. Available https://blockchain.info/en/pools. Accessed 1 July 2017 (Online)

Bitmain Antminer S7 Review: Is it Profitable to Buy? (Probably Not). Available https://www.buybitcoinworldwide.com/mining/hardware/antminer-s7/. Accessed 1 July 2017 (Online)

Spondoolies-Tech SP20 Jackson Review: Is it Profitable to Buy? Available https://www.buybitcoinworldwide.com/mining/hardware/spondoolies-sp20/. Accessed 1 July 2017 (Online)

Bitcoin Mining Calculator—BTCServ. Btcserv.net. Available https://btcserv.net/bitcoin-mining-calculator. Accessed 4 Nov 2016 (Online)

Milenković A, Milutinović V (2000) Cache injection: a novel technique for tolerating memory latency in bus-based SMPs. In: European conference on parallel processing. Springer, Berlin, pp. 558–566

Furht B, Milutinović V (1987) A survey of microprocessor architectures for memory management. Computer 20. Kovačević M, Diligenti M, Gori M, Milutinović V (2004) Visual adjacency multigraphs-a novel approach for a web page classification. In: Proceedings of SAWM04 workshop

Tomašević M, Milutinović V (1994) Hardware approaches to cache coherence in shared-memory multiprocessors. IEEE Micro 14:61–66

Grujić A, Tomašević M, Milutinović V (1996) A simulation study of hardware-oriented DSM approaches. IEEE Parallel Distrib Technol Syst Appl 4:74–83

Tomašević M, Milutinović V (1992) A simulation study of snoopy cache coherence protocols. In: System sciences, proceedings of the twenty-fifth Hawaii international conference, Hawaii, vol 1, pp 427–436

Tartalja I, Milutinović V (1997) The cache coherence problem in shared-memory multiprocessors: software solutions. IEEE Computer Society Press

Milutinović V (1997) Caching in distributed systems. IEEE Concurr 8:14–15

Milutinović V, Stenstrom P (1999) Special issue on distributed shared memory systems. Proc IEEE 87:399–404

Maxeler Technologies (2015) Multiscale dataflow programming

Milutinović V, Furht B, Obradović Z, Korolija N (2016) Advances in high performance computing and related issues. In: Mathematical problems in engineering

Korolija N, Popović J, Cvetanović M, Bojović M (2017) Dataflow-based parallelization of control-flow algorithms. In: Advances in computers, vol 104. Elsevier, pp 73–124

Trifunović N, Milutinović V, Salom J, Kos A (2015) Paradigm shift in big data supercomputing: dataflow vs. controlflow. J Big Data 2:4

Kos A, Tomažič S, Salom J, Trifunović N, Valero M, Milutinović V (2015) New benchmarking methodology and programming model for big data processing. Int J Distrib Sens Netw 11

Milutinović V, Salom J, Veljović D, Korolija N, Marković D, Petrović L (2017) Transforming applications from the control flow to the dataflow paradigm. In: DataFlow supercomputing essentials. Springer, Cham, pp 107–129

Milutinović V, Salom J, Veljović D, Korolija N, Marković D, Petrović L (2017) Discrepancy reduction between the topology of dataflow graph and the topology of FPGA structure. In: DataFlow supercomputing essentials. Springer, Cham, pp 19–66

Milutinović V (1989) Mapping of neural networks on the honeycomb architecture. Proc IEEE 77:1875–1878

Descriptions of SHA-256, SHA-384, and SHA-512. Available http://www.iwar.org.uk/comsec/resources/cipher/sha256-384-512.pdf. Accessed 1 July 2017 (Online)

Marek Palatinus (2016) Stratum Mining Proxy [Source Code]. https://github.com/slush0/stratum-mining-proxy

Meden R (2017) Mining bitcoins using maxeler data flow computer. M.S. thesis, Faculty of Electrical Engineering, University of Ljubljana, Ljubljana, Slovenia

p2pool Nodes. Poolnode.info. Available http://poolnode.info/. Accessed 4 Nov 2016 (Online)

Index

A

Acceleration, 86
Address, *see* bitcoin address
Algebra of graphs, 10
Algorithm, 57
Algorithm, *see* hashing algorithm
Amazon Dynamo, 4
Application-Specific Integrated Circuit (ASIC), 186, 187, 231, 242, 248–250, 252, 257–259, 282, 286, 289, 290, *see also* mining rig
Approximation, 60
Arithmetic logical unit, 229

B

BigData, 56
Bitcoin, 242–249, 251, 257, 258, 289
    address, 243, 245, 246, 252, 287
    miner, 242, 243, 247, 248, 257, 263–270, 280–283, 288
    mining, 242, 243, 247–249, 251, 252, 257, 258, 262, 280–282, 284–290
    network, 11, 242, 243, 246–253, 256–258, 270, 289
    protocol, 242, 243, 262
    transaction, 243–247
    wallet, 243–245, 247
Block, 4, 6, 8, 11, 12, 82, 95, 106, 111, 115, 116, 120–125, 134, 135, 139, 140, 142, 143, 183, 190, 200, 207, 243, 244, 247–252, 254–258, 263, 266, 272, 273, 275, 276, 278, 290
Blockchain, 242–244, 247–249, 251, 256, 257
Block header, 243, 250, 252, 254–256, 262, 263, 272–277, 279, 284

C

Central Processing Unit (CPU), 34, 42–45, 63, 65, 75, 77, 78, 82, 92–94, 96, 97, 134, 140, 143, 184–187, 202, 203, 222, 229, 233, 242, 248, 250, 258–263, 266, 267, 275, 280–285, 289
Class-based distribution, 22
Cluster, 86
Clustering, 27
Computational power, 248–250, 252, 289, *see also* hash rate
Control-flow, 41–44, 46, 49, 62, 91, 93, 95, 98, 102, 105, 111–113, 117, 120, 122, 124, 125, 127, 129, 229–232, 236, 239, 258
Cryptocurrency, 242, 243, 268, 289, 290

D

Dataflow, 63
    computing, 44, 46, 49, 56, 64, 79, 81, 96, 101, 134, 149, 183–185, 187, 194, 200–202, 212, 230, 231, 234, 236, 238, 239, 242, 258, 259, 261, 266, 276
    graph, 79, 81, 87, 91, 93, 95, 97, 102, 105, 113, 123, 126, 182, 184, 185, 194, 198, 201, 202, 206, 231, 232, 261, 263, 264, 267
Dataflow system, 8
Data server, 9
DFE, DFE board
    MAX2B, 243, 264, 283
    MAX5C, 243, 264, 266, 284, 286
Difficulty, *see* network difficulty
Distance function, 6, 28

© Springer Nature Switzerland AG 2019
V. Milutinovic and M. Kotlar (eds.), *Exploring the DataFlow Supercomputing Paradigm*, Computer Communications and Networks,
https://doi.org/10.1007/978-3-030-13803-5

E

Edge-based distribution, 23

Endianness

big-endian, 274, 279, 280

little-endian, 273, 274, 279, 280

Energy efficiency, 185, 186, 212, 257, 258, 285, 286, 290

Erlang, 8

Execution graph, 81

F

Field-Programmable Gate Array (FPGA), 57, 62, 82, 86, 92–97, 104, 105, 108, 134, 135, 142, 144–146, 171, 182, 183, 185–188, 199, 200, 202, 203, 206, 211, 212, 231, 232, 242, 248, 250, 260, 266, 289, 290,

Floating-point unit, 229

Fragment, 20, 27

Front server, 9

G

Getwork, 252, 262, 268, 270, 287, 288

Golden-section search, 57

Google bigtable, 4

Gradient descent method, 137

Graphical Processing Unit (GPU), 171, 187–189, 203, 242, 248, 250, 282, 284, 285, 289, 290

Graph-pattern, 10

H

Hash, 243–245, 249, 250, 252, 254–256, 263, 271, 274–279, *see also* SHA-256

Hashing algorithm, 253, 262, 267, 271, 272, *see also* SHA-256

Hash partitioning, 4

Hash rate, 248, 252, 253, 257, 258, 263, 266, 281, 284–288, *see also* computational power

I

Internet database systems, 5

K

Kernel, 43–45, 65, 77–79, 82, 97, 98, 101, 102, 105–108, 111, 117, 118, 120, 121, 125, 126, 143, 185, 186, 202,

203, 206–209, 211–213, 236, 237, 260–263, 265, 266, 275, 276, 280, 283

Keys

private, 244–246

public, 244–246

Knowledge graph, 4

L

List partitioning, 4

M

Manager, 10, 65, 66, 78, 79, 97, 98, 111, 120, 185–187, 202, 206, 207, 212, 213, 260, 262, 266

MAX2B, 264, 266, 267, 282, 284–289

MAX5C, 264, 266, 282, 284–290

MaxCompiler, 65, 78, 186, 203, 211, 212, 231, 264, 266

Maxeler, 58

Maxeler framework, 232

MaxIDE, 41, 47, 49, 212, 259, 264, 266, 268

Maximization, 67

MaxJ, 41, 42, 45, 65, 66, 77, 78, 97, 98, 121, 186, 203, 212, 232, 259, 266, 267, 280

Merkle root, 243, 244, 252–255

Merkle tree, 252–254

Midstate, 271, 274–277, 295, 297, 302, 303, 306–309

Miner, 242, 243, 247–249, 251, 252, 256–258, 263, 270, 275, 287–290, *see also* bitcoin miner

Minimization, 67

Minimum distances, 133

Mining, 242, 248, 249, 251, 252, 256, 258, 268, 270, 271, 281, 282, 284–286, 288–290, *see also* bitcoin mining

difficulty, 242, 248, 249, 251, 252, 257, 258, 288, *see also* network difficulty

node, 251, 256, 263, 268, 270, 274, 287, 288

pool, 243, 251–253, 256, 268, 270, 280, 281, 287, 288, *see also* P2Pool *and* SlushPool

process, 242, 243, 248, 251, 252, 257, 262, 289

profitability, 257, 258, 286, 290

profitability calculator, 258, 286

protocol, 252, 262, 268, 270, 287, *see also* Getwork

reward, 242, 243, 248, 251, 252, 256,
  258, 288, 289
rig, 242, 243, 248, 249, 252, 257–259,
  282, 285, 286, 288–290, *see also* ASIC

N
Network, 38
Network difficulty, 248–252, 257, 258
Network, *see* bitcoin network
Nonce, 252, 254–256, 262, 263, 265, 274–
  277, 279, 280, 284
Number of transistors, 230
Numerical method, 87

O
OpenSPL, 232
Optimization, 59

P
Parallel, 140
Physical algebra, 11
Power consumption, 49, 133, 135, 144, 183,
  185, 188, 189, 200, 230, 231, 239,
  257, 286
P2Pool, 287, 288
Predicate-based distribution, 21
Proof-of-work, 256, 290, *see also* SHA-256
Protocol
  bitcoin protocol, 242, 243, 262
  mining protocol, 252, 262, 268, 270, 287,
    see also Getwork

Q
Quadratic functions, 75
Query node, 13

R
Range partitioning, 4
RDF graph, 14

Resource Description Framework (RDF), 8
Round-robin partitioning, 4

S
Schema graph, 13, 15
Schema triple, 5, 15
Semantic distribution, 20
SHA-256, 253, 271–273, 277–279, *see also*
  hashing algorithm
Shared-nothing cluster, 8
Skeleton graph, 6, 24, 27
SlushPool, 288
SPARQL, 5
Spherical codes, 134
Stream, 11

T
Target
  threshold, 248–251, 255, 257, 263, 275,
    279
Transaction, *see* bitcoin transaction
Triple, 14
Triple pattern, 19
Triple-pattern localization, 30

U
Unimodality, 68

V
Vertex, 84

W
Wallet, *see* bitcoin wallet
Wikidata, 4

Y
YAGO, 4